Military, State, and Society in Israel

Theoretical & Comparative Perspectives

Military, State, and Society in Israel

Daniel Maman, Eyal Ben-Ari Zeev Rosenhek

Routledge
Taylor & Francis Group

LONDON AND NEW YORK

Preface

Paying Our Dues:
On the Intellectual Legacy of Moshe Lissak

This volume is dedicated to Professor Moshe Lissak. A major theme that has preoccupied Moshe throughout his career has been the study of the armed forces in general and of the Israel Defense Forces in particular. It was with this concern in mind that this volume was brought together by his friends, colleagues and former students. This collection covers examinations of many of the issues that Moshe first investigated, and the development of new avenues of research that have been emerging in the past decade.

Let us open this preface by relating the contributions and wider questions suggested by this volume to Moshe's scholarly and intellectual legacy. Our contention is twofold: first, through his work Moshe has been a key scholar to generate a consistent set of theoretical questions central to the agenda of the field linking the study of state, society and the military; second, many recent approaches that have developed new questions and perspectives on these themes have done so out of an engagement with his work. We do not deal with the micro-sociological process by which the study of "things military" in Israel has evolved nor directly examine Moshe's research on the armed forces of other societies. Rather, we will place his investigations and the analytical controversies they have raised within a broader theoretical and social context.

As a preliminary remark it is important to mention Moshe's very fruitful cooperative venture with Dan ("Dindush") Horowitz (Horowitz 1977, 1982). While Lissak began his study of the relations between the armed forces and society in the 1960s, his later work with Dan Horowitz (begun in the 1970s) provided some of the

i

most seminal formulations in regard to the character of the Israel Defense Forces (IDF), to the historical development of its social and political roles, as well as to its ties with other institutions in Israeli society. Without exaggeration, we would posit that it was Lissak and Horowitz who set the parameters for discussions of state, society and the military in Israel. Thus much of what we have to say in this introduction is also a serious acknowledgement to Dan Horowitz's contribution throughout the years. In this regard it is also important to note that both scholars have not only engaged in discussions with sociologists within and outside of Israel, but also of other disciplines such as history, political science, international relations, anthropology, law, and organization and administrative studies (Lissak 1984, 1990, 1993, 1995).

We would suggest that one of the most important features of Moshe's work has been the systematic application of theoretical insights, questions and frameworks from the sociology of the military to the Israeli case (Lissak 1984; Horowitz and Lissak 1996 [1988]; Peri and Lissak 1976). In line with a more general trend characterizing Israeli social science, he has sought to examine developments related to the IDF in terms of theoretical models developed elsewhere (primarily in America). More specifically, through personal and intellectual links, Moshe has consistently utilized, and contributed to, analytical concepts developed within the Inter-University Seminar on Armed Forces and Society (Burk 1993). Indeed, even a cursory review of the patterns of his acknowledgments and citations reveals his dependence on the work of such people as Morris Janowitz, Sam Sarkesian or Charles Moskos. Thus, for example, Moshe has examined a wide array of issues such as civilian surveillance of the military, the second career of army officers, the roles and interrelations between strategic elites, or centers and peripheries in the relations between military and society.

It is in this light that his work on other societies should be seen. Like many of the studies carried out by his teacher, Shmuel Eisenstadt, Moshe has constantly sought to place Israel within an explicit or implicit comparative perspective. In his earlier studies of such countries as Thailand, Indonesia, Pakistan or Burma, he deals primarily with the role of the armed forces in processes of nation-building (Lissak 1964a, 1964b, 1967, 1969, 1970, 1976). Here the governing notion has been an examination of the military's role expansion and its role in legitimating the new nation-state, mobilizing national resources and coping with the crises of modernization. In his later investigations with Dan Horowitz (Horowitz and Lissak 1989), comparisons were used in order to delineate different models found in the industrialized democracies (such as Britain) in terms of the gap between the civilian and military sectors of society. Here the aim was to sketch the "ideal type" of the Israeli polity and its relation with the armed forces.

But Moshe has not limited his efforts solely to the realm of academia. His work, while rooted in research and teaching, has constantly sought to grapple with the reality of Israeli experiences: with "nation-building," with the political role of the military, as indeed with the complex processes of the formation of Israeli identity. The book written with Horowitz, *Trouble in Utopia*, well underscores this point. In the past few years, Moshe, along with a number of other founding members of Israeli social sciences, has been singled out for criticism for the purported ideological mobilization at the base of their work. This charge has led to a rather heated debate among Israeli intellectuals (Peri 1996). Yet even in the heat of this debate, Lissak stands out as an intellectually honest and open scholar. For example, he appears in Kimmerling's (1992) acknowledgments in the latter's article called "Sociology, Ideology, and Nation-Building: The Palestinians and Their Meaning in Israeli Sociology," which attempts to uncover the ideological underpinnings of much of Israeli social scientific thinking. It is against this background that we begin to understand how Moshe, alone or with other scholars such as Dan Horowitz, has become a "founding father," a veritable "intellectual ancestor" for many younger scholars. As exemplified in this volume, later scholars invariably refer to his earlier work, sometimes acrimoniously, sometimes appreciatively.

Yet Moshe's theoretical perspective has not been static. While still rooted in the "armed forces and society" paradigm, much of his later work has developed out of an active dialogue with many of the contentions raised by scholars rooted in other approaches. For example, in their last major work together, Horowitz and Lissak (1989) developed a model that was much more conflictual than their previous one. While their former work "considered Israel an exceptionally consensual, consociational democracy, now Israel is considered an exceptionally diversified and strained polity" (Ram 1995: 61). It is out of this view that their notion of the "over burdened" polity of Israel emerged: a polity that is constantly grappling with the major cleavages that characterize Israeli society. Moshe's newer work thus is oriented much more to the diversity and pluralism of Israeli society and the ways in which power is spread among a multitude of groups, elites and institutions.

More recently, Lissak has developed his work in other directions. For example, in a cooperative project with Daniel Maman (who completed his Ph.D. under his supervision), he has asked questions about networks of military and other elites in Israel. This kind of conceptualization has allowed Lissak and Maman (1996) to add greater theoretical sophistication to what have become basic questions in the Israeli context: the relations between the civilian and military sectors and civilian supervision over the armed forces. This approach is more sophisticated than earlier formulations because it has added a new dimension to the stress on institutions: a focus on the micro structures and processes (such as networks of acquaintanceships, the promotion of personal interest and exchange of information) that bind or divide Israel's elite.

Finally, as mentioned before, Moshe has often participated in wider discussions and contentions. Recently, he has participated in the assaults, criticisms and counter-charges in what is known as the debate centering on the "New Historians" or "New Sociologists." The key argument here centers on the ideological mobilization of sociologists and historians who belong to the "mainstream" of Israeli social science and humanities (Kimmerling 1992; Lissak 1996; Peri 1996). Within the parameters of this debate, such questions as the place of war and conflict with Israel's Arab "neighbors," as the centrality of the military, are not only theoretically relevant, but take on meaning for a basic ideological commitment to Israeli society. According to the "new" scholars in their research and methods, older, mainstream or establishment social scientists served the needs and interests of the dominant national discourse. Moshe (Lissak 1996) has strongly countered that the "newer" scholars have been no less committed ideologically, and that the portrait of "older" scholars has not taken into account their commitment to scientific rigor, to self-reflection, and to their openness to alternative "readings" of Israeli society.

We do not wish to finesse the lines of conflict and dissensus among scholars of "things military" in Israeli society, but to stress that these discords can and should prod us to constantly think and rethink Israel and the place of the military in this society. It is in Moshe's spirit of constant engagement with differing theoretical approaches and viewpoints about Israel that we now turn to an introductory essay.

Acknowledgements

The conference which forms the basis for this volume was organized by the Department of Sociology and Anthropology and the Shaine Center for Social Research of the Hebrew University of Jerusalem. We wish to thank Meirav Ozeri-Ben-Ari and Hila Yoffe for warm, efficient and dedicated administrative assistance during preparations for and including the conference. Financial aid was awarded by the Faculty of Social Sciences of the Hebrew University, the Shaine Center for Social Research, the Leonard Davis Institute for International Relations, the Harvey L. Silbert Center for Israel Studies, the Harry S. Truman Research Institute for the Advancement of Peace, the Levi Eshkol Institute for Economic, Social and Political Research, and the Smart Family Foundation Communications Institute.

References

Burk, James 1993. "Morris Janowitz and the Origins of Sociological Research on Armed Forces and Society." *Armed Forces and Society* 19, 2: 167-85.

Horowitz, Dan 1982. "The Israel Defence Forces: A Civilianized Military in a Partially Militarized Society." In *Soldiers, Peasants and Bureaucrats*, edited by Roman Kolkowitz and Andrzej Korbonski, 77-105. London: Allen and Unwin.

Horowitz, Dan 1977. "Is Israel a Garrison State?" *The Jerusalem Quarterly* 4: 58-65.

Horowitz, Dan and Moshe Lissak 1996 [1988]. "Democracy and National Security in a Protracted Conflict." In *Democracy and National Conflict in Israel*, edited by Benjamin Neuberger and Ilan Ben-Ami, 73-113. Tel Aviv: The Open University. (Hebrew).

Horowitz, Dan and Moshe Lissak 1989. *Trouble in Utopia: The Overburdened Polity of Israel*. Albany: State University of New York Press.

Kimmerling, Baruch 1992. "Sociology, Ideology, and Nation-Building: The Palestinians and Their Meaning in Israeli Sociology." *American Sociological Review* 57: 446-60.

Lissak, Moshe 1996. "'Critical' Sociology and 'Establishment' Sociology in the Israeli Academic Community: Ideological Struggle or Academic Discourse?" *Israel Studies* 1, 1: 247-94.

Lissak, Moshe 1995. "The Civilian Component of Israel's Security Doctrine: The Evolution of Civil-Military Relations in the First Decade." In *Israel: The First Decade of Integration,* edited by Ilan Troen and Noah Lucas, 575-91. New York: State University of New York Press.

Lissak, Moshe 1993. "Civilian Components in National Security Doctrine." In *National Security and Democracy in Israel,* edited by Avner Yaniv, 55-80. Boulder, CO.: Lynne Rienner.

Lissak, Moshe 1990. "The Intifada and Israeli Society: An Historical and Sociological Perspective." In *The Seventh War — The Effects of the Intifada on Israeli Society,* edited by Reuven Gal, 17-37. Tel Aviv: Hakkibutz HaMeuchad. (Hebrew).

Lissak, Moshe (ed.) 1984. *The Israeli Society and Its Defence Establishment.* London: Frank Cass.

Lissak, Moshe 1976. *Military Roles in Modernization: Civil-Military Relations in Thailand and Burma.* Beverly Hills, CA: Sage.

Lissak, Moshe 1972. "The Israeli Defence Forces as an Agent of Socialization and Education: A Research in Role Expansion in a Democratic Society." In *The Perceived Role of the Military,* edited by M.R. van Gils, 325-40. Rotterdam: Rotterdam University Press.

Lissak, Moshe 1970. "Stages of Modernization and Patterns of Military Coups." *International Journal of Comparative Sociology.* 14, 1-2: 59-75.

Lissak, Moshe 1969. "The Military in Burma: Innovations and Frustrations." *Asian and African Studies* 5: 133-63.

Lissak, Moshe 1967. "Modernization and Role-Expansion of the Military in Developing Countries: A Comparative Analysis." *Comparative Studies in Society and History* 9, 3: 233-55.

Lissak, Moshe 1964a. "Selected Literature on Revolution and Coups D'Etat in the Developing Nations." In *The New Military: Changing Patterns of Organization,* edited by Morris Janowitz, 339-62. New York: The Russell Sage Foundation.

Lissak, Moshe 1964b. "Social Change, Mobilization and Exchange of Services Between the Military Establishment and the Civil Society: The Burmese Case." *Economic Development and Cultural Change* 13, 1: 1-19.

Lissak, Moshe and Daniel Maman 1996. "Israel." In *The Political Role of the Military,* edited by Constantine Panopoulos and Cynthia Watson, 223-33. Westport, CT: Greenwood Press.

Peri, Yoram 1996. "Is Israeli Society Really Militaristic?" *Zemanim* 56, 14: 94-113. (Hebrew).

Peri, Yoram and Moshe Lissak 1976. "Retired Officers in Israel and the Emergence of a New Elite." In *The Military and the Problem of Legitimacy,* edited by Gwyn Harries-Jenkins and Jacques van Doorn, 175-92. Beverly Hills, CA: Sage.

Ram, Uri 1995. *The Changing Agenda of Israeli Sociology.* Albany: State University of New York Press.

Military, State and Society in Israel: An Introductory Essay[*]

Eyal Ben-Ari, Zeev Rosenhek and Daniel Maman

This collection deals with the relations between the military, the state, and society in Israel. The past decade has seen the publication of a number of volumes devoted to issues centering on the place of war, security or military service in Israeli society. Yet these books have usually tended to focus on specific aspects of this triangle such as the social construction of war and national service (Lomsky-Feder and Ben-Ari 1999), the grand narratives of defense underlying Israeli views of security (Ezrahi 1997), or the role and place of the national-religious camp in the Israeli army (Cohen 1997). This volume, however, provides a broader perspective and makes three key contributions—theoretical, empirical and polemical—related both to the Israeli case and to wider debates about the place of war and the military in contemporary industrialized societies. In this introduction, we undertake the following tasks: to explain the contributions of this volume, to place it in its wider scholarly and intellectual context, and to introduce the specific papers.

The essays in this volume all proceed from an explicit recognition of the importance of theorizing the Israeli case. There are two reasons for this theoretical stress. First, it is only on the basis of the use of explicit analytical or interpretive frameworks that contemporary developments in Israel can be gauged historically and comparatively. And second, it is only on the basis of such frameworks that the contribution of the Israeli case to contemporary theorizing about such issues as war, the armed forces or security can be made clear. In this regard, Israel presents an interesting instance. While Israel has not figured in the formulation of general social theories about such issues as the development of modern or post-modern societies, its experience of continuous armed struggles and the centrality of its armed forces has been used to make some rather substantial contributions to those disciplines centering on military issues. Thus, for example, studies explicitly based on Israeli cases, data, and scholarship have been central to the development of expertise in such fields as applied psychology and psychotherapy (Breznitz 1983), the dynamics of small groups (Gal 1986; Greenbaum 1979; Shalit 1988; Shirom 1976) or models of leading and of leadership (Shamir and Ben-Ari, forthcoming).

The Israeli case is of no less importance in the development of more macro approaches to the study of "things military." Its potential contribution to general theory derives from the fact that, due to the central place of war in Israel's history and contemporary circumstances, it encapsulates in a very explicit manner many of the tensions in the relationships between the military, state and society found in other advanced industrial democracies. As such, it is an especially appropriate research site to examine questions raised in wider scholarship. In this regard, this volume stands at the core of contemporary debates between the two "master approaches" to the study of the relations between the military, society and the state: the "armed forces and society" school and the "state-making and war" perspective.

Essentially, the debate between these two perspectives has centered on the significance of the armed forces and their activities for contemporary societies. In general, the stress of the "armed forces and society" school has been on the social and political functions of the military, on the kinds of boundaries and arrangements that link or separate it from the civilian sector of a society, and on the various mechanisms by which it is controlled and supervised by (mainly) political elites (Janowitz 1971, 1976; Burk 1995). In contrast, and as attested to by its name, the "state-making and war" approach has emphasized the manner by which armed struggles and the necessary mobilization of human and other resources these entail have figured in the creation, consolidation and strengthening of states (Tilly 1985; Giddens 1985; Shaw 1988).

Such theoretical divergences are based on very different assumptions about modern democracies (Ben-Ari 1997; Ben-Ari and Lomsky-Feder 1999). First, while the "armed forces and society" perspective was developed in the theoretical context of structural-functionalism in the heyday of America's success in World War II and was essentially celebratory, the latter approach was formulated in the context of conflict-statist theories during the cold war and was highly critical of existent social and political arrangements. Second, the different frameworks called attention to distinct issues: the first to institutions and elites and to the inter-linkages between them, and the second to points of dissension, struggles and conflicts in and around the structures of the state. The third point is that on the basis of these diverging perspectives, the two approaches posed very different questions for research and analysis: one about the arrangements by which democracies can continue to function in the face of powerful armed forces; the other about how armed struggles figured in the manner by which structures of domination, especially states, were, and still are, created and maintained.

The guiding scholarly image of Israel according to the "armed forces and society school" is one of a non-militaristic society in which the military is "civilianized" and the civilian sector "militarized." The implication of these circumstances is that the more extreme consequences of both trends are offset (Horowitz and Lissak 1989). By contrast, the picture according to the "state-making and war" approach is that of a society characterized by a cultural militarism centered on a world view in which solutions to inter-statal problems are military in nature (Ben-Eliezer 1995a, 1997; Kimmerling 1993), and where the use of the military for nation-building purposes was an explicit aim of the elites from the state's formative years (Levy 1996). The confrontation between these two approaches has brought about a much more complex view of the role, centrality and consequence of the armed forces and of war. In general, as Ben-Ari and Lomsky-Feder (1999) suggest, the move has been from an analysis centered on Israel's unique status as a society that maintains democracy under conditions of protracted war and the centrality of the military, to more complex inquiries about this society as an instance of how democracy normalizes militarism, and how its armed forces have figured in the way the state has established its legitimacy and mobilized the population for collective aims (Ben-Eliezer 1997, 1998; Ehrlich 1987).

Contemporary Israel

This controversy should be viewed as part of more general developments in Israeli social sciences, which themselves derive from (and feed into) the transformations Israel has been undergoing in the past few decades. The most central of these changes have made Israel a society marked by deep tensions and contradictions. On the one hand, Israel is characterized by a certain decline in the acceptance of Zionist ideology, especially among the younger secular middle-classes, a strengthening of both consumerist and post-material values, greater acceptance of cultural pluralism and individualism, and deep-seated questionings of, and challenges to, the subordination of individual considerations to collective goals. These trends, which can be understood as the transformation of Israeli society from a "mobilized society" into a "normal" Western society, have resulted in a diminution in the state's ability to mobilize those groups who were the carriers of the classic Zionist project. On the other hand, however, social sectors that during Israel's formative years were relatively peripheral—for instance, the nationalist-religious and ultra-orthodox groups—have gained substantial political power and moved into many societal centers. This process has allowed them to challenge the previous hegemony and to present alternative models of society, ones much more oriented along "ethno-Jewish" lines. The very same trends that are seen by the former groups of the society as positive signs of "normalization"—growing individualism or cultural pluralism, for example—are understood by the latter as alarming signs of decadence and decline.

Contemporary Israel is thus the site of debates about, and interrogation of, many of the fundamental assumptions that have undergirded it as the Jewish nation-state: about the ethnic character of nationhood and statehood; about the role of the Jewish diaspora vis-a-vis Israel; the legitimacy of Jewish "ethnic pluralism"; the meaning of the Holocaust; the privatization of social life and the spread of consumerism; and the weakening of the centralized state as the agent of social transformation affecting housing, language, health, technology, production, dress, and child-rearing (Aronoff 1989; Ben-Ari and Bilu 1997; Dominguez 1989). More pertinent to the present volume, it appears that one important consequence of these internal conflicts and struggles has been a significant erosion in the almost sacred status once enjoyed by state institutions, and especially the military, among the majority of Jewish population.

A central assumption now being questioned by many groups is that of the centrality of the military to society and to definitions of "Israeli-hood" and full citizenship. Despite the hard-line taken by many of Israel's governments, many groups in contemporary Israeli society are no longer willing to grant the Israeli Defense Forces (IDF) its previous status of unquestioned professionalism and to view "state security" considerations as the only (or primary) criteria for national decision-making. In this context, new questions have arisen in and around a profusion of topics such as motivation for military service, the legal responsibility of commanders for casualties, the tension between private and public representations of commemoration, the links between conscripts' families and military authorities, or the official reasons given for suicides within the military (Ezrahi, 1997; Lomsky-Feder and Ben-Ari, forthcoming).

Interestingly, strong critical attitudes towards the state and the military appear not only among the groups representing a highly secularized "post-Zionist" perspective on contemporary Israel. They have also appeared—albeit articulated in inverse terms—among some of the most vocal right-wing sectors, especially settlers in the occupied territories. For instance, these groups had vigorously criticized the army for its involvement in the negotiations with the Palestinians during Rabin's government and for its acceptance of what they see as a policy of capitulation, or even treason in the face of problems centered on the country's very survival. In more general terms, the army is sometimes denunciated by some of these groups for having deserted its Jewish-national purpose.

On (Military) Theory and the Academy

In tandem with these socio-political transformations, Israeli social sciences have been experiencing important paradigmatic changes. The intensification of wider challenges to many of what were previously taken-for-granted notions characterizing the Zionist project and Israeli hegemony has allowed new interrogations of the state (and its military) to emerge within the academy and prompted younger scholars to take their inquiries into new, and often innovative directions. The newer theoretical orientations now emphasize struggles and conflicts, patterns of domination and exclusion, and the efforts that the Israeli state has invested in constantly producing and reproducing its legitimacy (Kimmerling 1992; Lustick 1988; Rosenhek 1998). Israel society, according to this broad coalition of newer approaches, is no longer seen solely as "the" Jewish society, nor as a system which is characterized by broad agreement or consensus over social goals and boundaries. It is now viewed in much more conflictual terms, as a society rent by deep divisions and constant struggles, and in which the very ground rules of public life are constantly negotiated and contested (Dominguez 1989; Ram 1989, 1995).

These broad paradigmatic shifts are linked, in turn, to a number of key developments—institutional, generational—in Israeli academia. To begin with, not only have Israeli universities expanded greatly in terms of numbers of students and faculty and fields of research, but new academic colleges have been established at an unprecedented rate in the past decades. These trends have contributed to an upsurge in the sheer amount of research now being carried out in relation to the study of Israel, and (perhaps more significantly) to a much more pluralistic scholarly arena in which a variety of theories and approaches are now used. In addition, the greater distance from the "heroic" stage of the Zionist project and the appearance of a new generation of scholars who have no direct experience in the saga of nation-building have also allowed the development of more critical approaches to the study of this society (Lissak 1996; Shafir 1996; Shalev 1996). Finally, some representatives of "marginal" social groups that have begun to obtain advanced degrees and gain access to faculty positions during the last years— such as women, "Oriental" (of Middle-Eastern origin) Jews and new immigrants—have introduced new perspectives that depart from the outlook that used to dominate mainstream Israeli academy and which was based on a male, Ashkenazi (of European origin) and veteran viewpoint.

Another factor has consistently encouraged the cultivation of critical approaches within Israeli social sciences: the strong connections with Western academic centers such as in America, Britain or France. Radical approaches developed in the West from the late sixties and seventies have been readily and regularly incorporated into Israeli social sciences and applied to the analysis of the Israeli case. This incorporation has been the result of the fact that Israeli academics read (and publish in) the journals and books of the West, participate in international forums in which intellectuals from these Western centers appear, and teach their students the theories and findings of these same people. This process of intellectual diffusion is further enhanced by the fact that large numbers of Israeli scholars have either obtained their Ph.D. degrees or spent significant stints (for sabbaticals or post-docs) abroad, especially in the USA and Britain (Ben-Yehuda 1997; Shamgar-Handelman 1996).

These general processes of change in Israeli society and in scholarly endeavors to examine it have had several important effects on the study of civil-military relations, the experiences of national service and war, or the internal structure and dynamics of the Israeli army. First, since the early 1980s there has been an enormous growth in the number of academic works dealing with these topics. Such has been this growth that in the past decade tens of English-language books and articles, and often comparable numbers of Hebrew-language titles, have been published each year (Ben-Ari 1998).

Second, this quantitative expansion has been accompanied by an increased diversification of subject matters. Most works published until the mid-1980s tended to focus either on the broad political aspects of civil-military relations (Lissak 1983, 1984; Peri 1977, 1981, 1985; Peri and Lissak 1976; Perlmutter 1968), or on the individual psychiatric effects of war experiences (Bar-Gal 1982; Breznitz 1983; Moses et al. 1976). Indeed, a significant part of contemporary research continues these lines of analysis (Etzioni-Halevy 1996; Lissak 1993; Peri 1993; Levy et al. 1993; Solomon 1990; Solomon et al. 1995). During the past two decades, however, scholars have begun to research such hitherto little studied issues as the economic links and effects of war or Israel's "military-industrial complex" (Kleiman 1985; Kleiman and Pedatsur 1991; Mintz, 1985), the links between military service and employment and stratification structures (Enoch and Yogev 1989; Maman and Lissak 1995), and the impact of security considerations on land use (Soffer and Minghi 1986). Other studies have focused on the internal organizational aspects of the IDF (Cohen 1995; Gur-Ze'ev 1997). Finally, a large number of scholars have examined the "cultural" place of the IDF and of wars in Israel through studies of central sites like Masada

(Ben-Yehuda 1999) or military cemeteries (Handelman and Shamgar-Handelman 1997), rites like Remembrance Day or Independence Day (Don-Yehiya 1988; Handelman and Katz 1995), or the means for effecting social memory (Sivan 1991).

Third, the diversification of subject matter has been accompanied by a strengthening tendency to examine the Israeli case using comparative research strategies. This propensity has strengthened a small number of earlier attempts to place Israel in a comparative perspective (Lissak 1967, 1969-70) and shows a significant departure from many previous works that have tended to emphasize the uniqueness of the Israeli case. To provide but two examples, while Horowitz and Lissak (1989) contrast Israel and Britain as two Weberian ideal types of civil-military relations, in a series of articles Ben-Eliezer (1995b, 1997, 1998) compares the Israeli case with France, Japan, or the Soviet Union in order to examine its potential for praetorianism or a military coup.

Fourth, along with the diversification of subject matter there has been a growing pluralization of theoretical orientations. Thus, unlike the period lasting until the end of the 1970s in which structural-functionalism provided the central paradigm for research into "things military" in Israel, today it is no longer possible to point to one such dominant perspective. One development has been the introduction of much more sophisticated theoretical and analytical tools to tackle broad, macro-level problems. Thus, for instance, Maman and Lissak (1997) have used an advanced social network approach to map out and explain the relations between military and other elites in Israel. Yet theoretical pluralization has manifested itself primarily in numerous studies that examine the micro level, and especially macro-micro links. On the micro-level, the stress is now placed less on various therapeutic or remedial approaches to "battle fatigue" than on tackling more analytical issues. For example, Liebes and Blum-Kulka (1994), Helman (1993) or Linn (1996) have used very different theoretical frameworks to examine how individual soldiers handle participation in unjust wars —Lebanon and the Intifada. Similarly, Lieblich (1989; Lieblich and Perlow 1988) uses propositions from developmental psychology in order to consider military service as a transition to adulthood.

Fifth, and finally, all of these trends are reflected in the kinds of questions now being asked. Contemporary studies tend to problematize previous taken-for-granted scholarly and popular notions about the IDF and the kinds of actions it participates in, and to consider the Israeli-Arab conflict not as an exogenous factor but rather as constitutive of Israeli society and state (Ehrlich 1987). At the risk of oversimplification, we suggest that there are two central themes common to many of these newer approaches to "things military" in contemporary Israel.

The first is the role of war and the military in the constitution of membership in Israeli society and polity and in the social construction of collective and individual identities. For many years, the stress in scholarly writings was on the "integrative" role of the IDF vis-a-vis diverse groups like immigrants (Azarya and Kimmerling 1984; Lissak 1972). This kind of emphasis has continued in some publications proceeding from assertions about the continued importance of the non-military roles of the IDF (Ashkenazy 1994). Other studies, like Shabtay's (1995, 1996) analysis of Ethiopian immigrants, while proceeding from the same premise of the IDF as a social integrator, have developed a much more dynamic picture of military service. But the overwhelming stress in much of the newer scholarship is on how the Israeli army has been and still is used (via recruitment, assignation, and retention of personnel) as the central mechanism for constructing different levels of inclusion and exclusion into society (Levy 1996; Kimmerling 1992, forthcoming; Rosenhek 1999). It is in this light that studies of the relations between gender and the military (Jerbi 1997; Levy-Schreiber and Ben-Ari, forthcoming; Sasson-Levy, forthcoming; Yuval-Davis 1985), or the incorporation of "minorities" (Frisch 1993; Peled 1998) and religious groups (Cohen 1997), should be seen. All of these studies show how military service serves both as a mechanism for building a complex hierarchy of social groups and as an indicator of this hierarchy.

The second theme common to many recent works centers on the ways in which the "sense of existential threat" to Israel has been created and used by the state and its representatives to gain and maintain legitimacy, to define standards for the distribution of resources, to shape public culture and the life worlds of individuals, and to construct the very agenda of Israeli social sciences (Ehrlich 1987). Thus, for example, critiquing Ben-Ari's (1989) earlier work, Paine (1992) has shown how despite carrying out policing roles during the Intifada, soldiers justified their action in military terms touch-

ing upon the very survival of the state. In other words, in compre-
hending the Intifada as a "normal" or "natural" military situation,
their activities were linked to the notion that military actions by
Israeli soldiers are related to the ultimate Zionist text: to the safety
and security of the country. Or, to take another example, Ezrahi
(1997) has carefully examined the central narratives of Zionism—
centering on defense—that govern the ways in which most Israelis
understand wars, political action, and the definition of enemies. Weiss
(1997) examined the ways in which the Israeli state has consistently
propagated an "ideology of bereavement" to sustain national boundaries, an ethos of sacrifice (allowing the continued mobilization and
retention of soldiers), and collective identities. Finally, Feige (1998)
demonstrates how military service continues to be used as a basis
for the creation of legitimacy by different groups.

It is against the background of these trends in the scholarly study
of the military in Israel—numerical growth, comparison, diversification, pluralization and critique—that the present volume, and the
essays comprising it, should be seen. While commencing from the
"master" debate between the two macro-sociological approaches to
the study of the military—the "armed forces and society" and the
"state-making and war" approaches—its theoretical dimensions are
not exhausted by this controversy. All the essays in this volume suggest how additional theoretical perspectives, not specifically developed in regard to the military, may be fruitfully used in order to
uncover and explain hitherto little researched aspects of war and the
armed forces. Among the variety of analytical tools used by the contributors to this volume are ideas about the role of the armed forces
in Israel's gender regime (Izraeli), the ways in which life narratives
normalize war (Lomsky-Feder), or the place of body-practices and
emotions in the construction of masculinity through military service (Ben-Ari). The use of such theories and interpretive schemes
may well signal a welcome "re-linking" of scholarly studies of
"things military" to some of the wider debates in contemporary social and human sciences, and offer examples of the insights to be
gained by explicitly placing issues related to war and the military
within new theoretical perspectives.

Similarly, many contributions to the volume apply various theoretical innovations marking recent scholarship of the military to the
Israeli case. In this sense, our volume is part of the wider interroga-

tion of war and the military that is now taking place in Israeli society. This interrogation centers on the manner by which the internal practices and arrangements of the military and its external representations and perceptions are being transformed. In this regard, however, while there is a very broad agreement between the different contributors to this volume, as well as other observers (Cohen 1998), regarding the decline in the importance of the military in Israel and with respect to the changes it is undergoing, they differ radically in the kinds of interpretations they attach to these changes. It is here that the theoretical perspectives adopted in the various essays are important. They are important because they provide prisms for understanding the very character of contemporary Israel as evincing a basic continuity of militarization and militarism, or its radical transformation into a "normal" industrialized society. In this sense, given the historical and present centrality of the military in this society, Israel provides a fruitful case for examining the place of the armed forces and more generally of security considerations in the development and transformation of contemporary industrialized societies.

The Articles

The two articles in the volume's first section serve to situate the Israeli case, and cases drawn from Israel, in its wider theoretical and comparative context. They show that, notwithstanding important distinctive features, the questions raised by the Israeli case are similar in nature to those topics upon which Western scholarship is currently focused. Furthermore, they also indicate how the study of this specific case can contribute to the theoretical understanding of contemporary changes in civil-military relations.

In his essay "Western-Type Civil-Military Relations Revisited," Bernard Boëne carefully locates his inquiry in the context of a general question that has been at the heart of much thinking about the military in any regime: how are the armed forces to be controlled and supervised? His specific focus is on how the changing social and historical circumstances of the industrialized democracies are now raising an array of new issues about the oversight and guidance of the military by established civilian (essentially political) frameworks and roles. He begins his analysis through the delineation of two previous models (or Weberian "ideal types") of civil-military relations. The first model, which was developed by Samuel Huntington (1957) and which was apt for the armed forces of the pre-World War II period, posited a strict separation between the political and military spheres of societies. The second model developed by Janowitz (1971, 1976), and further elaborated by Moskos (1975, 1976), was more suitable to the military establishments of the post-War period and the cold war. It, by contrast, stresses the integration of the military into society, and the convergence between military and civilian careers in terms of motivation, skills, incomes and lifestyles.

The third model, actually an extension of many of Janowitz's assertions, sketches out the major societal and political changes that the industrialized democracies have undergone in the past two decades and their implications for civil-military relations. Boëne outlines a complex model of change on multiple levels: macro-level transformations (the end of the cold war and the disappearance of mass threats, the role of non-governmental actors in the international arena, and the weakening of national symbols, for example); changes within military establishments (the move to greater openness to external civilian sectors, participation in multinational forces, or more politically sensitive roles for commanders); and the special necessity of new education for military leaders (for example, higher degrees or openness to cultural pluralism). (See Shamir and Ben-Ari, forthcoming). In this new model, older versions of military professionalism—based on being a "caste apart" or on an occupational orientation—are replaced by a much more labile version of political astuteness. It is a very "political" model because it underscores the kinds of special negotiations, bargaining and cross-pressures between civilians and military people that are now a considerable part of civil-military ties. Along with much of contemporary scholarship (see the special issue of *Armed Forces and Society*, 1998), Boëne's stress is very much on the political machinery that needs to be in place in order to assure that democracy can survive against the background of these changed circumstances.

Yet his analysis bears wider import in other respects. Theoretically speaking, Boëne cautions us, we must be careful to turn from a conceptualization based on singular models to formulating our contentions in terms of a variety of relationships between the military and civilian sectors. He suggests that while Janowitz' ideas about a movement between the two ideal types of civil-military relations has value, his assertions can be further developed if we conceptualize the issue in terms of plural, diverse arrangements that may coexist at any one specific point of time. In this regard, he cautions us as to the changing circumstances of different democracies during different historical periods, and to the fact that democratic institutions are not all equally conducive to effective civilian control. Boëne steers us away from a simple reductionism of civil-military relations either to a Huntingtionian or Janowitzian model. Indeed, the emphasis in his third model on the open and processual links between the military and civilian sectors reinforces this contention because it suggests the need to think in terms of change and transformation and a multiplicity of arrangements.

In a refreshing manner, Boëne's essay makes explicit a point that undergirds many analyses rooted in the "armed forces and society" approach: namely, that a major thrust of this perspective is a normative, prescriptive commitment to social betterment through an explication of the alternative arrangements by which democracies can create sound relations with the military (also Boëne 1990). For example, in many analyses belonging to this school, the lack of subordination of the military to civilian rulers is seen as an aberration while their full subservience is seen as the preferred state (Ben-Ari 1997). In this view, the "rational" (and therefore "normal") state of affairs is one where civilian politicians rule—oversee and govern—military commanders. The idea here is that the prerogatives that the armed forces take in interfering in politics (such as interfering in the management of foreign affairs) or the politicization of the military (as the outcrop of unclear boundaries between the civilian sectors and the armed forces) are deleterious to the democracies they serve. It is in this light that Boëne's sophisticated philosophizing should be read. He is careful to stress that one of the aims of theoretical essays like his is to suggest the ways of reaching the right balance that will determine the proper amount of power and influence accrued to the military establishment if it is to discharge its proper function without distorting the regime it is supposed to serve.

James Burk's essay, "From Wars of Independence to Democratic Peace: Comparing the Cases of Israel and the United States," examines the potential of different democracies for peace. He concurs with earlier findings that democratic states will tend to be more peaceful in their international relations than alternative kinds of government, but seeks to turn his analysis inwards, to the potential of democracies for internal peace. In theoretical terms he links three sets of variables: states, state strength (i.e., legitimacy) and the absence (or presence) of intra-state violence. Yet rather than positing a simple correlation between democratic regimes and internal peace, he points to some crucial differences between democracies which have usually been assumed to belong to one unitary category marked by common features. Through a focus on the American and Israeli cases, he suggests that the more unitary the democracy the more likely it is that the state's rule will be challenged by subordinate or marginal groups (Israel). Conversely, the more pluralist the democracy, the more accepted and secure will be the state's right to rule (the United States). The key to the development of these two different kinds of democracies, however, is the substance of the revolutionary ideology that "fired" the conflict for independence.

Burk's thesis is that the founding of democratic states through revolutionary wars of independence affect the ways in which a state's legitimacy—the degree to which its right to rule over a political community is accepted or not—is constructed. In this respect Weber's theory of the routinization of charisma is crucial. Why? Burk's point is that citizen-soldiers participating in wars or revolution carry meanings that extend far beyond their technical roles on the battlefields: they carry the charisma of the revolution in the sense of willingness and ability to confront the extraordinary and dangerous realm of war in the name of the new order that they are trying to establish. What is important is how the warriors' charisma — concentrated as it is in the revolutionary movement — is allocated (either concentrated or dispersed) through the bed of new political structures once the war draws to a close. When charisma is dispersed (to the central state, to all levels of government, or the individual citizens) as in America, the foundations of a pluralist democracy are established. When charisma is highly concentrated as in Israel, and allocated only to certain citizens and the central state, then a unitary democracy is founded: the unitary democracy of the Jews and especially the dominant Ashkenazi (of European origin) Jews. Sociologically speaking, the point here is the degree to which the ideology is tied to preexisting cleavages. Consequently, unitary democracies will tend

to be the most violent because opposition between groups is codified by the ideology of the war and later the political institutions.
Unlike the American revolutionary war, the Israeli war for independence was not moderated by the belligerents' hope to win the allegiance of the civilian population, but was rather a total war between
two civilian populations.

Burk's essay sheds light on a "puzzle" that has intrigued many
scholars about Israel: the purported existence of democratic ground
rules against a background in which the military is a central, if not
the central, social actor. As we saw, Horowitz and Lissak (1989;
Horowitz 1977, 1982) attributed this potential for democratization
in the face of a militarized Israel to the dual processes of
civilianization of the military and militarization of the civilian sector in a way that limits the extreme development of either.
Kimmerling (1993), by contrast, contends that various democratic
arrangements are no more than a facade covering Israel's militarism. What Burk's analysis does is to show that when placed in a
comparative context, the case of Israel is not one of "just any other"
democracy. As he shows, the revolutionary basis of the Israeli state
differs from that of other democratic states and it is this base that is
related to the centrality of armed struggles within and outside the country. What he seems to suggest is a theoretical innovation for the "armed
forces and society" school which brings it closer to the "state-making
and war" approach. It is an innovation because, as he stresses, war
should not be theorized as something that is external to society, but
as something inherent to the manner by which states are established.

While Burk's emphasis is on intra-state violence, an intriguing
aspect of his analysis which seems crucial for Israel, and has not
been explored in great detail, is the link between internal democratic processes and the management of external relations. The contention in the literature is that democratic countries establish institutions of self-government to resolve domestic conflicts peacefully,
and it is these "habits" that are then carried over into and begin to
characterize the realm of international affairs. If this is true, then
developments in the internal democracy and democratic arrangements of Israel seem crucial for the ways it will pursue peace or war
in the future.

The articles in the second section deal with issues central to civil-
military relations in Israel and elsewhere: civilian control of the

armed forces, the extent to which the military actively participates in political life, and the potential threats to democracy this participation might imply. Concentrating on current changes in Israeli politics, the character of the conflict with the Palestinians, and the place of military in society, the three papers touch, albeit from different theoretical perspectives, on the question of whether Israel is a militaristic society. At the same time, in the current context of the search for a political solution to the Israeli-Palestinian conflict, the contributions are also concerned with the potential role of the military in the "peace process" by asking whether that participation acts as an obstacle to the political resolution of the conflict.

In his contribution on "Civil-Military Relations in Israel in Crisis," Yoram Peri places his analysis in the context of a wider thesis about how Israel is now undergoing a process of decolonization — i.e., a process by which it is extricating itself from the occupied territories where it has been suppressing another national population which is in revolt. This starting point harks back to Boëne's contribution for it centers on the threats to, or crises of, Israeli democracy brought about by decolonization. Like Boëne, Peri contends that it is impossible to understand developments in Israel's civil-military relations without taking into account wider socio-political transformations. He identifies three major levels or dimensions to these turning points (all foreshadowing later essays): an internal crisis within the IDF centered on questions of identity and purpose; a question focused on the implications of new relations between senior military commanders and the political echelon above them for a stable democratic regime; and a crisis in the relations between the military and much wider sectors of civilian society converging on demographic changes and the collapse of the security meta-narrative of Israeli society.

In recent decades Israel has been steadily transformed into a society marked by greater individualism, democracy, civilian considerations and consumerism (Birenbaum-Carmeli 1994). On the one hand, these developments signal a greater potential for democratization both in terms of increased public debates about the IDF and the cultivation of various "watchdogs" overseeing it. These trends have led not only to contractions in the role of the IDF and lessened the intrusion of the military into civilian sectors, but have also brought about a more pervasive civilian presence in military life: for example, greater participation of parents of soldiers in the military (Katriel 1991), increased intrusion of the judiciary and legal norms into the armed forces, and greater scrutiny of the IDF by the media (Wolfsfeld 1997). On the other hand, the move to both consumerist and individualistic values has spelt a decline in motivation among the upper-middle class, and the recruitment of only parts of this group to elite combat and technologically advanced units. While this trend is welcome for the opportunities it has created for the participation of new groups in the military (women, Jews from Islamic countries, and some new immigrants), the decline in the mobilization of the elites of Israeli society has profound implications for citizenship and participatory democracy. What we may be witnessing to now is the influx, into the military, of new groups who have less commitment to democratic values and ground rules.

Uri Ben-Eliezer's essay is entitled "From Military Role-Expansion to Difficulties in Peace-Making: The Israel Defense Forces 50 Years On." In theoretical terms, he starts from a critique of the premise underlying many of the other essays that belong to what he terms the "civil-military relations" or what we have called the "armed forces and society" approach. His criticism centers on two central assumptions at the center of this approach: one, that there is, and that there can be, a separation between the civilian and military sectors of a society; and two, that militarism is a pathology or aberration in the normal state of functioning democracies. Ben-Eliezer contends that these assumptions are problematic. In its Israeli guise the civil-military relations approach has focused on the celebrated notion of the permeable boundaries between the military and society and on the civilianization of the military and the militarization of the civilian sector. As we saw earlier, this formulation contends that the dual influence between the two sectors leads to a balance between them. The IDF's role expansion, according to this view, is a functional mechanism to assure civilianism via nation-building tasks like education, "absorption of immigrants" or settlement of borders.

Ben-Eliezer, in contrast, anchors his analysis in a different theoretical perspective from those coming before him. Rather that using a functional—if sophisticated—approach that centers on questions about the survivability of social systems or democratic polities, he (following such scholars as Tilly 1985 or Shaw 1988) seeks to understand how the military is related to state-building. Consequently, rather than focusing on the centrality of the military in Israeli society (and there is widespread agreement about this centrality between scholars) and its relation to democratization, Ben-Eliezer suggests that the role-expansion of the military is not connected with modernization or nation-building, but with militarism and war. He contends that militarization has been an essential part of the creation of the Israeli state. The army (and war—Levy 1996) has been used to mobilize a population marked by deep divisions and by large waves of immigrant groups. According to him, the exercise of organized violence has been, and continues to be, used to create legitimacy, i.e., a belief in the justice and fairness of the state.

It is out of this perspective that we can begin to see militarism not as power that some groups have (or do not have) over others, but rather as a broad cultural and political phenomenon. Militarism, according to this view, is a way of thinking or interpretation which posits war as an effective, legitimate and necessary solution to inter-state problems and struggles (Ben-Eliezer 1998; Feige 1998; Kimmerling 1993). Militarism, as Giddens (1985) points out, is a proclivity in some societies for a section of the higher echelons to look for military solutions to political conflicts, and the readiness of large parts of society to accept such solutions. Viewed in this manner, the penetrable boundaries between the civilian and military sectors of Israel that Lissak and Horowitz discussed take on a different light. These permeable boundaries have allowed the Israeli state to create and maintain its legitimacy in the eyes of large parts of society. Similarly, Bar-Or's example (in this volume) of Prime Minister Rabin's attempt to integrate the military into the peace process as a means to garner legitimacy becomes an indicator of the militarism characterizing Israeli society itself.

Ben-Eliezer's paper raises a wider set of questions: how are we to interpret the continued use of the military for legitimation of state actions at the same time as there is lowered motivation for combat service among some groups and as others question the centrality of the military in Israeli society? Does this situation signal a rear-guard action on the part of the ruling elites against further erosion in their importance and status? The characteristics of the IDF and the changes it is undergoing as depicted by Ben-Eliezer are not different from those charted out by Peri: the decline in cultural militarism, the beginnings of a demarcation between society and army, and indications of contentions in and around the army. But where he radically departs from Peri is in his interpretation of where the threats to Israeli democracy lie. While Peri sees these threats as the outcome of a rupture, as a fundamental change in civil-military relations in Israel, Ben-Eliezer sees these dangers as emanating from the continued role expansion of the IDF. He contends that the army's continued support of the right-wing settlers in the occupied territories—in setting up local militias or integrating them into existing units—is an extension of the IDF's capacity and a continuation of militaristic attitudes and therefor a threat to Israeli democracy.

Along the lines we suggested earlier, the debate between Peri and Ben-Eliezer can be also read as part of the wider cultural debate now taking place in Israel. They can be read not only for their theoretical contentions but also for the polemical stance that they represent. This polemic transcends disagreements in the scholarly study of the military, for it is also linked to basic controversies concerning the understanding of the historical roots and contemporary character of Israeli society and polity. This kind of reading must make us much more aware of the social situatedness of the volume as a whole and is an issue we shall return to at the end of this introductory essay.

Stuart Cohen's tract, "Dimensions of Tension between Religion and Military Service in Contemporary Israel," carefully juxtaposes both macro and micro levels of analysis. Part of a wider set of investigations into the relations between different social groups and the military, Cohen's essay is an attempt to grapple with an issue that has become very important in the last few years: the specific kinds of tensions raised by religion, specifically from the national-religious camp, for the military in Israel. Two sources of tension are important in this respect: the allegiances of religious soldiers to Rabbinical authorities or military commanders; and the heightened participation of religious youths in the elite units of the IDF (often taking over from traditionally elite groups like Kibbutzniks) and the "danger" that they will enforce Jewish orthodox customs on their fellow soldiers. The background to these apprehensions are the different values and life-styles of the religiously observant and secular groups in Israel.

While acknowledging the dangerous potentials stemming from the national-religious sector, Cohen shows how such worries are exaggerated: for example, this camp is itself rent by deep divisions and as of yet there have not been any refusals of military commanders in the name of a higher Rabbinical authority. In addition, he shows that the long-term prospects of the "take-over" of the IDF by religious commanders is not a realistic option at this stage because religious establishments are themselves "greedy institutions" that posit demands no less serious than the army in terms of investment in careering. To put this point by way of example, a young religious officer contemplating commitment to a military career faces a difficult choice in terms of strong demands to return to religious establishments of higher learning.

It is in a similar light that another point that Cohen makes should be seen. In a number of points in his essay, Cohen demonstrates how, despite popular and scholarly imaginings to the contrary, the national-religious camp is not homogenous. It is heterogeneous both in the kinds of general worldviews that characterize it, and in the different kinds and degrees of commitment to military service that are found among its youths. Thus Cohen is careful to talk about patterns (in the plural) of service among religiously observant youths. This point is important because it underscores how groups within this camp show different kinds of responsibilities to Israeli democracy and to the kinds of civil-military relations that such commitments entail.

The third section deals with broad processes of identity formation and their relation to individuals' participation in the military arena in general, and in war in particular. The four articles in the section focus on the interface between macro features of state-making through war and the centrality of military experience, and micro processes of constituting social categories like citizens, soldiers and men. In this manner, all four contributions indicate how war and the military are not only instruments for state-making, but are also important factors in the formation of individual identities. Showing how these categories are recreated, and sometimes challenged, by individuals and social groups interacting with state agencies, the articles raise important questions related to the limits and potentials of human agency vis-a-vis state structures.

Dafna Izraeli's "Paradoxes of Women's Service in the Israel Defense Forces" deals with an issue that has become central to debates about the Israeli army in the last decade or so. While scholars have long noted the centrality of military service for the construction of gender relations in Israel (Kimmerling 1993: Yuval-Davis 1985), Izraeli's contribution represents the first full-scale attempt to systematically theorize this topic. Her analytical framework centers on the relation between the military and Israel's gender regime: the gendered division of labor and a gendered structure of power that both formally and informally sustains the taken-for-granted role of women as helpmeet to men. Izraeli shows how the military as a structure of power intensifies gender distinctions and then uses them as justifications for sustaining gender inequality. Furthermore, she uncovers the ways in which military service figures in the construction of images that explain, express and reinforce gender divisions, and in the patterns of daily interactions that reproduce patterns of dominance and submission between men and women (see also Jerbi 1997; Sasson-Levy, forthcoming). Finally, Izraeli examines the ways in which men accumulate different kinds of social and symbolic capital that grant them advantages in civilian life. In other words, she explains how through military service, men can gain social capital or valued resources (status, prestige, professional education) which they can then convert to their advantage in the civilian sector (like political success or access to jobs).

Izraeli's contribution opens up new questions (some of which are dealt with by Ben-Ari, Lomsky-Feder and Helman in this section) about the diverse ways in which the taken-for-granted nature of the tie between the military and gender inequality is constructed or how ideas of what being a man and being a woman become naturalized, invisible, and homogenized notions as they relate to the military and to military service (Levy-Schreiber and Ben-Ari, forthcoming). This is an important question because the relationship between gender and the military is a critical component of Israeli nation-building, as gender is a prime site for the naturalization of power; for the processes by which meanings get entrenched, become "taken-for-granted," and come to seem natural—through symbols, everyday rituals, discourse and practice (Yanagisako and Delaney 1995).

But when placed in its wider historical context Izraeli's chapter raises other questions. Is it only coincidental that it is only now when the IDF is contracting and when its social significance is waning that more and more roles which were previously the exclusive domain of men are opening up to women? For example, women have been allowed into the Border Police as front-line soldiers, into pilots' and naval commanders' courses, and are now allowed to participate in combat missions as doctors. But this greater access is occurring at precisely the time that the army is becoming socially less significant. In other words, despite greater access to combat roles (or other restrictively male military occupations), the ability of women to "convert" resources from their military service to civilian life is still limited, precisely because such resource conversion is becoming limited for men as well. One can wonder whether women, in their attempts to gain better positions in the army hierarchy, are not fighting a yesterday's war.

Eyal Ben-Ari's contribution, "Tests of Soldierhood, Trials of Manhood: Military Service and Male Ideals in Israel," continues Izraeli's exploration of the relations between gender and military service to deal with the social construction of manhood. Ben-Ari examines how military service is related both to ideals of manhood in (Jewish) Israeli society and to the construction of concrete standards and arrangements for the constitution, achievement and inculcation of male identity. Rather than proceeding from a focus on the experience of women or the relations between women and men as Izraeli does, Ben-Ari specifically "problematizes" the unmarked categories of men and masculinity as part of a set of relational ideas and concrete arrangements. Masculinity, he points out, is not a fixed construct or thing against which to explore women and notions about womanhood, but rather its constant construction is to be explained as well.

Ben-Ari chooses to approach these themes by looking at military service as a rite of passage to manhood and adulthood in Jewish-Israeli society. Yet he cautions us to beware of drawing a direct parallel between service in the Israeli army (or in the army of any other industrialized society) and rites of passage in tribal societies. In both kinds of societies such rites are charged with moving individuals from one male status to another. But in Israel this passage is uncertain. It is uncertain because the power of rituals is diffused in large-scale and pluralistic societies as Israel where some segments of the population—ultra-religious Jews or Palestinian citizens of the country, for instance—are distanced from central values and myths. It is also difficult because ideals of manhood are inextricably linked to specific organizational frameworks set up and maintained by the Israeli state. The point Ben-Ari makes is that this rite of passage is part of, and subordinated to, the overall logic of the military. In other words, the IDF uses ideals of manhood to motivate many of its recruits to join and participate in its elite combat units. Furthermore, as he argues, notions about manhood and their connection to military service are also related to national, state mandated goals: specifically to the idea that (young) men should be willing to sacrifice themselves for the nation-state.

But the analyses of Izraeli and Ben-Ari can be read in a different manner to raise a question about the reasons for the broad acceptance of mainstream typifications of the warrior image and of the military: Why are the Israeli armed forces used as signifiers of the unity of the country *despite* persistent inequalities and differences in the experiences of Israelis? We would suggest that the role of military ideals is to incorporate, to create a sense of a shared universe of meaning precisely because of the actual disparities and diversity of Israeli society (in terms of gender, ethnic affiliation, material conditions). The salience of these images of soldiers or of the IDF is one of shared vocabulary, not necessarily shared experiences or common achievements. Like ideologies and institutions everywhere, those of Israel "normalize." It is the thrust of ideological representation both to generalize and to naturalize, to claim for specific interests a natural universality. Much of the social and cultural dynamics related to the military in Israel may thus be read for the way these images have entailed distinctions that have reproduced and legitimated social differences, exclusionary practices, and the continued mobilization of resources for the military. Along these lines, the discourse on the "solidarity of warriors" (*akhvat lochamim*) accomplishes two things: it creates a language common to soldiers in general and to men who participate in combat duties in particular; and it blurs the differences within them and between them and other groups. Herein lies the complexity of meanings conveyed by military service: it radiates disparate but simultaneous messages of likeness and difference, of equality and inequality.

These two contributions further goad us to think about the need to re-read much of what has been written about war and the military in "gender neutral" terms as actually being studies about men. Here we would mobilize the central feminist insight that "the knowledge that is related to the gender base need not necessarily be specifically signaled as gendered topics" (Morgan 1994: 33-4). In other words, a gendered sociology of knowledge would focus not only on such ("obvious") military issues as rites-of-passage, sexual imagery of combat, women's emotionality or men's aggression but also on such matters as approved modes of administrative work, technical competence, or ideas of fun and leisure. Thus, along with current feminist contentions, we posit that even though it does not explicitly formulate its focus as such, a vast body of writing and research about supposedly neutral military issues is actually about men and masculinities. Could we not benefit then from re-reading (in the feminist sense) classic debates found within the sociology of the military about such issues as professionalization, career and promotion paths, or modes of recruitment as implicit examinations of central ideals about manhood and about ways to achieve these ideals?

Edna Lomsky-Feder's selection, "The Meaning of War Through Veterans' Eyes: A Phenomenological Analysis of Life Stories," deals with how war is represented within the personal narratives of veterans of the 1973 Yom Kippur War. Lomsky-Feder demonstrates how individuals interpret intense national events like wars by weaving them into their own life stories: those collection of events and experiences that individual narrators choose (consciously and unconsciously) to present as their personal story along a time axis. The group she studied is the hegemonic one—graduates of elite schools in Tel Aviv, Haifa and Jerusalem—and thus her findings reflect the hegemonic normalization of war by the governing elites. She finds that the veterans she interviewed do not represent their experience of war as traumatic but rather as a normal (that is an ordinary, common) experience.

The paper claims that just as war has been institutionalized and normalized into the macro-social arrangements in Israeli society (Kimmerling 1984, 1985), so too individuals synthesize experiences of armed conflicts into their personal biographies. Israeli soldiers do not represent the experience of war as a traumatic one, but rather as a normal—"natural" and expected—aspect of their lives. The two primary mechanism by which war is normalized are the creation of a narrative continuity between a soldier's previous knowledge and the experience of warfare, and the development of a notion by which war is simultaneously part and outside of his life. The importance of life stories is thus related to their being the primary means which people use to interpret their lives in a coherent and persuasive manner.

Lomsky-Feder's conclusions consequently contradict two kinds of common perceptions. First, they challenge Israeli beliefs about the lasting disturbances that the Yom Kippur War of 1973 is supposed to have created in Israel and among veterans of that conflict. Second, her findings about the normalization of war directly contradict much of the accepted popular and (especially) academic wisdom about the effects of combat and warfare. Put somewhat simply, this view sees war as inevitably (and universally) traumatic on both the social and individual levels. Hence, because current research usually proceeds from an assumption of war-as-trauma, it tends to foreclose other kinds of questions. Thus the tendency in much of the scholarly literature (and especially among psychologists) is not to reason about the possibility that participation in armed conflict may lead to the mobilization of personal definitions by which war is interpreted as an expected—albeit intense—experience in the lives of individual soldiers.

Sarit Helman's contribution, "Citizenship Regime, Identity and Peace Protest in Israel," examines the social movements that have emerged to question the centrality of the military and of the use of armed force to solve many of Israel's central problems. Such "peace" movements, as they are known in Israel, were first established in wake of the war of 1973, but have been especially active since Israel's debacle in Lebanon and the Intifada. Helman's thesis is that despite the long-term activity of these movements, they have been very limited in their potential for widespread mobilization of social groups and categories for two basic reasons. They are limited first because of their continued ideological stress on militaristic symbols—such as continued loyalty to the IDF or the military record of many of their leaders—that cannot transcend the basic assumptions underlying Israel's security. These movements are limited secondly because their support is derived from (and appeals to) only a certain ethnic and class fragment—the middle-class of European origin—and the fact that women have been allowed to play only a marginal role within them (Helman and Rapoport 1997; Sasson-Levy 1995). On a discoursive level, she thus tackles a contradiction that underlies the activities of such movements: their continued adherence to soldiering and motherhood as mobilizing frames at the same time as these frames are precisely the issues the movements are protesting against.

Helman's analysis centers on the concept of "citizenship regime." This regime is that set of arrangements and definitions that does more than simply incorporate social categories into citizenship. The power of a citizenship regime lies in the manner by which the state often uses previous existing ties for forming citizenship *and* exclusion from citizenship. Moreover, and perhaps more importantly, these modes of inclusion and exclusion enacted by the state are constitutive of political identities and frame the kinds of contestations and claims of social groups vis-a-vis the state. The point here is that the state structures the very bases of its opposition both in ideological terms and the coalitions of groups mobilized for collective action against it. Substantively, her contentions are that the Israeli state has nationalized, ethnicized and genderized citizenship through the formation of a hierarchical structure of citizenship based on war-making and the establishment and maintenance of the military. In short, through state action, the fighter's role—the ultimate expression of civic virtue—came to be "naturally" associated with Ashkenazi-Jewish males whereas Oriental Jews and females in general were relegated to the periphery of civic virtue.

It is this hierarchy of groups, in turn, that has channeled the kinds of opposition—in our case, peace movements—that have been mounted against the state. Consequently, those groups constituted as the prime carriers of civic virtue increasingly mobilized the fighter's role to challenge the policies of the state, and as the prime justification for claiming participation in security policies (Feige 1995). However, as she shows through the examination of four "peace" movements, the turn to the fighter's role as a basis for mobilization could generate consensus against the system only among the groups constituted as the carriers of civic virtue. By appealing to the association between military symbols and citizenship, these movements in effect continued to marginalize women, antagonized significant parts of Oriental Jews and alienated the Palestinian citizens of Israel.

Helman's study raises crucial questions about the future trajectory of change in Israeli society. A central element here involves the complex potential for the emergence of sustained protest against, and outside of, the state. On the one hand, it is true that in contemporary Israel, the middle and upper-middle classes are slowly moving towards creating a civil society. Yet, on the other hand, other groups—notably national-religious Jews and many Oriental Jews—are now advancing towards the state and its missions. Thus, Israeli militarism continues to be carried by groups that stressed the continued affinity between the military and citizenship and the persistence of graded hierarchies of inclusion and exclusion based on this affinity.

The fourth section refers to the social foundations of the notion of "national security" and their implications for Israel's security and foreign policies. The papers show how relations between institutions and "readings" of Israel's geo-strategic circumstances affect the contexts in which policies are formulated and implemented and their actual content. One basic question raised by the articles in this section centers on whether the institutional mechanisms and the strategic conceptions crystallized during the first 50 years of Israel's existence are still relevant in a changing post-cold war world.

Amir Bar-Or's contribution, "The Link Between the Government and the IDF During Israel's First 50 Years: The Shifting Role of the Defense Minister," deals with an issue central to questions about civilian control of the military. If Peri tends to look at the wider characteristics of Israel's society and polity, Bar-Or deals with a much more focused dimension of civil-military relations, that of interest groups and the nitty-gritty of power struggles. In his contribution Bar-Or examines how these micro-relations structure the ways in which strategic choices are formulated and decided upon. His contention is that a prime "site" for examining this aspect of civil-military relations in democratic societies are the interpersonal relationships between the Prime Minister, Defense Minister and the Chief of the General Staff. Of special importance in this respect is the role of the Defense Minister who acts—to use a term borrowed from organizational science—as a boundary-spanning role between the military and civilian (especially political) sectors. It is especially he (there have only been men in this position) that embodies the porous boundaries between these two sectors. Bar-Or traces the diverse historical contingencies underlying the actions of people who have acted in these capacities and the unsettled, often highly contentious processes by which military and security policies have developed.

To be sure, Bar-Or is well aware of the wider institutional context within which such relations are embedded: the Prime Minister's Office, the National Security Advisory Staff within the Ministry of Defense and the General Staff respectively. But his analysis underscores how the concrete, micro-level relations between the elites peopling these positions can influence the shaping of policy. His thesis is that despite any formal stipulations written into ordinances or laws, the relations between these three actors are marked by a great deal of ambiguity and room for maneuverability, and are determined by the personalities, experience in military matters, and political acumen that they have. In this way, Bar-Or adds a crucial dimension to this volume. While most other contributors stress systemic factors, he stresses the place of individual decisions and the ability of individuals to effect the course of events. Similarly, while he does not theorize the issue as such, his analysis focuses on the importance of individual agency and change and not on how the Israeli regime reproduces itself and its assumptions. Good examples of this role are Ariel Sharon's machinations in and around his role as Minister of Defense during the war in Lebanon, or (as alluded to earlier) Rabin's use of the military in legitimating the peace process.

Like Boëne, so Bar-Or provides an explicit normative or prescriptive view of the kinds of preconditions necessary for the full functioning of a democracy in Israel (Burk 1998; Desch 1998; Feaver 1996). One example he provides is the disaster of the "Grapes of Wrath" operation under then Prime Minister Peres. Bar-Or suggests that the presence of an independent National Security Advisory Staff capable of offering different frames, views and interpretations of intelligence reports and offering policy alternatives, could have conceivably not let this happen. But he also shows how the last failure of setting up such a framework—the latest in a long chain of aborted intents—resulted from the struggles between Prime Minister Netanyahu and Defense Minister Mordechai over their respective jurisdiction over security issues.

In his essay on "A New Concept of National Security Applied on Israel," Henning Sørensen continues a line of research applied by Moshe Lissak and Dan Horowitz to Israel (Horowitz and Lissak 1996;

Lissak 1995). Sørensen's essay is written with the explicit aim of generating discussion about the long-term prospects for peaceful resolution to the conflicts within which Israel is involved. Concretely, Sørensen explains how Israel's strategic options are limited by its view of national security and the ways in which its leaders define its external and internal security threats. In this sense, he places the case of Israel in a global context: in terms of how contemporary global contexts of armed struggles are changing. The industrial societies of today, according to him, are characterized by growing demands for legitimizing the use of violence on the part of the armed forces, calls for taking preventative (rather than offensive) military measures, and a continued need for states to correctly identify their security position so that they can prepare and act according to these changing circumstances. In order to delineate Israel's conception of national security, he sketches out a model based on a number of key variables: the position of a country vis-a-vis others in terms of enemies or allies; the threat perception of a nation which undergirds its military build-up; the time of intervention in conflict; and the goal of military involvement.

Written as it is by an outsider, Sørensen's contribution seems to have put his finger on what may be termed the essential conservatism of many policy makers in Israel. Instead of recognizing that threats to the country's existence have been reduced (terror is not an existential threat), and that the IDF can (arguably) curtail military expenditures, downsize the military, and remove conscription, the leaders of the country continue to act as a nation without allies and to rely on a heavy military buildup based on the view that only sufficient military power can eliminate or reduce attacks. These leaders tend, he concludes, to see security as a zero-sum game, where Israel's increased security implies decreased security for other countries. Along these lines, he finds that Israel is characterized by a gap. On the one hand, because of developments in Israel's international position, it is now less threatened from the outside and consequently has more room for maneuvering in foreign policy. On the other hand, however, a close examination of Israel reveals that it continues to be characterized by the threat perception of a nation with major enemies. It is this gap which stands in the way of the continuation of the peace process and a major change in the manner by which it conceives of terror organizations.

The final section signals a shift of focus from the previous papers. While most of the previous contributions center on the relations between the IDF and Israeli society and polity, the articles in this section focus on lines of continuity and trends of change in several aspects of the IDF's internal organizational structure. Yet these trends are analyzed as connected to wider social and political contexts. Reuven Gal's presentation, "The Israeli Defense Forces (IDF): A Conservative or an Adaptive Organization?" examines the internal reactive capacities and limits of the IDF vis-à-vis its external environment in terms of threat assessment, force structure, and training and education. The question Gal asks derives from placing the IDF in the context of changes in the military establishments of the industrialized countries of the world: why has the IDF, purportedly the epitome of a successful army, not adapted to its changed circumstances? In theoretical terms, his chapter represents an attempt to analyze the internal modes of thought of the IDF that characterize it as a conservative organization. His conclusion is that the IDF is characterized by conservative innovation. Gal's thesis is that the combination of a proclivity for tactical innovation and the constant pressure of demands for immediate reaction has led to a situation in which it is very difficult for the senior commanders to reflect on, to distance themselves from, their paradigms of thought.

Gal's paper raises an important question relevant to any military establishment, that of organizational learning. A commonly heard cliche states that militaries always prepare for the last war they have fought. Given the budgetary constraints and changed threat environments that armies around the world are facing, and given that these are very large and complex organizations, questions about their capacity for willed, intentional change seems especially important.

Louis Roniger's "Organizational Complexity, Trust and Deceit in the Israeli Air Force" develops some of the themes Gal raises in a different direction. The paper deals with the question of how the military ethos is translated into the concrete arrangements and culture of an organization. Through an examination of the uproar over a case of graft-taking by an Air Force general, Roniger uncovers a number of major assumptions about the ethos and basic codes of the IDF and especially its commanders. Hence, the "deviance" of this case underscores how the normal and normative expectations of the IDF were perceived to be deeply defiled. The moral condemnation this man received—he was called an "out-cast," "dirt," "betrayer," "traitor"—attest as much to the Air Force leaders' attempts to dissociate themselves from him as to the still relatively high level of trust granted to the IDF (and the shock of having this trust violated). Such condemnation can only be understood against the background of the still relatively strong notion of trust—beliefs in the goodness, ability, strength, honesty, and reliability—of the IDF (Meisels 1995; Lomsky-Feder 1998). Indeed, the fact that this general was promoted out of the administrative arm of the forces—that is, he was not an ex-pilot, a warrior—was seen as part of the reasons given for his (mis)conduct.

Throughout the paper Roniger indicates the importance of a cadre of "carriers" of the symbolic importance of the IDF: those military elites that figure as leaders not only in the instrumental sense of guiding the organization, but no less importantly in creating and recreating its public imagery. These informal circles are based on shared military experiences spanning recruitment, combat and joint duties (Maman and Lissak 1995; Lissak and Maman 1996). Like similar networks found among graduates of the *Palmach*, the British Army, the Paratroopers or the Armored Corps, the importance of these groups reverberates outward from the IDF to other realms of Israeli society. The point Roniger seems to be making is that, to follow Edelman (1976), these commanders are (beyond their formally assigned roles) also symbols of inspiration and reassurance. The military elite in Israel, then, concurrently symbolizes the defense of the nation (through remembrances of past glories and threats) and a variety of sentiments about professional standards, ethics, honesty, and non-partisanship. The commanders of the Air Force seem to "carry" these messages most fittingly: at the same time as they figure in the fame of the 1967 war and the heroic survival of the 1973 war, they are deemed leaders of the most professional, ordered, and technically advanced arm of the IDF.

Yet Roniger's composition documents something else: a continued questioning of previously undisputed assumptions held by state authorities and the majority of the population about military qualities and behavior. Many groups in contemporary Israeli society are no longer willing to grant the IDF its previous status of unquestioned professionalism and to accord "state security" considerations as the only (or primary) criteria for national decision-making.

Roniger's case should be seen as part of a much wider interrogation of the necessity of war and the military that is now taking place in Israeli society. We are now witness to reflections about such matters as the role of women in the army and in combat, the links between families and military authorities, the reasons for and handling of suicides by army authorities, or the leeway left to families to determine the symbols used in graves and military memorials.

The implications of Roniger's contribution lead us towards Moshe Lissak's epilogue to this volume, "Uniqueness and Normalization in Military-Government Relations in Israel." In this concluding article Lissak attempts to trace out the peculiarities of the Israeli case in terms of the ties that bind the military to various social and political entities. As he sees it, Israel is undergoing a process of "normalization." That is, the heavy stress on security—the "religion of security"—is gradually weakening and the country is beginning to take on similarities to other industrialized democracies.

Situating Our Volume

As we stated at the beginning of this essay, one of the contributions of this volume is related to what may be termed its "social situatedness," that is, its place in wider polemics about the character of contemporary Israel. During the past decade or so, a host of social scientific and humanistic studies have begun to critically explore mainstream Zionist assumptions at the base of many previous studies of Israel. These newer approaches—developed by scholars sometimes dubbed "the new historians" or "new sociologists"—have uncovered such biases at the base of studies of Israel as the equation of Israeli society with Jewish-Israeli society (Kimmerling 1992) and the concomitant neglect of Palestinian citizens (Rosenhek, 1998), or the stress on the "integration" of Oriental-Jewish new immigrants from the perspective of the dominant (European) groups (Ram 1995).

From the perspective of this volume the central question that these scholars have raised centers on, to follow Kimmerling (1993; see also Peri 1996), whether Israel is a militaristic society. Kimmerling's suggestions that Israel is characterized by a "civilian militarism"—i.e., by a stress on viewing the "world" through considerations of security and the resolution of inter-state problems through the use of military force—is questioned by Moshe Lissak in the epilogue to this volume. One perspective, which derives from the "armed forces and society" approach, contends that the militaristic turn in Israel is only evident in the past few years. The implication of this analysis is that while the civilian influences on the IDF were strong in the past, they are now under serious threat. The "state-making and war" approach, by contrast, contends that in reality Israel has always been a militaristic society and that it is only through the use of certain theoretical lenses that this is apparent. The contrast, then, is between a thesis of rupture in Israeli society in which the essential "civilianization" of the military is eroding and a contention about the continuity of a very basic militaristic worldview.

The essays in this volume extend and develop this debate and the diverging manner by which developments in the relations between war, state, the military and society in Israel are to be understood. To put this contrast by way of a number of questions: can we say that from the perspective of the "armed forces and society" school we are witness to the destruction of the clear institutional separation between the military and civilian elites and the potential for the "de-democratization" of Israel? From the perspective of the "state-making and war" approach, do changes in and around the IDF represent a potential for praetorianism, i.e., a military coup (Ben-Eliezer 1998), or (alternatively) the withering away of the state? Or, in what is an alternative to such questions, could the Israeli case represent the development of a new kind of military establishment and a novel model of its relations with external social and political groups?

Note

*   We wish to thank Noa Vaisman for her help in the research which led to this introduction, and to Edna Lomski-Feder for her comments on an earlier draft of the essay.

References

Ashkenazy, Daniella (ed.) 1994. *The Military in the Service of Society and Democracy: The Challenge of the Dual-Role Military.* Westport, CT: Greenwood Press.

Aronoff, Myron 1989. *Israeli Visions and Divisions: Cultural Change and Political Conflict.* New Brunswick, NJ: Transaction.

Azarya, Victor and Baruch Kimmerling 1984. "New Immigrants as a Special Group in the Israeli Armed Forces." In *Israeli Society and its Defense Establishment,* edited by Moshe Lissak, 128-48. London: Frank Cass.

Bar-Gal, David 1982. *Social Work as a Professional Career: Renewal vs. Erosion.* Paul Baerwald School of Social Work. The Hebrew University of Jerusalem. (Hebrew).

Ben-Ari, Eyal 1998. *Mastering Soldiers: Conflict, Emotions and the Enemy in an Israeli Military Unit.* Oxford: Berghahn Books.

Ben-Ari, Eyal 1997. "Review Essay: Samurai, Violence and State-Making: The Development of the Military and Militarism in Japan." *Bulletin of Concerned Asian Scientists* 29,2: 61-72.

Ben-Ari, Eyal 1989. "Masks and Soldiering: The Israeli Army and the Palestinian Uprising." *Cultural Anthropology* 4, 4: 372-89.

Ben-Ari, Eyal and Yoram Bilu 1997. "Introduction." In *Grasping Land: Space and Place in Contemporary Israeli Discourse and Experience,* edited by Eyal Ben-Ari and Yoram Bilu, 1-24. Albany: State University of New York Press.

Ben-Ari, Eyal and Edna Lomsky-Feder 1999. "Introductory Essay: Cultural Constructions of War and the Military in Israel." In *The Military and Militarism in Israeli Society,* edited by Edna Lomsky-Feder and Eyal Ben-Ari, 1-34. Albany: State University of New York Press.

Ben-Eliezer, Uri 1998. "Is a Military Coup Possible in Israel? Israel and French Algeria in Comparative Historical-Sociological Perspective." *Theory and Society* 27: 311-349.

Ben-Eliezer, Uri 1997. "Rethinking the Civil-Military Relations Paradigm: The Inverse Relation Between Militarism and Praetorianism Through the Example of Israel." *Comparative Political Studies* 30, 3: 356-74.

Ben-Eliezer, Uri 1995a. *The Emergence of Israeli Militarism 1936-1956.* Tel Aviv: Dvir. (Hebrew).

Ben-Eliezer, Uri 1995b. "A Nation-in-Arms: State, Nation, and Militarism in Israel's First Years." *Comparative Studies in Society and History* 37, 2: 264-85.

Ben-Yehuda, Nachman 1999. "The Masada Mythical Narrative and the Israeli Army." In The Military and Militarism in Israeli Society, edited by Edna Lomsky-Feder and Eyal Ben-Ari, 57-88. Albany: State University of New York Press.

Ben-Yehuda, Nachman 1997. "The Dominance of the External: Israeli Sociology." *Contemporary Sociology* 26, 3: 271-5.

Birenbaum-Carmeli, Dafna 1994. *A Good Place in the Middle: A Residential Area as a Means of Constructing Class Identity.* Ph.D. dissertation, Department of Sociology and Anthropology, Hebrew University of Jerusalem.

Boëne, Bernard 1990. "How Unique Should the Military Be? A Review of Representative Literature and Outline of Synthetic Formulation." *European Journal of Sociology* 31,1: 3-59.

Breznitz, S. (ed.) 1983. *Stress in Israel.* New York: Van Nostrand.

Burk James 1998. "The Logic of Crisis and Civil Military Relations Theory: A Comment on Desch, Feaver and Dauber." *Armed Forces and Society* 24, 3: 455-62.

Burk, James 1995. "Citizenship Status and Military Service: The Quest for Inclusion by Minorities and Conscientious Objectors." *Armed Forces and Society* 21, 4: 503-29.

Cohen, Eliot 1998. "An Appropriate Intellectual Challenge." *Haaretz,* September 20. (Hebrew).

Cohen, Stuart 1997. *The Scroll or the Sword? Dilemmas of Religion and Military Service in Israel.* Amsterdam: Harwood Academic.

Cohen, Stuart 1995. "The Peace Process and Its Impact on the Development of a "Slimmer and Smarter" Israel Defence Force." *Israeli* Affairs 1, 4: 1-21.

Desch, Michael C. 1998. "Soldiers, States and Structures: The End of the Cold War and Weakening of the U.S. Civilian Control." *Armed Forces and Society* 24, 3: 389-406.

Dominguez, Virginia R. 1989. *People as Subject, People as Object: Selfhood and Peoplehood in Contemporary Israel*. Madison: University of Wisconsin Press.

Don-Yehiya, Eliezer 1988. "Festivals and Political Culture: Independence Day Celebrations." *The Jerusalem Quarterly* 45: 61-84.

Edelman, Murray 1976. *The Symbolic Uses of Politics*. Urbana: University of Illinois Press.

Ehrlich, Avishai 1987. "Israel: Conflict, War and Social Change." In *The Sociology of War and Peace*, edited by Colin Creighton and Martin Shaw, 121-42. London: Macmillan.

Enoch, Yael and Abraham Yogev 1989. "Military-University Encounters and the Educational Plans of Israeli Officers." *Armed Forces and Society* 15, 3: 449-62.

Etzioni-Halevy, Eva 1996. "Civil-Military Relations and Democracy: The Case of the Military-Political Elites' Connection in Israel." *Armed Forces and Society* 22, 3: 401-17.

Ezrahi, Yaron 1997. *Rubber Bullets: Power and Conscience in Modern Israel*. Berkeley: University of California Press.

Feaver, Peter D. 1996. "The Civil-Military Problematique: Huntington, Janowitz and the Question of Civilian Control." *Armed Forces and Society* 23, 2: 149-78.

Feige, Michael 1998. "Peace Now and the Legitimation Crisis of "Civil Militarism." *Israel Studies* 3, 1: 85-111.

Feige, Michael 1995. *Social Movements, Hegemony, and Political Myth: A Comparative Study of Gush Emunim and Peace Now Ideologies*. Ph.D. dissertation, Department of Sociology and Anthropology, Hebrew University of Jerusalem.

Frisch, Hillel 1993. "The Druze Minority in the Israeli Military: Traditionalizing an Ethnic Police Role." *Armed Forces and Society* 20, 1: 51-68.

Gal, Reuven 1986. *A Portrait of the Israeli Soldier*. New York: Greenwood Press.

Giddens, Anthony 1985. *The Nation-State and Violence*. Cambridge: Polity

Greenbaum, Charles W. 1979. "The Small Group Under the Gun: Uses of Small Groups in Battle Conditions." *Journal of Applied Behavioral Science* 15: 392-405.

Gur-Ze'ev, Ilan 1997. "Total Quality Management and Power/Knowledge Dialectic in the Israeli Army." *Journal of Thought* (Spring): 9-36.

Handelman, Don and Elihu Katz 1995. "State Ceremonies of Israel: Remembrance Day and Independence Day." In *Israeli Judaism: The Sociology of Religion in Israel*, edited by Shlomo Deshen, Charles Liebman and Moshe Shokeid, 75-85. New Brunswick, NJ: Transaction.

Handelman, Don and Lea Shamgar-Handelman 1997. "The Presence of Absence: The Memorialism of National Death in Israel." In *Grasping Land: Space and Place in Contemporary Israeli Discourse and Experience*, edited by Yoram Bilu and Eyal Ben-Ari. Albany: State University of New York Press.

Helman, Sara 1993: *Conscientious Objection to Military Service as an Attempt to Redefine the Content of Citizenship*. Ph.D. dissertation, Department of Sociology and Anthropology, Hebrew University of Jerusalem.

Helman, Sara and Tamar Rapoport 1997. "Women in Black: Challenging Israel's Gender and Socio-Political Order." *British Journal of Sociology*, 48, 4: 681-700.

Horowitz, Dan 1982. "The Israel Defence Forces: A Civilianized Military in a Partially Militarized Society." In *Soldiers, Peasants and Bureaucrats*, edited by Roman Kolkowitz and Andrzej Korbonski, 77-105. London: Allen and Unwin.

Horowitz, Dan 1977. "Is Israel a Garrison State?" *The Jerusalem Quarterly* 4: 58-65.

Horowitz, Dan and Baruch Kimmerling 1974. "Some Social Implications of Military Service and the Reserves System in Israel." *Archives European de Sociologie* 15: 262-76.

Horowitz, Dan and Moshe Lissak 1996 [1988]. "Democracy and National Security in a Protracted Conflict." In *Democracy and National Conflict in Israel*, edited by Benjamin Neuberger and Ilan Ben-Ami, 73-113. Tel Aviv: The Open University. (Hebrew).

Horowitz, Dan and Moshe Lissak 1989. *Trouble in Utopia: The Overburdened Polity of Israel*. Albany: SUNY Press.

Huntington, Samuel P. 1957. *The Soldier and the State*. Cambridge, MA: Harvard University Press.

Janowitz, Morris. 1976. "Military Institutions and Citizenship in Western Societies." *Armed Forces and Society* 2, 2: 185-203.

Janowitz, Morris 1971. *The Professional Soldier: A Social and Political Portrait*. New York: Free Press.

Jerbi, Iris 1997. *The Double Price: The Status of Women in Israeli Society and Women's Service in the Military*. Tel Aviv: Ramot. (Hebrew).

Katriel, Tamar 1991. *Communal Webs: Communication and Culture in Contemporary Israel*. Albany: State University of New York Press.

Kimmerling, Baruch Forthcoming. "Political Subcultures and Civilian Militarism in a Settler-Immigrant Society." In *Concerned With Security: Learning From Israel's Experience*, edited by Daniel Bar-Tal, David Jacobson and Aharon Kleiman. Stamford, CT: JAI Press.

Kimmerling, Baruch 1993. "Patterns of Militarism in Israel." *European Journal of Sociology* 34: 196-223.

Kimmerling, Baruch 1992. "Sociology, Ideology, and Nation-Building: The Palestinians and Their Meaning in Israeli Sociology." *American Sociological Review* 57: 446-60.

Kimmerling, Baruch 1985. *The Interrupted System*. New Brunswick, NJ: Transaction Books.

Kimmerling, Baruch 1984. "Making Conflict a Routine: Cumulative Effects of the Arab-Jewish Conflict Upon Israeli Society." In *Israeli Society and its Defense Establishment*, edited by Moshe Lissak, 13-45. London: Frank Cass.

Kleiman, Aharon 1985. *Israel's Global Reach: Arms Sales as Diplomacy*. Washington, DC: Pergamon-Brassey's.

Kleiman, Aharon and Reuven Pedatzur 1991. *Rearming Israel: Defense Procurement Through the 1990s*. Boulder, CO: Westview.

Levy, Amichai, Zahava Solomon, Michel Grank, and Moshe Kotler 1993. *Combat Reactions in Israel's Wars: 1948-1982*. Tel-Aviv: IDF, Chief Medical Officers Office. (Hebrew).

Levy, Yagil 1996. "War Policy, Interethnic Relations, and the Internal Expansion of the State: Israel 1948-1956." *Teoria VeBikoret* 8: 203-24. (Hebrew).

Levy-Schreiber, Edna and Eyal Ben-Ari forthcoming. "Body-Building, Character-Building and Nation-Building: Gender and Military Service in Israel." *Studies in Contemporary Judaism*.

Liebes, Tamar and Shoshana Blum-Kulka 1994. "Managing a Moral Dilemma: Israeli Soldiers in the Intifada." *Armed Forces and Society* 27, 1: 45-68.

Lieblich, Amia 1989. *Transition to Adulthood Military Service: The Israeli Case*. Albany: State University of New York Press.

Lieblich, Amia and Meir Perlow 1988. "Transition to Adulthood during Military Service." *The Jerusalem Quarterly* 47: 40-76.

Linn, Ruth 1996. *Conscience at War: The Israeli Soldier as Moral Critic.* Albany: State University of New York Press.

Lissak, Moshe 1996. "'Critical' Sociology and 'Establishment' Sociology in the Israeli Academic Community: Ideological Struggle or Academic Discourse?" *Israel Studies* 1, 1: 247-94.

Lissak, Moshe 1995. "The Civilian Component of Israel's Security Doctrine: The Evolution of Civil-Military Relations in the First Decade." In *Israel: The First Decade of Integration*, edited by Ilan Troen and Noah Lucas, 575-91. New York: State University of New York Press.

Lissak, Moshe, 1993. "Civilian Components in National Security Doctrine." In *National Security and Democracy in Israel*, edited by Avner Yaniv, 55-80. Boulder, CO.: Lynne Rienner.

Lissak, Moshe (ed.) 1984. *The Israeli Society and Its Defence Establishment.* London: Frank Cass.

Lissak, Moshe 1983. "Paradoxes of Israeli Civil-Military Relations: An Introduction." *The Journal of Strategic Studies* 6, 3: 1-12.

Lissak, Moshe 1972. "The Israeli Defence Forces as an Agent of Socialization and Education: A Research in Role Expansion in a Democratic Society." In *The Perceived Role of The Military*, edited by M.R. van Gils, 325-40. Rotterdam: Rotterdam University Press.

Lissak, Moshe 1969-70. "Center and Periphery in Developing Countries and Prototypes of Military Elites." *Studies in Comparative International Development* 5, 7: 139-50.

Lissak, Moshe 1967. "Modernization and Role-Expansion of the Military in Developing Countries: A Comparative Analysis." *Comparative Studies in Society and History* 9, 3: 233-55.

Lissak, Moshe and Daniel Maman 1996. "Israel." In *The Political Role of the Military*, edited by Constantine Panopoulos and Cynthia Watson, 223-33. Westport, CT: Greenwood Press.

Lomsky-Feder, Edna 1998. *As if There Was No War: The Life Stories of Israeli Men.* Jerusalem: Magnes. (Hebrew).

Lomsky-Feder, Edna and Eyal Ben-Ari 1999. "From 'The People in Uniform' to 'Different Uniforms for the People': Professionalism, Diversity and the Israel Defence Forces." In *Managing Diversity in the Armed Forces*, edited by Joseph Soeters and Jan van der Meulen. Tilburg: Tilburg University Press.

Lomsky-Feder, Edna and Eyal Ben-Ari (eds.) 1999. *The Military and Militarism in Israeli Society.* Albany: State University of New York Press.

Lustick, Ian 1988. "The Voice of a Sociologist, The Task of a Historian: The Limits of a Paradigm." In Books on Israel, edited by Ian Lustick, 9-16. Albany: State University of New York Press.

Maman, Daniel and Moshe Lissak 1997. "Military-Civilian Elite Networks in Israel: A Case in Boundary Structure." In A *Restless Mind: Essays in Honor of Amos Perlmutter*, edited by Benjamin Frankel, 49-79. London: Frank Cass.

Maman, Daniel and Moshe Lissak 1995. "The Impact of Social Networks on the Occupational Patterns of Retired Officers: The Case of Israel." *Forum International* 9: 279-308.

Meisels, Ofra 1995. "An Army in the Process of Liberation." Paper presented at the conference: 'An Army in Light of History: The IDF and Israeli Society.' Hebrew University of Jerusalem. June. (Hebrew).

Mintz, Alex 1985. "Military-Industrial Linkages in Israel." *Armed Forces and Society* 12, 1: 9-27.

Morgan, David H.J. 1994. "Theater of War: Combat, the Military and Masculinities." In *Theorizing Masculinities*, edited by Harry Brod and Michael Kaufman, 165-82. Thousand Oaks, CA: Sage.

Moses, Rafael, David Bargal, Joseph Calev, Avner Falk, Hai HaLevi, Yacob Lerner, Mili Mass, Shabtai Noy, Batia Perla, Meir Winokur 1976. "A Rear Unit for the Treatment of Combat Reactions in the Wake of the Yom Kippur War." *Psychiatry* 39: 153-62.

Moskos, Charles 1976. "The Military." *Annual Review of Sociology*, 2: 225-77.

Moskos, Charles 1975. "The American Combat Soldier in Vietnam." *Journal of Social Issues* 31, 4: 25-37.

Paine, Robert 1992. "Anthropology Beyond the Routine: Cultural Alternatives for the Handling of the Unexpected." *International Journal of Moral and Social Studies* 7, 3: 183-203.

Peled, Alon 1998. *A Question of Loyalty: Military Manpower Policy in Multiethnic States*. Ithaca, NY: Cornell University Press.

Peri, Yoram 1996. "Is Israeli Society Really Militaristic?" *Zemanim* 56, 14: 94-113. (Hebrew).

Peri, Yoram 1993. "The Arab-Israeli Conflict and Israeli Democracy." In *Israeli Democracy Under Stress*, edited by Ehud Sprinzak and Larry Diamond, 343-57. Boulder, CO: Lynne Rienner.

Peri, Yoram 1985. "Civilian Control During a Protracted War." In *Politics and Society in Israel*, edited by Ernest Krausz, 362-84. Albany: State University of New York Press.

Peri, Yoram 1981. "Political-Military Partnership in Israel." *International Political Science Review* 2, 3: 303-15.

Peri, Yoram 1977. "Ideological Portrait of the Israeli Military Elite." *The Jerusalem Quarterly* 3: 29-39.

Peri, Yoram and Moshe Lissak 1976. "Retired Officers in Israel and the Emergence of a New Elite." In *The Military and the Problem of Legitimacy*, edited by Gwyn Harries-Jenkins and Jacques van Doorn, 175-92. Beverly Hills, CA: Sage.

Perlmutter, Amos 1968. "The Israeli Army in Politics: The Persistence of the Civilian over the Military." *World Politics* 20: 606-43.

Ram, Uri 1995. *The Changing Agenda of Israeli Sociology*. Albany: State University of New York Press.

Ram, Uri 1989. "Civic Discourse in Israeli Sociological Thought." *International Journal of Politics, Culture and Society* 3, 2: 255-72.

Rosenhek, Zeev 1999. "The Exclusionary Logic of the Welfare State: Palestinian Citizens in the Israeli Welfare State." *International Sociology*, 4, 2: 195-215.

Rosenhek, Zeev 1998. "New Developments in the Sociology of Palestinian Citizens of Israel: An Analytical Review." *Ethnic and Racial Studies* 21, 3: 558-78.

Sasson-Levy, Orna forthcoming. "Subversion within Oppression: Constituting Gender Identities among Female Soldiers in 'Male' Roles." In *Hear My Voice: Representations of Women in Israeli Culture*, edited by Yael Atzmon. Tel Aviv: Hakkibutz HaMeuchad. (Hebrew).

Sasson-Levy, Orna 1995. *Radical Rhetoric, Conformist Practices: Theory and Praxis in an Israeli Movement*. Shaine Working Papers No. 1. Department of Sociology and Anthropology. Hebrew University of Jerusalem.

Shabtay, Malka 1999. *Best Brother: The Identity Journey of Ethiopian Immigrants.* Tel Aviv: Cherikover. (Hebrew).

Shabtay, Malka 1996. "The Melting Pot Works from the Outside: The Dynamics of Identification with Ethiopian Culture: A View from the End of the Military Service." *Society and Welfare* 17, 2: 199-216. (Hebrew).

Shabtay, Malka 1995. "The Experience of Ethiopian Jewish Soldiers in the Israeli Army: The Process of Identity Formulation within the Military Context." *Israel Social Science Research* 10, 2: 69-80.

Shafir, Gershon 1996. "Israeli Society: A Counterview." *Israeli Studies* 1, 2: 189-213.

Shalev, Michael 1996. "Time for Theory: Critical Notes on Lissak and Sternhell." *Israeli Studies* 1, 2: 170-88.

Shalit, Ben 1988. *The Psychology of Conflict and Combat.* New York: Praeger.

Shamgar-Handelman, Lea 1996. "Family Sociology in a Small Academic Community: Family Research and Theory in Israel." *Marriage and Family Review* 23, 1-2: 377-416.

Shamgar-Handelman, Lea 1993. "The Social Status of War Widow." In *Women in Israel*, edited by Yael Azmon and Dafna Izrael, 35-50. New Brunswick, NJ: Transaction.

Shamir, Boas and Eyal Ben-Ari forthcoming. "Challenges of Military Leadership in Changing Armies." *Journal of Political and Military Sociology.*

Shaw, Martin 1988. *Dialectics of War: An Essay in the Social Theory of Total War and Peace.* London: Pluto Press.

Shirom, Arie 1976. "On Some Correlates of Combat Performance." *Administrative Science Quarterly* 21: 419-32.

Sivan, Emmanuel 1991. *The 1948 Generation: Myth, Profile and Memory.* Tel Aviv: Ministry of Defense. (Hebrew).

Soffer, Arnon and Julian V. Minghi 1986. "Israel's Security Landscape: The Impact of Military Considerations on Land Uses." *The Professional Geographer* 38, 1: 28-41.

Solomon, Zahava 1990. "Does the War Stop When the Shooting Stops? The Psychological Toll of War." *Journal of Applied Social Psychology* 20/21: 1733-45.

Solomon, Zahava, Karni Ginzburg, Yuval Neria, and Abraham Ohry 1995. "Coping With War Captivity: The Role of Sensation Seeking." *European Journal of Personality* 9: 57-70.

Tilly, Charles (ed.) 1985. *The Formation of National States in Western Europe.* Princeton, NJ: Princeton University Press.

Weiss, Meira 1997. "Bereavement, Commemoration, and Collective Identity in Contemporary Israeli Society." *The Anthropological Quarterly* 70, 2: 91-101.

Wolfsfeld, Gadi 1997. *Media and Political Conflict: News from the Middle East.* Cambridge: Cambridge University Press.

Yanagisako, Sylvia and Carol Delaney 1995. "Naturalizing Power." In *Naturalizing Power: Essays in Feminist Cultural Analysis*, edited by Sylvia Yanagisako and Carol Delaney, 1-24. London: Routledge.

Yuval-Davis, Nira 1985. "Front and Rear: the Sexual Division of Labour in the Israeli Army." *Feminist Studies* 11, 3: 649-76.

Part 1

Theoretical and Comparative Perspectives

1

Western-Type Civil-Military Relations Revisited

Bernard Boëne

Introduction

The issue of how the specialists in legitimate organized violence ought to and do relate to the society they are supposed to protect and to those who rule it, is at least as old as political thought itself: one finds it raised in Book III of Plato's *Republic*. It was long dealt with in terms which restricted it to one single question: how to secure the full subordination of *de facto* military power to legitimate, sovereign political power? This strongly suggests that such a problem is not specific to liberal democracies alone: that it in fact derives from the nature of things.[1]

This was Clausewitz's message when he emphasized the intrinsically political nature of any resort to force, i.e., the notion that military force cannot be an end in itself, and that what counts is the political landscape which its use, or even its mere existence, are apt to fashion. Such theoretical justification of civil control in the domain of external relations finds its internal counterpart in the nature of sovereign political power as the ultimate arbitrator among competing values, interests or manifestations of power within the polity. It follows that armed forces, standing for security-related interests and values that vie for attention and resources with those of economic prosperity and sociopolitical harmony, are in essence but one tool among others in statesmen's hands; it follows also that their possible transformation into autonomous political actors can only be regarded as an aberration fraught, in many historical cases, with dire consequences.

That such has been the case across periods, places and regimes, and not only under conditions prevailing in liberal democracies, is convincingly borne out by the fact that absolute monarchs, dictators or totalitarian leaders have been quite as anxious as democratic rulers, if not actually more, to enforce civil control over armed force,[2] and that all regimes resort for that purpose to means which may differ in detail and implementation but not in principle. All aim to stop those entrusted with the task of defending, or promoting the interests of, the sovereign community through force of arms from turning against it and coercing it into submission instead of serving it loyally by obeying its legitimate masters. As a result, any examination of the armory of civilian control in democracies is bound to start with classical methods of universal validity before it can concentrate on the differences in manner which separate them from less liberal regimes.

This, however, can no longer be the whole picture. Restriction of the civil-military problematique to issues of subordination and control was typical of periods or countries in which armed forces were simpler, hence more transparent to external controllers; freer in action, i.e., less dependent upon the parent society's material and symbolic support; and, given the slowness of communications, more difficult to supervise from a distance. Such periods have long since gone by, and in only precious few countries do such primitive conditions still apply today. All through the twentieth century, security imperatives, the very size of armed forces, the amount of technology they utilize, the potential demographic, economic, social and environmental consequences of their action, have all bestowed on them a more central and permanent place in modern society, as well as transformed them into complex organizations that are more opaque to outside controls, and more sensitive to public opinion when it comes to recruiting, budgeting, internal norms and dynamics, or even missions. Conversely, increasingly rapid communications have lessened or removed the need for a priori methods of distant control.

Complexity, dependence, legitimacy constraints and easier communications go far to explain how, beside the traditional issue of subordination *which remains fundamental,* another problem has emerged: that of coordinating military power and the various facets of civilian power within the state apparatus, as well as of mobilizing popular energies and backing in support of sovereign action. The problem is then less to restrict military autonomy or avert military coups—a pathology which now rarely affects the more complex societies—than to determine the proper amount of power and influence which should be accruing to the military establishment if it is to discharge its function without distorting the regime it is supposed to serve. In other words, the question resides in the type of principles which should govern the relationship between the military specialist and the statesman, by nature a generalist,[3] and in the mechanisms most likely to optimize the balance between functional imperative—military effectiveness—and sociopolitical imperative: the harmonious integration of the military into society and state, which (in combination with perceptions of external threats) conditions its legitimacy. There again, the issue by far exceeds the bounds of liberal democracies. It is raised in broadly similar terms in most complex societies. But democracy clearly places special demands, which shall be underlined in due course, on civil-military systems. The following will be predicated on the thesis that democratic solutions to the problem as posed in contemporary societies are not fixed once and for all, give or take differences in manner, as when it comes to controlling the armed forces, but that the underpinnings of an ideal balance change over time as a function of trends affecting both external and internal contexts.

In brief, this paper's main line of argument will move from a review of the classical methods of civil control to an examination of factors in modern civil-military coordination. The treatment will essentially be normative, even though history—in ideal-typical form—will be used: during the course of the last century the same values and imperatives have led to different outcomes according to the state of society and international relations. In the interests of conciseness and economy, it will leave out caesarism, praetorianism, garrison states and military-industrial complexes, i.e., the various pathologies, ancient or contemporary, of civil-military relations, the circumstances which are apt to generate them, the consequences they entail. The reference here will be to an ideal, stable, functioning democracy, deeply-rooted in political culture and institutions.

Figure 1.1

Controls Over the Military

	Objective	Subjective
External	*Direct control* *(subordination)* Precautionary and retrospective surveillance.	*Indirect control* *(subordination* and *coordination)* Amount of public support and societal regard, a priori and a posteriori.
Internal	*Indirect control* *(subordination* and *coordination)* A priori trust + a posteriori surveillance.	*Direct control* (subordination) Trust based on precautionary measures (recruitment screened, pledges of fidelity, shared values, common interests).

Classical Methods of Civil Control Over Armed Forces

Principles of Direct Control

To secure the loyalty and subordination of soldiers (in the broader sense), all political regimes traditionally resort to an array of direct methods comprising *objective/external* and *subjective/internal* means. The latter, by nature precautionary, depend on screening personnel and manipulating symbols so as to maximize the likelihood of obedience and trust; the former, applied either a priori (precautionary measures) or a posteriori (retrospective), rely on institutional coercion, the power to allocate resources, and the utilization of divisions and balance-of-power relationships among the various components of the state security apparatus. Adding indirect methods (*objective/internal* and *subjective/external*: to be introduced later), the following 2x2 table, shown in Figure 1.1, summarizes the goals and principles of such controls:

Thus, the constitutional clause which makes the sovereign (in title or actuality) commander in chief of the armed forces, control of budget resources, periodic rotation of units (and prohibition of unauthorized movements: "cantonnement géographique"), the restriction of military personnel's civil liberties and rights (association, speech, vote, eligibility, strike: "cantonnement juridique"), the functional separation of military from police forces, or that other application of the *divide et impera* principle which consists in encouraging inter-service rivalries, the promise of severe punitive sanctions for violating institutional restraints, temporary or permanent watchdog committees, all comprise the most frequent basis on which external objective political control rests.

It is usually supplemented in standing forces by administrative controls, directed at the middle and lower levels, less easily supervised than the upper echelons. Such controls normally include the power of evaluation and sanction of individual performance, unambiguous command chains clearly establishing responsibilities at all levels, and a coherent body of regulations prescribing sterner discipline and stronger commitment to organizational goals than are usually to be found outside the services. Finally, judicial control of offenses against the law of the land rounds out the system of restraints imposed on soldiers. This last type is generally weaker than the previous two in that civilian judges often lack the resources necessary to enforce their rulings upon the military, and their jurisdiction not infrequently stops at the threshold of a specific body of martial law enforced by military courts.

Subjective/internal control, on the other hand, consists in reserving access to the instruments of legitimate organized violence for those in whom the regime feels trust is not misplaced. When it comes to military leadership positions at the top, the power of appointment and dismissal is generally sufficient. The problem raised in that regard by mid-ranking personnel and the rank and file is more complex, since their loyalty cannot be verified on a personal basis. Temptation is therefore strong, especially if manpower requirements are low, to bar from the military those who are deemed potentially disloyal. To prevent the military establishment from becoming a state within a state, exhibiting ideals at variance with dominant civilian attitudes or pursuing sectional interests in conflict with perceptions of the common weal, governments can fill the ranks with citizen-soldiers, in the form of either conscripts or short-term volunteers. Less likely to develop a separate ethos than their long-career superiors, such citizen-soldiers act as a potential guard, in time of crisis, against any seditious intent or behavior on the part of their military leaders. Selective promotion can be used to the same effect among cadres. In more normal times, or when force level requirements are high, simple emphasis on patriotic socialization, stressing central sociopolitical values and loyalty to existing institutions (sovereign, constitution, nation, official ideology, often in the guise of an oath of allegiance) is usually enough. Should patriotism be weak or suspect (as was the case in West Germany after World War II), civic and political indoctrination, as well as machinery to weed out unsavory orientations, are resorted to. Finally, compensation for the real or potential hardships and constraints of military life (risk to life and limb, open-ended terms of service, formal and informal demands placed on individuals), when it takes the form of substantial material or symbolic privileges, adds self-interest to the reasons soldiers have to remain loyal to their political masters.

Such practices are so basic and habitual as to become almost invisible. They nonetheless remain in force in the constitutions, laws, regulations, organizational cultures and personnel management norms of most states, and continue to shape behavior and attitudes.

Democratic Norms of Civil Control

Having recognized that democracies face the same fundamental civil-military problems as other, less liberal regimes of the past or present, and that some of the solutions to which they have recourse do not differ greatly in principle from those used under non-democratic conditions, the next question is what distinguishes them from absolute monarchies, dictatorships or totalitarian states.

The most obvious answer is that liberal democracy is a form of government based on popular consent and political compromise. Executive and legislative holders of sovereign power are directly or indirectly elected by the people, and subject, like all citizens, to the rule of law. The arbitrariness inherent in any exercise of power thus finds itself formally limited. It follows that where dictatorial and totalitarian regimes rely on personal (often charismatic) ties, political discrimination, secrecy, brute power relationships, police surveillance or even terror, democracies allow persuasion through values or self-interest, the authority of impersonal rules, and public opinion pressures to play a prominent role in civil-military relations.

Democratic control of military behavior is less abrupt, and less dangerous for those concerned, since beyond the special demands imposed on them by their legal status, their rights as citizens are formally guaranteed. The *subjective/external* form of indirect control through the amount of public support and social honor enjoyed by the military, though present and (in some circumstances, not least war) critical under any regime, figures prominently in democracies since public opinion is more central to the way political games are played in them, and status differentials cannot be imposed from above.

The second major difference is that while totalitarianism equates loyalty with ideological conformity (if not positive fervor and zeal), and dictatorships insist on active political allegiance, democracies only imply partisan neutrality on the part of uniformed personnel, at least when acting in an official capacity. The need for such apolitical, nonpartisan behavior, without which democratic political pluralism would soon be impaired, accounts for the restrictions, of varying scope according to time and place, applied to soldiers' civil liberties. Yet, this principle finds its limits in the no less necessary adherence to the tenet of political pluralism itself, for the risk exists (as was observed time and again in the West during parts of the nineteenth and twentieth centuries) that military professionals may construe partisan neutrality less as remaining *outside* than as standing *above* politics. The latter conception, with its implicit assumption of a supra-constitutional status for the military ("governments, regimes pass, the armed forces remain"), betrays lack of respect for that central character of democratic life: the elected representative of the people, and may lead the top military leadership to picture itself as arbitrator or last recourse, thereby sanctioning the ruin of democratic principles.[4]

This brings us to a third difference. There are more natural tensions, both structural and ideological, between a democratically organized society and its military institutions than between the latter and illiberal regimes. Civilian pluralism contrasts with centralized military command; middle-class predominance, with the quasi-aristocratic status of officer corps within the social structure of armed forces; civilian values—freedom, equality, prosperity—with values such as authority, cohesion, discipline and honor, which derive naturally from the performance of truly military (i.e., combat) roles. The democratic need for openness and freedom of information contrasts sharply with the military necessity for (or, as the case may be, infatuation with) secrecy. The list of such contrasts could be lengthened at will.

The various manners in which these tensions are resolved characterize each polity and period: more on that later. But success in resolving them requires legitimacy and stability, hence strong institutions. Democracies, while on the face of it more vulnerable than less delicate regimes, have historically proved more resilient than them if the long view is taken.[5] It is so despite the fact that they are less able to control the factors which govern the military's informal influence upon society. But no doubt, the absence of aggressive militarism normally (though not always actually) associated with democracy contributes to stability, just as—all things equal—prosperity and the protection of individual rights serve to foster democratic loyalties among military and civilian alike.

If indeed one regards military coups as the most serious setback civil control may suffer, and if one takes seriously the conclusion reached by political scientists after decades of comparative studies—to wit, that it is not so much force of arms as power vacuums which generate military intervention in politics[6] —, then the fact that contemporary democracies are spared that phenomenon seems to confirm their superiority. While totalitarian regimes proved capable of firmly controlling their militaries, they did so at prohibitive costs which led to their ruin: internal coercion, external aggression, and eventual exhaustion of social energies through war or economic ineffectiveness.

The Diversity of Democratic Political Institutions and its Impact on Civil Control

It should be noted, however, that democratic institutions are diverse, and not equally conducive to effective civilian control, notably of the external/objective variety. Some make it easier, others more problematic. Thus, political systems based on a structural division of sovereignty, such as a strict separation of executive and legislative powers, or federalism, experience a major hindrance to the subordination of the military.[7] One can easily see why: in such systems, military elites can always play off President against Congress, or individual states' rights against federal authorities, in order to preserve more or less substantial margins of *de facto* autonomy. This weakness is all the more likely as the Legislative, aware that its means of control are more limited than the Administration's, can ally itself with the military, or sections of it, to counter designs it disapproves of in the Executive's agenda, or again to ensure the reelection of parliamentarians in constituencies where defense expenditures have a particular economic impact. U.S. history supplies the best illustrations of that type of difficulty, offset in that case by a strong democratic culture. Checks and balances undoubtedly have many positive qualities, but they have proved less adaptable to the needs of civil control than the political systems of West European unitary states and parliamentary democracies. The machinery of a Cabinet politically accountable to Parliament and whose continued existence is vulnerable to a vote of no confidence, but which is backed by a majority induced to act responsibly by the prospect of dissolution, preserves the unity of sovereign power, and thereby inhibits the development of interstitial autonomy among government bureaucracies more effectively than a system based on dual control of the same.

Likewise, the articulation of political and military power within the Executive branch is not indifferent. Without going into details, an articulation which does not encourage overlap and confusion of roles (e.g., by appointing a soldier as Minister or Secretary of Defense), or which establishes clear lines of authority while it makes due allowance for the different requirements of political and military functions (as is the case with the *balanced pattern* advocated by Samuel Huntington), is superior to others.[8]

Lastly, and to name but a few, other considerations may impinge on the quality and scope of civilian control: the legal status governing military secrecy, the modalities (detailed or not) of parliamentary budget reviews, the importance attached to long-term planning (which may tie the hands of subsequent leaders), the tacit or open alliances among military and civilian (notably economic) groups— all come into play. It does not take great analytical skills to assert that even where democracy enjoys a long tradition and favorable conditions, optimal civilian control is not guaranteed and must be vigilantly attended to.

The Modern Conditions of Civil-Military Coordination

The issue of optimal conditions, in the twentieth century, leads to that of coordination. It was raised first when military art, which from time immemorial had been, as Napoleon put it, *"un art simple et tout d'exécution,"* became a more complex expertise based on a corpus of military and administrative science. The onset of rationalization (technology, management) profoundly changed the name of the game. Armed forces mostly made up of combatants among whom division of labor was minimal were succeeded by military establishments in which role specialization increased dramatically and active combat functions were soon reserved for a minority.[9] From that moment on, traditional methods of civilian control turned out to be insufficient. Facing career specialists who alone could master the intricacies of their trade, controllers felt at a disadvantage. More importantly, the levels of human and material resources mobilized meant that military dependence upon society and the state was on the increase. In such a context, traditional controls, aiming to contain armed force within politically acceptable limits, ran the risk of impairing military effectiveness: the chances of success in war or in deterring foes, at a time when that translated into chances of societal survival. It became necessary to place a higher amount of trust in a body of military experts on whom civilian control would bear less heavily, but who, in exchange for the broader autonomy thus granted, would be required to acknowledge and internalize the necessity for, and legitimacy of, subordination to civilian political leadership. The emphasis thereby shifted from direct (objective/external and subjective/internal) forms of surveillance, to an *internal/ objective* method of indirect control: the last cell so far left untouched in the 2x2 table offered earlier.

Professional Norms as Indirect Civil Control

To analyze the logic of the new situation, the sociological concept of professionalism[10] was applied to officer corps. A formal parallel was drawn between officers on the one hand, doctors, lawyers, clergymen or professors—i.e., the classical professions—on the other. Leaving conceptual refinements aside, three main dimensions define professionalism: expertise based on higher (initial and continuing) education, in a socially important domain; an ethic of service and responsibility to society which, based on higher values or principles, partly removes specific types of expertise from the purview of market forces; a sense of corporate unity, and delegation of public authority in setting and enforcing professional standards. In other words, professionals form groups of experts for whom self- and mutual control partly substitutes for direct political control, on the understanding that only experts can supervise the work of experts, and that in upholding higher values they partly renounce self-interest (pursuit of power or wealth), and deserve the trust placed in, and the societal regard (prestige) accruing to them.

Applied to officership, the concept of professionalism has proved immensely valuable, from both a normative and an analytical standpoint. It made it possible to think in terms of military autonomy and civil-military coordination, while preserving the principle of military subordination to the civil power. Yet, it has not given rise to *one* best model of democratic civil-military relations, endowed with universal validity. Two dominant conceptions based on professionalism have emerged, associated with the names of American scholars (Samuel Huntington of Harvard, Morris Janowitz of Chicago) who have left their mark on the field of civil-military studies. The relevance of these conceptions has varied across time and cultural traditions as a function of internal and external contexts. It is possible to construct, derived from them, two ideal-types of equilibria in armed force/ state/ society relations, applicable respectively to pre-nuclear times and the Cold War.[11] There is some doubt, and a measure of controversy, as to whether military professionalism is still a relevant notion to analyze the post-Cold War situation, and whether a different configuration is in the making. This certainly calls for the revisiting of the subject announced in the title.

Threats of Total War, Charisma of the Nation-State: Radical Professionalism

The Huntington model of civil-military relations is related to mass armed forces and the possibility of total war, in the pre-nuclear sense.[12] In the West, it materialized historically, in more or less pure forms, between the late nineteenth and mid-twentieth centuries, i.e., until such time as strategic thought ceased to think in terms of a

Third World War which would have resembled World War II on a larger and more destructive scale. It remains relevant in some other parts of the world to this day.

This model is marked by a strict separation of political and military spheres of activity. Such clear division of labor stems from the fact that the two domains do not yet overlap much: foes are clearly identified, the threat on the borders (or strategic stakes) is massive, and in the days of industrialized mass armies, strategic plans are too complex to become an easy bone of political contention. Thus, the distribution of roles assigns politicians in office the definition of ultimate goals, risks and expected utility, the assessment of political opportunities and expedient methods, and the mobilization of resources—human, material and symbolic—necessary to achieve the objectives set; it reserves for the military leadership strategic planning, the organization of armed forces, and operational command in wartime. Put differently, the statesman ideally fixes the missions, allocates the means (after consulting with generals), and leaves implementation to the professionals: he refrains from interfering in a domain where he is incompetent. Conversely, the military officer forgoes any considerations alien to his jurisdiction, which would warp his expert judgment. Politics, in particular, is beyond the scope of his professional competence (though *policy* may not entirely be where it overlaps with strategy, if he recognizes that the former takes precedence over the latter). He is thus subordinated to sovereign political power, which his professional ethic leads him to accept and even desire; but he enjoys professional autonomy and a priori trust, and he knows he will be judged by his effectiveness in action alone. Such is the meaning of internal objective control. As corporate prestige is the main form of gratification enjoyed by professionals, the fear of losing society's respect provides a powerful incentive for professional officers to adhere to the central norms of military professionalism—honor (the willingness to fight when ordered to do so, regardless of personal consequences) and loyalty.

The mass armed force, geared to total war, provides the structural framework for such professional autonomy. A true mirror of the larger (male) society, it comprises all the support functions it needs to wage war, and is in practice largely independent of its civilian environment as soon as its legitimacy and the resources it requires are secured. Such a configuration faithfully expresses liberal aspirations for a strict distinction between civilian and military society, wartime and peacetime norms.

Under such circumstances, the military are in a position to cultivate their collective identity and live apart from the parent society. They can do so at little or no cost since strategic insecurity and the strength of national sentiment are generally enough to legitimate the existence and ways of armed forces. The nation-state's charisma is then at its zenith. As a form of sociopolitical organization, it not only provides security against external threats, but also promotes democratic development through mobilization of social energies aiming at rational control of collective destiny.[13] The prestige of officers, who better than any other group embody public service ideals, is at its height. Their merits include the sacrifices they incur in the service of society, but also the transformation of armed forces into large-scale bureaucracies, which at that time pass for the most effective form of organization, and from which many public administrations and large private firms will in due course borrow innovations and norms.[14] Even antimilitarist liberals are forced to admit that, in democracies, war and armies have at least the advantage of fostering citizen ideals and cooperative social action in the cause of democratic utility. Though suspect, the conservative penchant of officer corps, counterbalanced by the social and political representativeness of conscript mass armies, is readily overlooked on such grounds.

The substantive conservatism of officers[15] stems in part from their social origins. An influential minority among them is drawn from the ranks of traditional elites: the landed gentry,[16] then threatened with social decline, whose sons find in a military career an honorable way of keeping up their position in society. The majority, of more humble extraction, are promoted through meritocratic procedures: they tend to adhere fully to the ideology and specific norms of an institution which provides a then unique avenue of upward mobility. But such conservatism is also related to the possibility (and over the long haul, likelihood) of total war, resulting in an exacerbation of functional values, poor in democratic content (as underlined earlier), and thereby widening the gap between the military leadership's ethos and dominant civilian attitudes.

Thus, during the period considered, many factors conspired to turn career officers into professionals who derived a substantial amount of legitimacy and prestige from the assiduous cultivation of their military identity, and were therefore encouraged to stand apart from society by idealizing their duties and unique characteristics well beyond the sole requirements of military effectiveness.[17] That is what, using the label coined by one of Huntington's commentators (Larson 1974), can be termed *radical professionalism.*

Such a model of civil-military equilibrium is not indeed without its faults. The strict functional separation of politics and war may bring military leaders to forget that political goals and meaning are paramount in any military conflict, and to wage war as if it were an end in itself.[18] In another connection, the apolitical behavior required of uniformed personnel by democracy can be given a conservative twist which, if relayed in public opinion by strong nationalist currents, may turn the military services into a symbolic rallying place for antidemocratic forces, and lead to militarism. Cases in point abounded in the interwar period, and in some instances produced monstrous regimes. Yet, the model proved reasonably functional and stable in the stronger democracies.

Cold War, "Secularized State": Pragmatic Professionalism

A second model, analyzed—and in many respects predicted—by Janowitz (1960), became dominant in the last three decades of the Cold War. Basic traits of the previous model are no longer part of the picture. Though the military threat remains massive, risks of total war are powerfully diminished (but not altogether abolished) by mutual nuclear deterrence. The charisma of the nation-state, retrospectively held responsible for past cataclysmic wars, is gnawed away by a rising tide of hedonistic and competitive individualism, born of increased specialization of roles, economic prosperity and the personal social security afforded by welfare. The military officer no longer is the sole standard-bearer, enjoying high recognition, of public service ideals—now entering a period of steep decline. These were precisely the underpinnings of legitimacy and prestige which had previously enabled "radical professionals" to remain functionally and symbolically central while living on the social margins.

Hence, everything incites military cadres to renounce their erstwhile splendid isolation, and to become as fully integrated into the larger society as they can. To the extent that they succeed in emulating other groups in terms of skills, income and lifestyles, this is the source of a new status which, though lower than previously (but all professionals or public servants experience the same relative decline), minimizes loss of societal regard.

The major role played by technology obliges them to adopt a more liberal leadership style, leaving room for initiative among subordinates, who now have to be motivated rather than disciplined. It introduces massive doses of instrumental rationality in the form of technical or management skills, which they not infrequently acquire at civilian universities. Transferable skills make for external reference groups and easier career switches to private firms or public administration. The rank structure, formerly pyramidal, takes on a double diamond shape not unlike that in civilian capital-intensive organizations, while technical specialization of roles results in a broader social base of recruitment at all levels. Military family lifestyles, which used to be highly divergent from those on the outside, now tend to resemble them: married cadres' residence is now mostly in the civilian sector, while rising proportions of military wives, gainfully employed, are unable (or unwilling) to assume the informal support and social role they traditionally volunteered within the service community. As a consequence, the ubiquitous sense of sacrifice and decorum gives way to a balance struck between soldier duties and citizen rights: between those special demands of service life that remain functionally warranted, and other private, off-duty aspects, over which institutional control is no longer tolerated.

In summary, the new external and internal context substantially erodes the objective and subjective gap between armed forces and society, thereby reducing (though again not annihilating) the "natural" tensions between military-functional and democratic values. Such (relative) loss of identity and status leads to a more pragmatic ethos. Professional officers know, and more or less readily accept,[19] that under the conditions set by nuclear deterrence, real military action will be scarce while training and readiness standards have to be kept high, lest friendly weaknesses provoke the enemy into attempting a strategic surprise attack. They are aware that in the eyes of public opinion they are legitimate only insofar as they can guarantee maximum security at least cost to blood and treasure (not to mention the population's anxiety levels in the face of nuclear risks).

Legitimacy, now one of their primary concerns, is also conditioned by the degree to which the military's composition is representative of society's mainstream.

The other facet of such *pragmatic professionalism* (Larson 1974; Segal 1986)[20] resides in the changes affecting the political-military nexus. The main difference in that regard is a much weakened division of labor between political and military leaders. The former now routinely intervene, sometimes down to the lower echelons, in what hitherto was the latter's province. The reason behind such major change is twofold. Use of military force now operates within narrow legitimacy constraints. This is in part a consequence of the role played by nuclear weapons, and the ambivalent feelings it generates: the public fears the intrinsic risks entailed by the mere existence of such mass destruction armories, yet is aware, if only dimly, that they act as powerful inhibitors of major war between the two sides of the East-West stand-off. But these legitimacy constraints also have to do in part, towards the end of a century in which democracies were involved in two world wars followed by unpopular limited conflicts, with the long-term erosion of sacred national egoisms and enthusiasm for military solutions to problems created by the clash of national interests. Hence, politicians in office can no longer ignore matters such as military strategy, organization, readiness and operations, which they were heretofore happy to leave entirely to the military leadership. After all, the Chief Executive would bear personal responsibility for the ultimate decision to press the nuclear button. Under mutual deterrence, as authors like Raymond Aron or Thomas Schelling noted at the beginning of the period considered, crisis management and political brinkmanship substitute for large-scale hostilities. So that the distinction between peacetime and wartime becomes blurred, and spheres of military and political competence overlap.[21] Secondly, for fear of possible escalation, what little amount of military action is called for takes place at the periphery of areas of major strategic interest, with limited objectives, resources and—ideally—duration. Means of instant communications enable politicians, intent on keeping political costs of such limited operations as low as possible, to influence events in the field directly if they so choose by issuing orders in real time to their few actors (which they could not possibly do in such detail if there were hundreds of divisions or equivalent formations engaged in action at once, as in the world wars). As a result, the military's operational autonomy finds itself potentially curtailed as no doubt never before.

The degree of dependence upon the larger society increases somewhat in proportion to the gradual decline of mass armed forces allowed by strategic requirements and imposed by sociopolitical developments (to say nothing of economic imperatives). However slow, reductions in manpower levels imply that support functions are in greater numbers either contracted out or entrusted to civilian public employees, thereby intensifying functional civil-military exchanges at the grass-roots level (just as residence outside of military installations increases the chances of civil-military intercourse and sociability at the family level).

These same reasons explain why the military find it impossible to restrict themselves to purely military considerations, and broaden their perspectives to the political, economic and social conditions and consequences, both domestic and foreign, of military action. Likewise, the fact that the legitimacy of armed services or the use of force is now conditional rather than automatic obliges the military establishment to equip itself with public relations management capacities. These bear on public debates when it comes to defining missions and resource levels, or even assessing opportunities. In other words, the military turns itself into an administrative pressure group while the mixing of political and military considerations gives the governmental decision-making process the "fusionist" appearance of a negotiation between civilian and uniformed leaders.

However, such military influence is by no means boundless. Political leaders normally have the last word; other interests and values than those related to security can make themselves heard; and in most democratic countries, "fire-alarm" communities of civilian scholars and journalists exist, in which security issues are debated. When the free play of these forces is not inhibited, a balance is struck which, despite pressures from "military-industrial complexes," stands a fair chance of being satisfactory. This balance is based in part on the internal objective control mechanisms dominant in the previous model. But it depends in part also upon the internalization by officer corps of (a) the pluralist logic of political institutions and (b) the limits inherent in the use of force, i.e., upon subjective control. Put differently, the trust placed in a responsible management of legitimate organized violence by the military has its basis in a professional ethic which obliges officers to recognize the need for and acceptance of subordination to politics, partisan neutrality and, in this context, reduced operational autonomy; on the other hand, the military's broader political influence is deemed acceptable only by reason of their spontaneous convergence with society, which leads them to share its dominant values and makes their base of recruitment more representative.

The characteristic trait of that model and period is that the substantive conservatism so prevalent among officers in the previous ones is considerably toned down.[22] So much so that those democracies which had long retained conscription as a guard against possible sedition on the part of officer corps suspected of flirting with far-right ideologies no longer feel the need to do so. Technology, both nuclear and conventional, having reduced manpower requirements, conscription declines: in some countries where it was not deeply-rooted, the draft is abolished; in others, the length of conscript service is curtailed, and exemptions are granted more liberally. In a context marked by weakening citizenship ideals, this is not without raising problems of equity, which a U.S. commission of the 1960s aptly summed up by asking *Who Shall Serve When Not All Serve?* The social representativeness of those who bear arms, at both leadership and rank and file levels, becomes central to the legitimacy of military service. But while it is necessary to democratize the recruitment base of the upper ranks, the requirement regarding the lower echelons is to attract the sons (and in the new all-volunteer forces, daughters) of the privileged classes who tend to legally escape service under arms altogether (Moskos 1988).

The Post-Cold War Context, External and Internal

Since 1990, after the watershed symbolized by the fall of the Berlin wall, we seem to have entered a new era in western democratic civil-military relations. (There is in fact reason to believe that similar conditions would apply elsewhere to other advanced countries with democratic institutions and culture, were old, long-stalemated conflicts to come to a definitive conclusion.) Though it may be too early to regard the emerging configuration as final, a brave attempt at characterizing it ideal-typically may now be in order.

The great novelty, of course, is that the massive military threat of Cold War days is gone, and unless a new major strategic cleavage emerges, calling for force reconstitution on a large scale, the missions assigned to armed forces will be concentrated in the low to medium intensity range. The very success of the Gulf War means that revisionist regional powers, or "rogue states," are deterred from emulating Saddam Hussein's aggressive designs. More frequent than formerly, military action on the part of democracies is likely to assume the shape of multifunctional, multinational interventions, often improvised at the last minute, in which their forces act as third parties on the strength of explicit or implicit international mandates. The goal of such missions (e.g., Kosovo, East Timor) is to terminate or freeze violent local disorders which the international community deems intolerable, either because, in a realist perspective, they threaten the viability of a status quo in which democracies now have a vested interest, or because, in somewhat more idealist terms, public opinion tends to impose upon states the same imperatives of conscience as bear on individuals.

Today's international order is unipolar. But it is marked, on the part of the only superpower left, by the temptation of splendid isolation or, at best, by some reluctance to live up to all the consequences of the responsibilities that come with superpower status. It is thus likely to be imperfect and unstable, at least at the periphery of areas where military breaches of the democratic peace have become unthinkable.

Another new development lies in the fact that nation-states are no longer the only actors in the international arena: scores of non-governmental organizations and sub-state groups[23] have invaded it. Satellites, real-time news coverage and the Internet have removed the barriers which kept local moods and states of national opinion fairly separate, with a resulting trend toward *domestification* of international relations. Terrorism, too, has turned global, and now affects countries which it had hitherto spared. The linkage between external and internal security is much closer than it used to be.

The political equation of military force grows more complex as legitimacy constraints deepen or multiply. "Peace dividend" pressures have lowered manpower levels and defense budgets, sometimes dramatically.[24] Together with their declining functional relevance in the new context, increasing dislike of nuclear weapons leads powers of the first rank to de-emphasize them (though they continue to carry political weight and structure world order). Mass armies are now the rare exception (linked to local situations in peripheral areas) rather than the rule, and volunteer recruitment replaces the draft wholly or in part as the normal way of meeting low manpower requirements. But it often proves more difficult under the new circumstances to *man the equipment* than it was to *equip the forces* a few decades earlier (Segal 1989).[25] With fewer troops available, there is a need to concentrate the energies and training of uniformed personnel on operational roles, and to delegate service support roles to civilians, thus increasing functional dependency upon society to the point where units can no longer operate in action without direct civilian input, hence without the legitimacy required for moral support.

National symbols seem to count for less in a situation where nation-states are, as Daniel Bell once put it, "too small for the big problems of life, too big for the small problems," and are led to devolve authority and competence to supranational and local echelons. Some countries so reduce their defense resources, and are so eager to place them under allied command in multinational formations, as to renounce in practice any military expression of their national sovereignty (see Klein 1998).[26] In such a context, patriotism is less easily invoked, national citizenship ideals and readiness for disinterested public service, much weakened by several decades of erosion, prove harder to revive.

Military professionals tend to worry about the direction of trends which seem to call into question so much of what they cherish. Yet, their public image and prestige have rather distinctly improved in the new context. In many democratic countries, their new missions in support of peace and suffering humanity are deemed noble by large segments of public opinion (which not so long ago remained suspicious of military action in the cause of national interests).[27] The basis of their legitimacy is shifting again. Typical of complex affluent societies where diffusion of veto power has shortened social distances, grass-roots political action by minorities has given rise over the years to multiculturalism, which makes mainstream identities more uncertain, and tends to regard any attempt at reviving citizenship or strengthening integration as a new form of majoritarian tyranny. Under such circumstances, all groups claim a right to cultivate their identity. In advanced democracies, at once egalitarian and libertarian (some would say *postmodern),* identities are chosen rather than ascribed, and largely based on lifestyles, values, and agendas, thus undermining old normative systems. Public institutions therefore increasingly rely for their legitimacy on *cultural* diversity rather than on *social* representativeness.[28] In what a French sociologist has called the "age of tribes," in which no value (apart from life, tolerance and equality of dignity) stands above others, armed forces come to be legitimized as a tribe among tribes, free to cultivate military-specific values and lifestyles *if* by so doing they renounce old "school-of-the-nation" concepts, do not deny internal diversity or equal treatment, and avoid spilling too much (friendly to begin with, but probably to be generalized later to *any)* blood.[29]

Heightened prestige, missions combining traditional *Realpolitik* goals with *Idealpolitik* impulses, and that are unforeseeable as to their specifications and location (and therefore cannot be planned long in advance), multinational dimensions, real-time media coverage, legitimacy constraints that are mitigated in some respects but become heavier in others: such are the basic traits of the new context in which military institutions now move.

The Future of Professionalism and Civil Control

Renewed Intellectual Debate

Can pragmatic professionalism weather the change and survive in the post-Cold War era? In theory, all things equal, it should be able to, since it came about in the first place, from the 1960s onward, as a result of rising complexity, de-emphasis of force in international relations (for fear of escalation risks and Armageddon), and the need for the military to factor in non-military considerations. Now, precisely, complexity continues to rise, and success in action as in institutional survival often implies that officers take on new, more politically sensitive role models (Moskos et al. 2000) such as *soldier-diplomat, soldier-scholar* or *soldier-communicator,* while they retain, for worst-case scenarios, that of *soldier-soldier.* Use of force itself, somewhat more likely than at the height of the Cold War, and still the cutting edge of military action (though power ges-

ticulation does achieve results), is by no means given free rein. Indeed, it is resorted to mostly in peace support operations, hardly
conducive to absolutist military values—even if soldiers sometimes
chafe under unrealistic rules of engagement imposed by international
bureaucrats. And new layers of intricacy and paradox in the ends-
means nexus, whether present ("military-technological revolution")
or still to come ("information warfare") point to situations where
the direct experience of violence among soldiers will be even less
than it has been in the last decades. Added complexity makes detailed, intrusive formal monitoring by civilian leaders more problematic, which should normally heighten the value of internal controls, both objective and subjective, on which pragmatic
professionalism rests. Integration of military professionals into the
society and culture, which makes it possible, remains (with the possible exception of the United States: see Ricks 1997) a major feature in the new configuration.

However, the viability of professionalism itself, civilian or military, is now in doubt. Hailed from the 1940s through the 1960s as a
solution to the predicament of modern societies, professionalization
of occupations is now regarded as creating more problems than it
solves. The reason behind this negative change is that the classical
professions have reached a state of crisis. Mystical beliefs in the
virtues of competition and market forces, which became dominant
after Keynesian policies and welfare excesses brought discredit upon
themselves, have undermined the structural and normative underpinnings of the trust and authority professionals were once accorded.
Their monopoly on delimited spheres of competence is under attack, and often opened to competition from unchartered practitio-
ners; their expertise is contested, sometimes in court, by clients
whose higher education levels and consumerist attitudes place them
in a position to denounce failings and inconsistencies. Concern for
economic efficiency, fed by the need to contain public deficits (or
rising insurance premiums) and reduce taxation levels, brings many
professionals under strict external controls effected by accountants
or financial analysts ("managerialism"). Democratization of higher
education and the rise of affluence have broadened both demand
and supply of professional services, thus making them banal and
lowering the status of those who provide them. Bereft of their former
prestige, professionals are tempted to rank self-interest above ser-

vice. In short, two decades of ascending neoliberal capitalism, by reinforcing the effects of what Karl Mannheim (1940) identified as the "fundamental democratization" of industrial societies, have powerfully contributed to a devaluation of professionalism (Dandeker 1994).

This temptation to substitute self-interest for normative involvement had been noted in western military establishments even before its effects had time to come to full fruition in civilian societies. It was notably the case in the United States in the wake of the Vietnam trauma.[30] Some authors (Perlmutter 1977; Betts 1977) were led to relativize the benefits to expect of military professionalism. This in essence was the message, with a subtext warning against the deleterious effects of such trends on military effectiveness, delivered by Charles Moskos (1977) when he first offered his celebrated thesis positing a shift from "institution to occupation" in military establishments.

This general crisis of professionalism has generated a reexamination of old civil-military issues along lines which no longer accord that notion the central place it was accustomed to receiving in earlier classic treatments. But, as the fact that those who advance the newer theories are part of a younger generation strongly suggests, other factors are involved. One is the recognition that postCold War developments have probably ushered us into a new era; another is that the older theories were western-centered and lacked relevance in other parts of the globe, hence the degree of generality a good theory should have. So that the time seems ripe for a change of paradigm.

The first attempt at renewal, which goes by the name of "concordance theory," starts from the commendable premise that due allowance should be made for cultural and historical diversity. Where one would expect ideal-types of historical-cultural configurations trying to cover most known civil-military contexts, models of relationships and outcomes across time and space, it singles out three actors (political elites, the military establishment, the citizenry) and four indicators (composition of the officer corps, political decision-making process, recruitment method, military style), and goes on to posit that when the three actors are in agreement on the four indicators, military intervention is unlikely. The theory is then in a position to explain why coups do not happen when predicted by the likes of

Huntington, or why they do occur in spite of predictions to the contrary by the old theorists (Schiff 1995). One wonders how such a simplistic causal model can account for so complicated a subject as comparative civil-military relations (unless they are restricted to the conditions leading or not leading to military coups, which is only part of the subject). The new "theory" seems especially off the mark when, in its desire not to superimpose particular (i.e., western) values and histories upon nations, it rejects the postulate that civilian institutions must control the military (Schiff 1996). One suspects confusion as to the meaning and status of civil/civilian control: the logical subordination, grounded in both strategic and government theory, of military to sovereign political power, which applies (as underlined by this writer from the outset) even when the two functions are discharged by groups that are socially indistinguishable.

A much more interesting and intellectually significant attempt at reformulation is the Agency Model offered by such authors as Deborah Avant (1994, 1997) and Peter Feaver (1996b, 1997). The Principal-Agent framework, already used by political scientists to analyze relations between Congress or President and civilian bureaucracies, is fruitfully applied to civil-military relations. Both groups of institutional actors are faced with a binary choice: intrusive monitoring or delegation for the political leadership, compliance or non-compliance ("work" or "shirk") for the military establishment. The four cells of the ensuing 2x2 table are then examined in terms of the parameters (context, symbolic or material costs and capacity for initiative, sanctions, etc.), provided by empirical/ historical evidence, and of the equilibria to which the actors' "strategic" interactions are apt to lead. This deductive, game-theoretic approach, unlike "concordance theory," takes the necessity for civil dominance as its point of departure, and goes on to explore the factors that shape the changing patterns of civil control. Sensitive to the seemingly tautological nature of the "professionalism-equals-civil-control" thesis common to both Huntington and Janowitz, it argues, with an eye to economy of effort, in favor of a unified theory covering all historical/cultural configurations by combining values-based with interest-based factors and external controls. This does indeed hold the promise of a finer-grained institutional analysis of the "pulling and hauling" that takes place between political and military elites even in normal times,[31] as well as greater applicability to a variety of contexts. However, its agnosticism as to what is acceptable or not in view of the civil-military problematique and (in the case of liberal political regimes) democratic theory raises problems.

Given the inescapable justifications for civil control, the field of civil-military relations is intrinsically normative, hence prescriptive, and a theory which does not offer criteria for distinguishing between the normal and the pathological falls short of expectations.[32] That the classics of the 1950s defined away the problem by labeling "professional" those officer corps which accept civil control while they cultivate functional virtues (separating them to varying extents from society at large), and by seeking the proof of professionalism in the absence of any undue military influence or intervention in politics, is a charge that falls flat when one realizes that both Huntington and Janowitz sought to define *the conditions under which the equation is not a tautology.* Their respective prescriptions flowed from there: an *ideology* sympathetic to, or at least tolerant of, the military virtues for the former, a better *social and cultural integration* of officers and soldiers in general into the parent society for the latter. Both recognize that the crucial variables involved in fostering a professional ethos are symbolic in nature: the prestige which enabled radical professionals to remain central while living apart and cultivating a separate identity, or the sense of moral worth and self-esteem to which Janowitz referred the meaningfulness of pragmatic professionalism.

Toward a Partial Return to Radical Professionalism?

The thesis advanced here posits not only that military professionalism is still relevant from a theoretical/normative viewpoint for successfully combining military effectiveness and civil control in the new context, but also, on the empirical level, that the military is no longer alienated from society (though in some countries it may be alienated from civilian *elites* who no longer support established institutions).

In the two decades just elapsed, most western armed forces have recovered from the difficult situation which had led Charles Moskos to hypothesize a rise in occupationalism. That difficult period witnessed the painful adjustment from the conditions which had fostered the dominance of radical professionalism to those which, as predicted by Janowitz, opened the way to a new, "pragmatic" equilibrium. Moskos (1986, 1988) recognized, after a full decade of comparative studies, that his I/O thesis had overstated the case. The post-Cold War context has confirmed, and accelerated, a recovery that is all the more remarkable as it contrasts with the continuing crisis which, in varying degrees, affects most civilian professions.

Though its pace has varied from one country to the next, the change has consisted of a distinct rise in the legitimacy of armed forces. They are now (as opinion surveys indicate) among the most respected public institutions. Interestingly, improved public images and approval ratings appear less sensitive than they were only ten or fifteen years ago to blunders or scandals, some of them serious, such as have affected the military establishments of Canada, Italy, Germany, Britain or the United States in the last few months.[33] Acceptance of the military's place in society seems less conditional than in the previous period. Nor are the reasons behind such a change particularly mysterious. The benefits derived from positive public perceptions of new missions in the cause of peace and humanitarian concerns have already been mentioned. But there are others. For one thing, the burden of defense on the citizen, in terms of either taxation levels or conscript service is almost everywhere substantially lighter than under the Cold War. For another, disinterested, non-partisan military elites compare rather favorably with civilian counterparts who, because they seem bent on pursuing personal advantage rather than the common good (politicians), because the institutions they direct are seen as dysfunctional (judiciary), because they no longer appear to believe in the values or traditions they are supposed to uphold (teachers, clergy, journalists), or sacrifice the long-term social benefits of the many for the short-term economic gains of a privileged few (corporate leaders), suffer from a lack of public confidence. Technological sophistication, organizational innovations imposed by unpredictable missions and fiscal retrenchment, the high level of education and training required by their functions, higher than average pay, and falling numbers (at a time when civilian elites go through a process of dilution) all conspire to heighten the status of service members.

This translates into more self-assurance and less reluctance to make the military voice heard in public debates when corporate interests are at stake. In some countries, as support roles tend to be contracted out and uniformed personnel to concentrate on the management of violence more restrictively defined, organizational identity tends to harden. Yet, that trend is powerfully held in check by functional dependency. If the services stray too far from societal norms (or what's left of them), before long recruitment and retention suffer, the mechanics of transfer to civilian employment upon leaving the forces become problematic, and social/cultural corrections have to be introduced urgently: reliance on volunteers, which helps maintain standards of excellence, is also the Achilles' heel of today's military. When manpower levels are so depleted as to allow the ranks to be filled by single-mindedly promilitary volunteers alienated from their society, who think of themselves as members of an elite institution superior to anything civilian life has to offer, and of their domestic role as guardians of the true culture (as seems to be the case, according to some journalistic accounts, in the United States today: Ricks 1997), then such arrogance is likely to turn opinion leaders off the military and trigger campaigns against them. Today's western societies are characterized, after three decades of rising individualism, by a desire for a return to a more consensual order, but also by unprecedented levels of irreverence. The need to combat the social disorders which threaten to make post-modern tolerance difficult to sustain is symbolically in tune with the success latter-day military institutions have met with internally in mastering such social ailments as racism, drugs or deviant violence, and more generally with the orderly image they exude. Yet, the only thought of militaries leading their countries culturally other than by example would no doubt soon make them the butt of thickly laid-on irreverence, thus depriving them of much of their current influence and legitimacy.

There are, however, no such automatic stabilizing mechanisms in the new context as regards politics, and here may reside some novelty. To military self-confidence, and the temptation to harden values, must be added career interests to vindicate publicly, a higher degree of political know-how, and the greater political latitude afforded by the new missions. All of this promises to make the politics of civil-military relations more lively than it had been for some time.

Budgets, force levels and definition of missions are more hotly contested, especially in a period when drawdowns are the order of the day, than in the days of mass armies made up of civilians in uniform. Whereas conscripts were eager to be demobilized, volunteers fear "reductions in force" simply because they mean loss of jobs (and perks not available to civilians), fewer platoons, battalions or brigades to command, fewer ships to sail or aircraft to fly, i.e., less desirable career prospects. Likewise, the military are more likely to resist what they see as the transformation of the armed services into social laboratories where cultural innovations (women in combat positions, open homosexual behavior, etc.) are tried out without much regard for their sensitivities or for potential consequences, which they wish to avoid, on functional effectiveness. As politicians in countries which have opted for all-volunteer forces tend over time to have hardly any first-hand experience of military life, therefore little interest in and appreciation of it, misunderstandings and mutual resentment are more probable than they previously were.

This comes at a time when officers are much better equipped to deal with politicians because their professional education and training now routinely include courses in law, economics and management, and the social sciences, not least political science. The proportion of senior officers holding graduate or even doctoral degrees before they assume higher command is rising, simply because a blurred division of labor at the political-military interface calls for such skills. Military leverage is not negligible in situations where politicians in office govern from the political center (probably the only option available in societies fragmented into cultural communities entering shifting coalitions), and who are vulnerable to blasts from both left and right, not to mention voter defection.

Last but not least, peace support operations involve the military in politics, whether they like it or not: on the scene (relations with native authorities), at home (due to real-time coverage and its effects on public opinion and policy) and at the international level (intricacies of coalition diplomacy, contradictions between national and multinational chains of command, etc.). The line that separates politics from military action in operations other than war, tightly interwoven with civil affairs, is even more blurred than on more conventional missions. This gives officers increased interstitial autonomy which closer (objective) political control cannot hope to reduce significantly, or places them in situations where they have to decide for themselves how to resolve dilemmas of divided loyalties. Bosnia alone has provided an abundant supply of cases in point: a British general defending UN action against negative comments from conservatives in the U.S. Congress; a French flag rank officer ignoring or defying the authority of the UN Secretary General; French and British senior officers in Sarajevo assuming stances at variance with the official positions defended by their countries at NATO headquarters;[34] a French general repatriated after expressing doubts about the applicability of the Dayton accords (although he was only giving voice to unofficial misgivings in Paris). Assessing the feasibility of having known war criminals arrested by NATO troops without incurring casualties clearly illustrates the kind of political/military role overlap one has in mind. Recalling his time in Somalia, an Italian general candidly acknowledged he had had to arbitrate between orders from New York and Rome.

All of this results in a new military assertiveness on the national scene. The clearest case is that of U.S. Army General Colin Powell, then chairman of the Joint Chiefs, who, basking in the glory of success in the Gulf War, presumed to spell out what the missions of the military ought to be, state conditions for the use of force, and reject the idea of trends going against the grain of an entrenched military self-image (Powell 1992).[35] Later, he briefly considered riding the wave of popularity to run for the presidency as a candidate of either party, but then declined (thereby inadvertently belittling the political process). This is only one episode in a larger series of incidents which recognized academic authorities (Weigley 1993 ; Kohn 1994 ; Snider and Carlton-Carew 1995 ; Feaver 1996a) have labeled a crisis in American civil-military relations—one that can be traced all the way to military perceptions of civilian betrayal in Vietnam, and attests long memories.

West European countries have been spared so far the bitterness of open conflict, but there, too, examples can be found of political-military tension. In the Netherlands and Belgium, chiefs of staff routinely threaten to resign in protest over drawdown decisions the military disapproves of; in Germany, the Bundeswehr chief was reprimanded by his Defense minister for anticipating the government's lead in providing guidance for German participation in peace support operations; in France, an interview given two years ago to *Le Monde* by the incoming Army chief of staff (with prior clearance from his minister) made front-page headlines—deservedly so: in it, the President was warned that the reforms he had personally initiated the previous year would not be implemented if the Prime Minister's recent defense budget decisions were carried over to FY 1999. This as widely interpreted as an attempt, the first of its kind, to drive a wedge, in a period of cohabitation, between a conservative President and a leftist Cabinet.[36]

In short, we have clearly entered a new era: one which, if a diagnosis may be ventured, retains some features of the previous period's equilibrium based on pragmatic professionalism (missions placing restrictions on the use of force, blurred division of political and military roles, social integration of military families, etc.), but in some respects (prestige, growing assertiveness) takes us back to the radical professionalism of earlier times and the problems it sometimes raised.

Conclusions

Whether any talk of crisis, real or potential, is warranted is an open question: answers are apt to vary from country to country. The political aspects are no doubt the more troubling: while nothing egregious appears in the offing, the "pulling and hauling" has gained in intensity, publicity and acrimony.

Study of these processes, which take place in gray areas where safe guidelines are not easily defined, is called for. These are areas that the classics, having drawn the boundaries and spelled out the basic principles, left pretty much uncharted. This is where the contribution of new theory along the lines of the Principal-Agent framework can be invaluable.[37]

But in the end, the solutions must needs be normative, and follow Janowitz's prescription, to which Huntington (1977) at least partly rallied: keep the armed forces *distinct, but not distant* from the society they protect; avoid a separate military ethos by fostering meaningful exchanges across the civil-military interface so as to ensure shared values and mutual understanding; help soldiers stay clear of exaggerated notions of what force and the military virtues can achieve; but in return for loyalty, make sure if you can that the conditions for the military's effectiveness, self-respect and public appreciation are met. Pragmatic professionalism is probably still the best answer around.

Notes

1. This is a revised and updated version of an expanded lecture script published earlier in French: Boëne, 1996.
2. It is symptomatic of military regimes that they subject their armed forces to closer political surveillance than do most civilian regimes. This is so because their legitimacy is generally made weaker by lack of popular consent, and the military establishment is, or may become, politically divided: if use of force to seize political power has benefited some individuals or factions within it, others can think of following suit to serve their own interests. Third world countries have provided numerous illustrations of that process over the past half-century.
3. As seen from that angle, civil-military relations are but a specific case (marked by a monopoly on legitimate organized violence, and the functional values which derive from it) of the broader relationship between politics and administration. Long-term trends have over a century brought about a shift from a "separatist" concept, of broadly Weberian inspiration (though sometimes ahead of Weber's published writings: cf. Goodnow 1900), reserving ultimate ends for politicians and faithful, objective implementation for the (trained, rational) civil servant, to a "fusionist" concept that, on the contrary, recognizes the symbiotic linkages between the two categories of actors in a decision-making process which the diffusion of power induced by complexity transforms into a quasi-negotiation. It is possible, as will be seen later, to locate an intermediate stage (cf. Friedrich and Mason 1940) in that conceptual evolution: that which leaves the civil servant broad autonomy, subject to self- and internal controls. That intermediate stage corresponds with the notion of *professionalism,* which for over forty years has supplied a basis to the classical theory of civil-military relations.
4. The necessity for partisan neutrality of the military while on duty may, in some democratic traditions, clash with the respect for political pluralism. Thus, in the Federal Republic of Germany after rearmament (1955-56), particular care was taken to ensure that active-duty service members were in a position to act as "citizens in uniform": contrary to French or British practice (which deems active political participation incompatible with military status), their right to belong to political parties and to hold elective office was recognized and made effective as long as it did not hinder proper discharge of their military duties. The origins of such norms are to be found in the will to avoid a possible return to the weaknesses (and eventual ruin) of the Weimar Republic in which an apolitical military barely masked antidemocratic orientations. It remains that party membership and elective office generate the risk of military politicization, and can only be recommended where democratic tradition and institutions are stable and strong. The ideal balance between the two imperatives is easy to identify neither in theory nor in practice.
5. Such is at least the dominant feeling as the century draws to a close. That other periods (1930s, late 1970s) did not warrant the same optimism cautions us against believing too smugly that democracy will always keep the upper hand in the end.
6. Political scientists in the field of military studies tend to give pride of place to the systemic level of analysis, and to look for the reasons inherent in the political structure which incite the military to intervene in politics or exert undue influence. They therefore stress pull factors: the weaknesses of the

fallen regime which made praetorianism possible. Sociologists, on the other hand, take into consideration the modalities of the military's integration into the larger society, as well as push factors: institutional conditions within the military establishment which may provide the motives for an intervention. In other words, while the former are satisfied with sufficient conditions (which can be so described only if a natural attraction to power is hyper-realistically hypothesized), the latter broaden the perspective to include necessary conditions (which they find in sub-systemic factors: intermediate variables which inhibit or strengthen the will to intervene).

7. Huntington (1957: 177) recognized as much: "The separation of powers is a perpetual invitation, if not an irresistible force, drawing military leaders into political conflicts."

8. Huntington (1957) synthesizes the various possible modes of articulating political and military leadership roles within the Executive branch as follows:

Balanced pattern	Coordinate pattern	Vertical pattern
Chief Executive ⬇ Minister/Secretary ⬇ ⬇ Military chief (ops) Bureau chiefs (adm)	Chief Executive ⬇ ⬇ Military chief (ops) Minister/Secretary ⬇ Bureau chiefs (adm)	Chief Executive ⬇ Minister/Secretary ⬇ Military chief ⬇ Bureau chiefs (adm)

9. Available historical statistics show that circa 1850, land as well as naval forces were comprised of some 90 percent pure combatants, i.e., military actors without civilian equivalents. A century later, for reasons related to the impact of technology, the proportion had come down to 1/4. Today, in most advanced military establishments, it is as low as 15-20 percent of individual soldiers. In terms of units, the teeth-to-tail ratio is approximately double that figure (which goes to show that even in combat outfits, non-combatants make up a large share of the total manpower).

10. The first sociologists to have elaborated on that concept were British: Carr-Saunders and Wilson (1933). They were massively followed by American scholars in the wake of the 1947 Taft-Hartley Act. The first to apply the concept to officership were also British (or scholars closely related to the British tradition): Lewis (1939) and Elias (1950). There again, their posterity was to flourish in America as from the mid-1950s onward.

11. In other words, instead of viewing Huntington's and Janowitz's theses as conflicting interpretations (as has long been customary in the field: see Larson 1974), the present writer holds them to be equally valid, but applicable to different historical contexts: the charisma-of-the-nation-state/total-war/mass-armed-force era for the former, the Cold War/secularized-state/decline-of-citizenship era for the latter.

12. Huntington (1957: 32-34) relates the rise of the military profession to industrialization and increased division of labor, the growth of, and rivalry among, nation-states, and the advent of democratic ideals. However, he derived therefrom an institutional theory of civil-military relations of purported *non-dated, non-localized* normative validity: insensitive to the variability of historical-cultural contexts. This is the main weakness (which Huntington later cor-

rected) in what remains an invaluable contribution to the field. It is difficult to see why, if military professionalism is a product of history, new historical conditions could not change professional norms, or do away with professionalism altogether. The thesis offered here is that Huntington described and analyzed the logic which prevailed in the West roughly from the late 1860s to the early 1950s, and later in other parts of the world. Yet he applied to it a narrow political science approach. To take but one (central) example, he ascribed officer corps' ideological conservatism, characteristic of that period, to the sole requirements of their military role. A wider sociological treatment would (will here) inquire *also* into social origins as a possible determinant of attitudes.

13. Such conditions were not realized in the United States until the Wilson administration (temporarily), later the Roosevelt through Johnson presidencies (i.e., until the early 1970s, when a return to a more conservative market philosophy was initiated). The novelty of mobilizing social energies in the name of a political (external and internal) agenda may explain the rash of American scholarly books, which soon emerged as classics, on civil-military relations: Smith (1951), Mills (1956), and Huntington (1957). The advent in the 1950s of a strategic practice self-consciously based on mutual nuclear deterrence, which these authors did not take into consideration, generated a time-lag between their scholarly perceptions and the military realities of the day. It was not until the next decade, with the publication of yet another classic of the field (Janowitz, 1960), that this intellectual time-warp came to an end.

14. An interesting illustration of such turn-of-the-century beliefs among sociologists is to be found in Park (1900).

15. Ideological military conservatism translated into both political and functional conservatism, which raised problems at a time when democratization was on the march in the polity, and modernization proceeded apace within the armed forces. The period referred to here was fertile in controversies between military traditionalists and modernists, notably centering upon the symbolic status of horsemanship when reformers advocated the substitution of armor for cavalry units. Modernists owed their eventual victory, if at all, to the approach of war.

16. This trend is visible in most European countries, especially in France under the nascent Third Republic, when the lesser nobility who, for political and social reasons, had mostly stayed away from the services during the Revolution, Napoleonic times and the July Monarchy, began to stream back to careers upon which nationalist fervor now bestowed new luster. The military's professional worldview was now closer to their own social ethos. Provided the meaning of "gentry" is extended to include small town elites in the rural hinterland, the same proposition seems to apply (as Janowitz and others have shown) to America for much of the same period.

17. Military personnel were, literally or figuratively, in uniform 24 hours a day, 365 days a year: there was hardly any distinction between on- and off-duty life. For cadres, few aspects of supposedly private behavior (marriage, residence, education of children, social life, etc.) escaped the influence of formal or informal military control. For the rank and file, the armed forces assumed the characteristics of what Erving Goffman later called "total institutions."

18. It can give rise to an erroneous interpretation of Clausewitz, equating his "continuation of politics by other means" with military dominance in government during wartime (cf. German experience in World War I, U.S. experience

in World War II, and the MacArthur-Truman controversy). Taken one step further, such a conception may lead to a reversal of Clausewitz, along the lines of Ludendorff's *Der totale Krieg* (1935) which, echoed by the likes of Nicholas Spykman in America, turned politics into the continuation of war between two periods of open hostilities.

19. The frustration hypothesized by Janowitz (1960) and others (Moreigne 1971) in the 1960s or early 1970s was plausible, but mostly failed to materialize in open behavior. It certainly was little in evidence in West European militaries. (In the United States, most professional frustration was in connection with Vietnam and the lasting impressions it left of the dire consequences of civilian micromanagement.) The only interpretation one can think of for such lack of frustrated expression on the part of officers is either that frustration was actually nonexistent, perhaps due to the fact that exacting readiness requirements were enough to exhaust energies, or that legitimacy considerations prevented them from complaining: under Cold War conditions, war, however limited, carried risks out of proportion with possible gains; more generally, democracies do not start little wars to please their frustrated militaries.

20. Janowitz's formulation of a pragmatic approach to soldiering, based on a balance between functional and sociopolitical imperatives (which used to add up where they now partly run in opposite directions), has proved better adapted empirically than the hypothesis (first put forward in Moskos 1977) of a shift to occupationalism among the military.

21. Relations with allies (notably within NATO) work on both levels, and serve to deepen the trend.

22. Military conservatism now tends to be *formal* (preservation of the status quo, irrespective of its ideological contents). The functional repercussion is that innovation is no longer hindered, and becomes routinized.

23. Local wars now involve factions rather than states: in 1994, out of a total of some thirty military conflicts worldwide, *none* pitted one recognized state against another.

24. Budgets tend to decline in lesser proportions than do manpower levels, since democracies (as they are wont to do) compensate for low service strengths by resorting to technology. Fixed costs and loss of economies of scale are also involved.

25. This was noted by Segal (1989) for U.S. forces in the 1980s. The trend is more in evidence today, especially in Europe. The British Army is suffering manpower shortages even though its strength has gone down by fully 1/3 in the last decade. In Italy, plans to professionalize the forces are hindered by lack of prospective candidates. In France, planning for the future AVF has fixed manpower requirements at an all-time low in anticipation of similar difficulties.

26. See Klein (1998). The Bundeswehr officially advertises itself as the prototype of a "post-national" force. The armed services of the Benelux countries are now almost entirely subsumed under multilateral commands. Even a country such as France, which had prided itself upon its relative autonomy vis-à-vis the Alliance in defense matters, is now a party to numerous European units or endeavors, and is moving (however slowly) closer to NATO's integrated military structure. The United States itself recognizes that there is precious little it could do in the long run without allied cooperation and support.

27. Such is undoubtedly the case in many European countries and former British dominions. The case is harder to decide in the United States where public

enthusiasm is counterbalanced by fear of casualties, sizable reluctance to play the role of world gendarmes, and the military's lack of taste for peacekeeping.

28. Representativeness, not so long ago, was assessed in a strict mathematical sense (though the multiplication of criteria: socioeconomic background, race, gender, ethnicity, etc., complicated the issue to the point of rendering it operationally unmanageable). In a tightly integrated society in which service under arms was a widely shared experience, equity (hence legitimacy) implied that the risks involved should not be imposed disproportionately on disadvantaged categories of citizens, either by law (conscription) or through selective socioeconomic mechanisms (all-volunteer forces). The entity which served as baseline for measuring representativeness was the polity as a whole (an ideal easier to approach under draft conditions than in all-volunteer formats, because the latter tend to reflect the compositional make-up of the working population rather than the citizenry). The new imperative of cultural diversity raises fewer problems in that regard. All that is required is for the armed services to recognize as legitimate new values whose vehicles are active minorities in society (feminists, gays, ethnics, etc.) which, imitating the American Black movement of the 1950s and 1960s, now reject the domination and lack of legitimacy to which their old, ascribed underdog status hitherto subjected them. It is enough for these minorities that the enviable positions that were closed to them now be opened on merit, i.e., free of discrimination, as for instance in the case of military females and the combat roles from which they are usually still barred. It is most likely that only a few women would volunteer for such roles: low numbers would suffice to represent *all* women. In other words, under conditions of free choice, cultural diversity and low force levels, representativeness of the military becomes qualitative and symbolic rather than mathematical. The entity to which reference is made is not the polity, but a mosaic of groups enjoying a recognized status in society.

29. These new legitimacy constraints are not easy to satisfy. Achievement of a zero-casualty objective is impossible to guarantee. As for internal diversity, it runs counter to the conclusions of classic social science works on the determinants of primary and secondary unit cohesion. "The effects of few variables on group cohesion have been so thoroughly studied as that of group homogeneity. Similarity of race, ethnicity, occupation and age, for example, contributes to group cohesion primarily through normative integration. Individuals of similar backgrounds tend to share (to a greater degree than individuals of dissimilar backgrounds) common attitudes and values. Background also contributes to interpersonal attraction, however, because individuals tend to like people similar to themselves in values and attitudes. Further, homogeneity influences more remote relationships because individuals tend to believe, whether true or not, that people with backgrounds similar to their own also share attitudes similar to their own. Many armies (...) have taken advantage of this in organizing units in regional, racial or ethnic background," wrote Wesbrook (1980). Yet it is true that while primary and secondary cohesion is more difficult to achieve in units and small groups that are culturally diverse, it is not altogether impossible: only longer to promote, all things equal. Also, effectiveness in action can rely on emulation or competition to a greater extent than in homogeneous units.

30. However, the range of options offered by Cold War conditions left little room for the military as a group to allow material or symbolic self-interest, or "absolutist" values (for instance, the belief in the desirability of strategic victory,

in MacArthur's sense), to prevail at the expense of the pragmatism required by a situation eminently dangerous for all parties concerned. The phenomenon thus remained within the acceptable limits of the games played by an administrative pressure group in a democracy.

31. This is where the classics left room, between empirical analysis and broad normative prescriptions, for theoretical developments which are certainly welcome.

32. Agency theory is perfectly compatible with the normative dimension. All that is required is that description and analysis of "is" be supplemented with the limits imposed by "ought," so that aberrant or deviant institutional behavior (military "shirking" beyond those limits) be identified as such.

33. These recent incidents and scandals have involved the cover-up (or neglect) of isolated murders of civilians while on overseas peace support duty (Canada, Italy), overt neo-Nazi behavior (Germany), racist overtones in hazing (Britain), and sexual harassment (United States). It should noted, however, that failure to deliver on popular expectations of effectiveness in protecting human rights or averting atrocities on the part of belligerents in peace support operations may lead to dramatic loss of legitimacy: public questioning of the raison d'être of armed forces, as was the case in the Netherlands following the Srebenica massacre, which Dutch troops deployed to Bosnia had been unable to stop in 1995.

34. Cf. "Who's in Charge in Bosnia ? NATO and UN Fight it Out," *International Herald Tribune,* 3 October 1994.

35. It must be said in his defense, however, that he was to a large extent merely restating the principles of the 1984 Weinberger Doctrine. The forceful way in which he chose to phrase that reaffirmation probably owed much to the pulling and hauling between Pentagon and State Department, in view of the latter's post-Cold War interventionist leanings.

36. "Le chef d'état-major de l'armée de terre interpelle Lionel Jospin sur la baisse des crédits militaires," *Le Monde,* dimanche 16-lundi 17 novembre 1997, p. 1. Little notice seems to have been taken of the fact that the socialist Minister of Defense felt obliged to authorize publication of the interview, which confirms that the military wield more clout in their relationship with politicians than at any time since the beginnings of the Fifth Republic.

37. A causal model consisting of a few deductive propositions generating game-theoretic equilibria as a function of parameters governing interaction among principals and agents, to be checked against the empirical/ historical record, can be a powerful tool. Yet Feaver (1996b) is wrong to consider that the Weberian approach used by the Janowitzian school (inductively derived ideal-types) does not lend itself to hypothesis testing: causal adequacy of hypotheses derived from ideal-types to empirical findings is a primary concern in Weberian methodology. Cross-fertilization of the two approaches is both possible and useful.

References

Avant, Deborah 1997. "The Principles of Agency : New Institutionalism and Civil-Military Relations." Paper presented at the Biennial IUS Conference, Baltimore, MD, 24-26 October.

Avant, Deborah 1994. *Political Institutions and Military Change.* Ithaca, NY: Cornell University Press.

Betts, Richard 1977. *Soldiers, Statesmen and Cold War Crises*. Cambridge, MA: Harvard University Press.

Boëne, Bernard 1996. "Les rapports armée-Etat-société dans les démocraties libérales." *Revue Tocqueville*, XVII, 1: 53-81.

Carr-Saunders, Alexander and Paul Wilson 1933. *The Professions*. Oxford: Clarendon Press.

Dandeker, Christopher 1994. "A Farewell to Arms?" In *The Military in New Times*, edited by Burk, James, 117-39. Boulder, CO: Westview Press.

Elias, Norbert 1950. "Studies in the Genesis of the Naval Profession." *British Journal of Sociology*, 1: 291-309.

Feaver, Peter 1997. "An Agency Theory Explanation of American Civil-Military Relations during the Cold War." Paper delivered at the Biennial IUS Conference, Baltimore, MD, 24-26 October 1997.

Feaver, Peter 1996a. "An American Crisis in Civil-Military Relations?" *Tocqueville Review*, 17, 1: 159-82.

Feaver, Peter 1996b. "The Civil-Military Problematique: Huntington, Janowitz and the Question of Civilian Control." *Armed Forces and Society*, 23, 2: 149-78.

Friedrich, Carl and E.S. Mason 1940. *Public Policy*. Cambridge, MA.: Harvard University Press.

Goodnow, Frank 1900. *Politics and Administration*. New York: Macmillan.

Huntington, Samuel P. 1977. "The Soldier and the State in the 1970s." In *Civil-Military Relations*, edited by Andrew Goodpaster and Samuel P. Huntington, 5-28. Washington, DC: American Enterprise Institute.

Huntington, Samuel P. 1957. *The Soldier and the State: The Theory and Politics of Civil-Military Relations*. Cambridge, MA: Harvard University Press.

Janowitz, Morris 1960. *The Professional Soldier: A Social and Political Portrait*. Glencoe, IL: Free Press.

Klein, Paul 1998. "Vers des armées post-nationales?" In *Les Armées en Europe*, edited by Bernard Boëne and Christopher Dandeker, 161-9. Paris: La Découverte.

Kohn, Richard 1994. "Out of Control: The Crisis in Civil-Military Relations." *National Interest*, 35: 3-17.

Larson, Arthur 1974. "Military Professionalism and Civil Control: A Comparative Analysis of Two Interpretations." *Journal of Political and Military Sociology*, 2, 1: 57-72.

Lewis, Michael 1939. *England's Sea Officers: The Story of the Naval Profession*. London: Allen & Unwin.

Mannheim, Karl 1940. *Man and Society in an Age of Reconstruction*. New York: Harcourt, Brace.

Mills, C. Wright 1956. *The Power Elite*. New York: Oxford University Press.

Moreigne, J.P.1971. "Officiers, pour quel office?" *Défense nationale*, mai: 718-27.

Moskos, Charles 1988. *A Call to Civic Service*. New York: Free Press.

Moskos, Charles 1986. "Institutional/Occupational Trends in Armed Forces: An Update." *Armed Forces and Society*, 12, 3: 377-82.

Moskos, Charles 1977. "From Institution to Occupation: Trends in Military Organization." *Armed Forces and Society*, 4, 1: 41-50.

Moskos, Charles and Frank Wood (eds.) 1988. *The Military: More Than Just a Job?* Washington, DC: Pergamon-Brassey's.

Moskos, Charles, John Williams and David Segal (eds.) 2000. *The Post-Modern Military*, New York: Oxford University Press.

Park, Robert E. 1900. "The German Army: The Most Perfect Military Organization in the World." *Munsey's Magazine*, 24: 376-95.

Perlmutter, Amos 1977. *The Military and Politics in Modern Times*. New Haven: Yale University Press.

Powell, Colin 1992. "U.S. Forces: The Challenges Ahead." *Foreign Affairs*, 71, 5: 32-45.

Ricks, Thomas E. 1997. "The Widening Gap Between the Military and Society." *The Atlantic Monthly,* July: 66-78.

Schiff, Rebecca 1996. "Concordance Theory : A Response to Recent Criticism." *Armed Forces and Society*, 23, 2: 277-83.

Schiff, Rebecca 1995. "Civil-Military Relations Reconsidered: A Theory of Concordance." *Armed Forces and Society*, 22, 1: 7-24.

Segal, David 1989. *Recruiting for Uncle Sam*. Lawrence: University Press of Kansas.

Segal, David 1986. "Measuring the Institutional/Occupational Change Thesis." *Armed Forces and Society*, 12, 3: 351-75.

Smith, Louis 1951. *American Democracy and Military Power*. Chicago: University of Chicago Press.

Snider, Don and Miranda Carlton-Carew (eds.) 1995. *U.S. Civil-Military Relations: In Crisis or in Transition?* Washington, DC: Center for Strategic and International Studies.

Weigley, Russell F. 1993 "The American Military and the Principle of Civilian Control from McClellan to Powell." *Journal of Military History*, 57, 5: 27-58.

Wesbrook, Stephen D. 1980. "The Potential for Military Disintegration." In *Combat Effectiveness*, edited by Sam Sarkesian, 244-78. Beverly Hills, CA: Sage.

2

From Wars of Independence to Democratic Peace: Comparing the Cases of Israel and the United States*

James Burk

Introduction

Perhaps the grandest hope in the tradition of modern liberal thought is that the world might some day live in peace. What distinguished this hope from a utopian aspiration or a messianic dream was the belief that the means for realizing it were within our grasp. The key to establishing "perpetual peace," as Immanuel Kant (1983) thought over two centuries ago, was to ensure that every nation was also (what we now call) a democratic republic.[1] The citizens of democratic republics who had to bear the costs of war would be far more reluctant to assume war's risks than a despot who could dispatch a country to war without having personally to pay the costs imposed. So simply put, Kant's idea seems naïve (Thompson 1994). Yet, in recent years a rich literature has developed that confirms the essential insight that there is a close and positive connection between democracy and peace (Dixon 1994; Russett 1993). Democracy, of course, has not spread to every nation. Nor do we enjoy perpetual peace. Nevertheless, we possess enough experience with democracies to be relatively confident that they are unlikely to go to war against one another, however likely they may be to wage war on other states. There is good evidence to warrant hope for a democratic world order in which collective violence is much less common than it is at present.

Should we ask why that is so, Kant's argument based on the willingness of citizens to bear the costs of war appears to be (if anything) too cynical. He reduced matters too quickly to self–interested calculations of material loss and gain when normative factors are more important. Using evidence drawn from contemporary and ancient times, Bruce Russett (1993) argues persuasively that democratic government depends to a considerable degree on the institutionalization of peaceful norms of conflict resolution. These norms do not only shape internal politics. Democratic leaders expect, Russett writes,

> to resolve conflicts by compromise and nonviolence, respecting the rights and continued existence of opponents. . . . Therefore, democracies will follow norms of peaceful conflict resolution with other democracies, and will expect other democracies to do the same with them. (1993: 35)

His conclusion is not an isolated one. William J. Dixon (1994) also found that democracies settled their differences relying on institutionalized norms of peaceful conflict resolution. When he distinguished various phases of conflict (from the least to the most serious) and degrees of democratization among states, he found that as conflicting states were more democratic, they were also more likely to settle their differences peacefully in an early (less serious) phase of the conflict. His findings held even after controlling for alternative explanations that considered whether states belonged to common alliances, relied on third–party mediators, or had prior experience of military conflict with one another. Ironically, peaceful relations among democracies may even be a source of power in international politics. In his recent overview of Cold War history, John Lewis Gaddis (1997) argues that one (albeit inadvertent) reason for NATO's success over the Warsaw Pact was that the principal members were democracies. Although the United States was the leading power in the alliance, its leaders carried their democratic habits with them into negotiations with their allies. This gave European NATO members a larger influence over policy and higher morale than did their Warsaw Pact neighbors who labored under the tutelage of an authoritarian Soviet regime.

No one knows with certainty whether the new wave of democratization is evidence that international relations are moving away from a realist model toward one that conforms more closely to our understanding of democratic community (Diamond and Plattner 1993; Wendt 1994). There are reasons to be wary. The present wave of democratization cannot be presumed to continue without interruption. Democracy depends on a double identification of leading elites with one another and of this unified elite with ordinary citizens. Current trends in economic globalization at a minimum weaken the connections among economic and other elites and the dependence of economic elites on the ordinary citizens of any one nation. Economic interdependence may have a perverse effect on the trend toward democracy. Other threats to democratization, less subtle than this, can easily be imagined. If we believe, however, that some movement toward a democratic peace is possible, we are assuming a projection of domestic constitutional arrangements onto the world stage. Democratic countries establish institutions of self–government to resolve domestic conflicts peacefully, according to the rule of law. The habit of living under these arrangements is then carried over into and begins to characterize the realm of international affairs. The assumption is warranted, though as we shall see it requires clarification.

It is far more difficult than in the past to maintain a sharp distinction between domestic and international peace. Indeed over the last fifty years, major outbreaks of collective violence within states have become far more common and in fact are more common than wars between states. Consider as examples the civil wars in Lebanon, Somalia, Bosnia and elsewhere in Africa and Eastern Europe, the too numerous and too frequent terrorist attacks, and the determined violent opposition between groups in Northern Ireland, Iraq, and Israel. These conflicts are examples (by no means all inclusive) of what Kalevi J. Holsti (1996) calls "wars of the third kind." We may wonder why they have proliferated in the current era. Obviously, the capacity of states to maintain the rule of law, to limit the scope of collective violence within their borders is highly variable. But what explains the variation? Holsti's inquiry into this matter suggests that intra–state violence (like inter-state violence) is least likely among the most stable democratic countries. The reason why is that democracies are also "strong states."

State strength can be largely conceived in instrumental and material terms. But what Holsti emphasizes instead is the importance of legitimacy, defined simply as the state's uncontested right to rule over the political community.[2] That legitimacy rests he argues on "authority, consent, and loyalty to the idea(s) of the state" and on the degree to which "groups and communities within the polity accept and tolerate each other" (1996: 84–90). Democratic states are more likely than others are to meet the test of legitimacy and to be strong. Their citizens are less likely to believe they must resort to collective violence to achieve economic, social or political objectives.

In this paper, I try to clarify these ideas about democracy, state strength, and the absence of intra–state violence. So far, the discussion treats democracies as a homogeneous class of states, varying in degree, but not in kind. Yet we know that democracies are not all cut from the same cloth and we may suspect that all kinds of democracy are not equally immune to intra–state violence or immune only to the degree that they are democratic. It is reasonable to ask whether all are constituted equally in ways that their citizens accept as legitimate and that effectively establish and maintain peace within their borders. If not, what are the critical differences between them? Why are some democracies more likely than others are to keep the peace? The questions are too broad to answer completely, but some progress might be made. Limiting the task undertaken here, I want specifically to examine how the founding of democratic states through revolutionary wars of independence affects our answer to these questions. Much depends, I suggest, on the substance of the ideological movement that led to the establishment of the democratic state and the way that substance was embodied in the constitution of the state, making it a democracy of a particular kind.

Translation from a revolutionary ideology to a working constitution is no simple matter. Revolutionary movements are one kind of charismatic movement. To go from revolution to constitution then is a special case of the routinization of charisma. When focusing on this problem, Max Weber (1978) argued as if charisma was always extraordinary and could never last. For him, studying the routinization of charisma was largely a matter of how to choose the next generation of leaders who themselves would not be charismatic.

He thought that how the new leaders were selected would help us know whether the new society would survive and if so whether it would tend toward traditional or rational legal forms. The charisma itself would expire. I want to treat the matter more broadly, building on Edward Shils's (1975: 134) observation that charisma has a normal as well as an extraordinary form, that charisma is a real, continuous and effective force in the ordinary life of society. The process of routinization entails more than solving the problem of succession. It is in essence the problem of writing (or settling on) a constitution, of establishing the rules, norms, offices, etc. that embody the charismatic qualities, however attenuated, on which ordinary life depends. To understand how likely democracies are to establish and maintain domestic peace we need to ask how charisma, intense and concentrated in a revolutionary movement mobilized for war, is subsequently routinized and allocated through the constitution of the state.

To develop this theme, I will compare two important cases of democracies, the United States and Israel. While the two states were founded in different eras and under radically different circumstances, they are alike in certain critical respects. They both were founded as a result of wars for independence waged by revolutionary leaders who were charismatic and ideologically informed. They are not democracies like that of Germany or Japan or the many new born democracies in Eastern Europe whose forms of government were externally imposed after war as a condition for peace. Nor are they democracies like France or Britain that could rely for a sense of unity on a long history of continuous settlement in one region. The United States and Israel are immigrant societies. At their founding, both had explicitly to confront difficult questions about who in their fledging republics would count as citizens, why, and to what degree. Most important, in both cases, they are democracies established by revolutionary wars of independence. This distinguishes them from other major cases of revolution—not only in Russia, China and Iran, but also in England and France—that did not end by establishing democratic states, but whose immediate aftermath was to create authoritarian rule or at best a conservative restoration regime.[3] Before examining the cases, however, it is important to state in greater detail the logic of the argument that links revolutionary wars of independence to the prospects for a domestic, democratic peace.

Revolutionary Wars of Independence and
the Constitution of Democracy

The literature on revolutions is broad and deep, far beyond in scope what is possible or necessary to review for present purposes. We are not interested here in the origins of revolution, the factors affecting the success of revolutionary wars or even the outcomes of revolutionary war. Our ultimate interest is to know how the prospects for domestic "wars of a third kind" vary within democratic regimes and our hypothesis is that these prospects are connected with the way democratic revolutions are settled, particularly with the constitutional settlement that allocates revolutionary charisma throughout society. Consequently, we are assuming as settled what remains an outstanding problem, namely that the revolutionary struggle will end with the establishment of a democratic state. Yet our problem is large enough. To solve it, I suggest that we begin by examining the role of the citizen soldier in revolutionary movements. As Jack A. Goldstone (1991: 435–6) has observed, armies of the revolution embody the revolutionary ideology and "their eventual dominance is not merely a triumph of their own strength but also a final victory of the revolution." If we want to know what that victory entails we have to know something about the soldiers who made the fight.

Soldiers in revolutionary movements are unique. Recall Amos Perlmutter's (1977) distinction between professional, praetorian, and revolutionary soldiers. In essence, his typology rests on Max Weber's (1978) theory of legitimacy. He associates a type of soldier with each one of Weber's three forms of legitimate domination. Professional soldiers represent the prototypical bureaucrat of rational–legal domination. They are rational experts employed as state officials and operating within a legal framework for the benefit of the nation–state. Praetorian soldiers represent prototypical traditional domination. They exercise power to benefit the regime, ethnic group, tribe or other status group that they decide best represents the interests of the country; not surprisingly they value political loyalty and coup–making skills above professional expertise in the management of violence. Finally, revolutionary soldiers represent prototypical charismatic domination mobilized in party movements, opposing the status quo and embodying a law unto itself. Unlike professional soldiers, revolutionary soldiers do not wish to distance themselves from the larger society that their movement would lead. On the contrary, they are integrated with the revolutionary movement and with that movement's identification with society. But, like all charismatics, their "impact is greatest during the period of transition to the revolutionary regime. Once institutionalized, order and a more normal regime are needed" (Perlmutter 1977: 209). In other words, as charisma is routinized, the revolutionary soldier moves toward either professionalism or praetorianism.

Where does the citizen soldier fit within this typology? The phrase is not one that Perlmutter uses. What I shall argue, without attempting slavishly to follow Perlmutter's distinctions, is that citizen soldiers are what he would call revolutionary soldiers, that they are representatives of a charismatic movement and are themselves possessors of charismatic authority. If true, this gives citizen soldiers a social and political importance that extends beyond (and does not necessarily depend on) their technical competence as soldiers on the field of battle. Nonetheless, their charismatic possession (attributed to them by others and perhaps self–attributed as well) rests on their willingness and ability to confront and master the extraordinary and dangerous realm of war. Moreover, they do so not simply in their own names or for their own sakes, but in defense of the society they belong to and help constitute. It is an essential element of their charismatic status, therefore, that their identity is not exclusively that of a soldier. Citizen soldiers must be seen to be successful in their pacific civilian roles as well; their civilian status as settled members of the community confirms their charismatic accomplishment as soldiers: they are living in a democratic order whose survival depends on their sacrifice. This, perhaps, helps to explain the militancy with which citizen soldiers oppose deployments for purposes other than home defense even when material considerations about their contribution to the local economy or the physical security of their families do not apply. Their status as citizen soldiers is a local accomplishment meaningful only as tied to the political order they defend. Sent abroad, the citizen soldier becomes just another soldier, either professional or praetorian depending on the particular historical circumstance. Even in victory, the connection between accomplishment far from home and the survival of the soldier's own democracy may be hard to draw.

While charismatic, citizen soldiers are not self–mobilizing, even for home defense. They are recruited into service and trained as part of a revolutionary, broadly democratic social movement that justifies the existence of the state. The movement itself is charismatic and might well be called a constitutional movement. Its purpose is to establish the foundation and framework within which a particular democratic society can flourish. It carries within itself an idea of justice and a rule of law, defining how people ought to relate with one another. It embodies a direct connection with the transcendental realm of ideals. And when citizens are mobilized as soldiers in its cause, they become bearers of its charisma. How that charisma is socially organized remains to be said. Of course, it fluctuates in intensity. Speaking generally, it is concentrated most intensely when the constitutional bases of society are directly threatened by violent means and it is most dispersed when threats are distant or hard to imagine (Shils 1975, esp. chap. 7). But how this fluctuation affects the connection between citizen soldiers and democratic society is a matter of great importance that no one to my knowledge has adequately explored.[4]

We can start by noting that, in all democratic societies, sovereignty is supposed to reside somehow in the people. Yet this is a relatively vague and unsatisfactory formulation. In practice, democracies vary considerably in how their sovereignty is organized. I contend that the particular organization of sovereignty results from the way charisma is allocated and routinized when societies are established (or reestablished) through war. The warrior's charisma concentrated in democratic constitutional movements is allocated—either concentrated or dispersed—through the bed of political structures that the movement establishes as war draws to a close. In substance, the practical ideology of the victorious revolutionary movement—the ideology that actually motivated citizen soldiers to mobilize and fight—sets parameters around what these structures can be (Goldstone 1991: 416–58). And the allocation is routinized institutionally as citizen soldiers agree to uphold the new political arrangements whether directly through the force of arms or indirectly by acquiescence or tacit consent.

An important factor affecting the character of the new democratic order is whether the allocation of charisma is concentrated or dispersed. When charisma is widely dispersed, allocated to the central state, to lower–level governments, to intermediary institutions, and to individual citizens, then the foundation for a pluralist democratic society is laid. What that means specifically has been well described by Ernest Gellner's (1994) discussion of a civil society. At a minimum, it refers to a society with two traits: (a) some measure of political–coercive centralization whose power is countervailed by a highly pluralistic social and economic organization, and (b) a "modular" culture that endows individuals with common organizational skills allowing them easily to associate in cooperative pursuit of life's goods. In such a society, the interests of the community, the citizens, and the institutions they form are balanced, with none ultimately dominating the others. Underlying this arrangement is a belief that everyone possesses the capacity for self–government and a good society is one that respects and nurtures that capacity by making it difficult for any individual or group to govern over all.

When charisma remains highly concentrated, allocated only to individual citizens and their central government (a minimalist definition of democracy), then the foundation is laid for a majoritarian democratic order. What that means specifically has been well–described by Tocqueville (1945) in his analysis of the dangers of the unlimited power of the people and the tyranny of the majority. "Tocqueville's fear," Adrian Oldfield (1990: 121) notes, "was that people would actually choose to live under a system . . . which concentrated all political decision–making at the centre." And when that occurs, now quoting Tocqueville, "the state reaches the extreme limits of its power all at once, while private persons *allow themselves* to sink . . . to the lowest degree of weakness"(original emphasis). Underlying this arrangement is a belief that the country is itself a moral enterprise, embodying the will and aspirations of the people, and that securing its well–being is the first responsibility of everyone. Democracy it may be. But it is democracy in which the balance between individual powers and interests and the interests and powers of the community are tilted very much in favor of the community. Or, perhaps I should say, they are tilted very much in favor of the dominant elite exercising power through the central state. Tocqueville (1945: vol 2, 301) believed—perhaps because of the French Revolution—that such an outcome was most likely to occur:

> at the close of a long and bloody revolution, which, after having wrested property from the hands of its former possessors, has shaken all belief and filled the

nation with fierce hatreds, conflicting interests, and contending factions. The
love of public tranquillity becomes at such times an indiscriminate passion . . ..

Whether this prediction is always borne out is a matter that requires further study. It is a plausible hypothesis to emphasize the
intensity and the character of the conflict. I would add that how
intense the conflict is depends on how closely the revolutionary ideology is tied to the pre–existing structure of social cleavages. If
belligerents in a war for independence were already members of
opposing social factions (ethnic, religious, economic, etc.) and their
opposition was codified in the ideology that justified the war, then
the intensity of the conflict would be exacerbated. The effect would
be greater if it were reinforced by social cleavages among bordering
states. In such a circumstance, the victors, however committed to
democratic ideals, would be more likely to concentrate in their own
hands control over the democracy they established.

Based on these expectations, majoritarian democracies in which
charisma is highly concentrated are unlikely to be the most peaceful
ones. Beliefs are shaken. The nation is filled with fierce hatreds,
conflicting interests, and contending factions. And the newly established state is unwilling to share its sovereignty. Belief in the legitimacy of the state's right to rule is, under these conditions, most
likely to be confined to the dominant, victorious elite. The capacity
to adjudicate and peacefully settle conflicts between the dominant
and subordinate groups is likely to be weak. In short, the state is
weakened, despite its centralization, and more prone either to endure wars of a third kind or from its love of public tranquillity, to
preempt them by adopting undemocratic means of coercive control.
Ironically, pluralist democracies, the product of a less intense
struggle, may be in a less precarious position despite the greater
dispersion of charisma, that is, of the sovereign power to rule.

Which kind of democratic order will citizen soldiers help to establish and maintain in the immediate aftermath of a revolutionary
war for independence? The answer, I believe, depends on the intensity of conflict, to include consideration of the relation of the conflict to pre–existing social cleavages, and on the substance of the
revolutionary ideology used to mobilize and motivate citizen soldiers to fight. These factors, closely related, affect how charisma is
allocated in the political settlement that follows the revolution. They
affect whether the revolutionary bearers of charisma in democratic

constitutional movements will found a pluralist or a majoritarian democratic order, an order that will be less or more likely to endure Holsti's wars of a third kind. To illustrate these contingencies and the differences they make, let us briefly compare certain features of America's Revolutionary War and the Israeli War for Independence. In both cases, citizen soldiers played a decisive role in the political (and, in Israel's case, in the military) outcome. In both cases, the revolutionary wars ended with the establishment of stable and enduring democracies. But the cases differed in the character of the conflict and in the substance of their revolutionary ideologies. These differences led in one case to the establishment of a pluralist democracy and in the other to a majoritarian democracy, with the predicted consequences for the prospects of a democratic peace.

America's Revolutionary War and Pluralist Democracy

Once we think of war as an intense charismatic movement, our attention turns to consider what kind of social order the movement would constitute if it prevailed. It is a way of thinking about "war aims" less in terms of strategic objectives and more in terms of normative commitments. Although it could be said that both the American and the Israeli wars for independence were fought for "liberty" —what modern war was not?—in fact their normative commitments can be sharply distinguished from one another. The differences between them show up in the intensity and character of the conflicts, in the motives and conduct of the citizen soldiers who fought, and in the political settlements (the allocation of charisma) made at each war's conclusion.

The American Revolutionary War was committed to break colonial ties to the British crown, to establish the independence of the colonies as sovereign states bound in union together, and to transform colonial subjects into citizens of a republic. The conflict was over allegiance. Was allegiance due to a King whose government failed to respect the subjects' elemental rights of petition and representation before imposing laws on them? The revolutionaries who signed the Declaration of Independence thought not. Colonists had to decide whether they wished to remain loyal subjects of the King or to become faithful citizens of the new republican states that were founded under the terms of that declaration and banded together through the Continental Congress. The choice was real. A South Carolinian physician and historian noted at the time, the difference between subject and citizen was "immense"; the subject is "under the power of another" and must "look up to a master," while the

> citizen is an unit of a mass of free people, who, collectively, possess sovereignty. . . . Each citizen of a free state contains within himself, by nature and the Constitution, as much of the common sovereignty as another. (Quoted in Wood 1992: 169)

In addition to this immense difference, there was also the practical reality that how one decided had consequences for the security of one's property and ultimately for the security of one's family and one's life.[5]

Nevertheless, it cannot be said that the war was as violent as it might have been. On the contrary, it was fought with considerable restraint on both sides. In John Shy's (1990: 219) well–known formula, the conflict was a "triangular" war in which two armed forces contended less against each other than for support and control—for the allegiance—of the civilian population. In such a war, excessive use of force would be counter–productive. Another issue that Shy does not address was equally important. The revolutionaries challenged the right of King George (and his parliament) to rule over them and so aimed to overthrow the pre–existing aristocratic elite, with consequences for social leveling that were perhaps unexpected by some (Wood 1992). But otherwise the conflict was not organized around other social cleavages. Neither religious nor ethnic nor other collective identities that have so often intensified the fighting of wars were salient here.[6] It was not a conflict of one people against another.

For revolutionary patriots, use of coercion against their countrymen was especially difficult. The revolution was authorized by majority votes in representative assemblies and the principle of majority rule (they said) made the decision binding on the minority who were either indifferent, apathetic, or otherwise disagreed with it. Nevertheless, while open disaffection with the majority decision could not be tolerated, neither could coercion be used to make this so–called minority compliant. Resort to coercion would violate a fundamental principle over which the revolution was being fought, namely, that legitimate government rested on consent and consent had to be freely given. The difficulty was overcome largely by legal means that encouraged the unwilling and apathetic to serve as citizen soldiers in the new state militias. Following the Declaration of Independence, the new states passed treason laws to take effect as of a certain date, allowing individuals time to choose sides. Whoever continued to live in the state after that date presumably had chosen to be a citizen (women and children excepted) and would be liable to the charge of treason if caught giving aid to the enemy. Whether remaining in place reflected a positive or free choice could be debated, of course; not everyone had resources to move. Nevertheless, even after these laws were passed, executions for treason were rare as convicted traitors were often given pardons if they agreed to enlist as soldiers.[7]

Probably such reluctant citizen soldiers added little to the military's effective fighting force. The main point lies elsewhere. Effective or not on the battlefield, the militia "played the role of political teacher" (Shy 1990: 236). While many (including those who were not traitors) were reluctant to serve in the state militias and were even more reluctant to serve in the Continental Army, they did serve and in relatively large numbers. Morris Janowitz reports that as many as 100,000 men—between 40 and 50 percent of all men of military age—fought for the revolution (Janowitz 1983: 36). It was the surest way of demonstrating one's loyalty and loyalty demonstrations were required at a time when failure to take sides was at least to put one's property at risk. Once in the militia, even citizens who fought no battles carried out raids and other acts that branded them as rebels and identified their fates with the fate of the revolution. Collectively, they enforced the new republican law and order wherever the British army did not impose, which was a vast expanse of territory. In short, once the revolution was declared, those who chose the patriot's side and entered militia service were offered an intense course of political education in the constitution of a democratic republic and its vision of society (Shy 1990: 236–43). They came by experience to know that they contained within themselves as much of the common sovereignty as any other and that made them defensive of their rights, even against their fledging government.[8]

Once the war ended, it was time formally to address the constitutional question the temerity of citizen soldiers had only raised. How would this common sovereignty be shared among citizen soldiers, the states, and the Continental Congress of the United States that waged the war? The answer was to establish a pluralist democracy in which sovereignty was widely shared. After a few years under the weak Articles of Confederation, which concentrated virtually all power in the states, leaders of the revolution gathered to write the Constitution that established a stronger federal system by dividing power more equally between the central government and the states. Worth recalling is that this idea was strongly opposed at the time. The "anti–federalists" worried that the new republic would be too large to win the loyalty of the people, that the newly empowered central government would be tempted to use its military capacity to raise a standing army to enforce its laws, and that the Constitution failed to define and guarantee the rights of citizens (McDonald 1985: 266–7, 285; Storing 1981). The writers of the Federalist papers tried, with some success, to assuage these fears, including fears that a standing army might displace the citizen solider and undo the revolution.[9] Nevertheless, in the end, ratification of the Constitution was won on the promise, subsequently kept, to amend the Constitution to include a bill of rights. Those amendments protected the pluralist character of the new republic by guaranteeing, among others, the freedoms of speech, religion, assembly and (not least) the right of citizens to bear arms as soldiers in a well–regulated militia.

It is commonplace now to note that the new American republic failed fully to realize its own pluralist democratic dreams. Large segments of the population were excluded from the circle of citizenship and so denied any share in government. They were not equal participants in the allocation of charisma. Women were the largest of the excluded groups. For a brief period after the revolution, women in New Jersey could vote. But this was exceptional and did not last (Norton 1980). Black slaves were recruited to fight on both sides of the conflict and promised their freedom in return, a promised sometimes kept (Quarles 1973). But slavery remained a scourge on American democracy, as has its legacy even after slavery was put finally to an end. Native Americans fought mainly for the British, against the Americans, rightly perceiving that the expansion of the American nation was a dire threat to them. Sending an expedition into Iroquois country in 1779, George Washington said simply that he wanted the Iroquois "destroyed" (Millet and Maslowski 1984: 73–5). Native Americans did not seek to become citizens after the revolution was won. Nor would their application for citizenship have been welcomed. There were limits to pluralism. Nevertheless, American democracy was much more pluralist than the colonies had been. For example, during or soon after the revolution, many Northern states moved to abolish slavery altogether, while many Southern states barred further imports of slaves (Jameson 1926). Perhaps more important, the revolution provided an enduring political vocabulary that idealized and legitimated a pluralist liberal democracy. Only once, during the American Civil War, was there a violent domestic uprising that challenged the right of the pluralist state to rule. The end of that tragic conflict was to expand the constitutional basis for an inclusive, pluralist regime. To be sure, in practice, dominant groups used their dominance to win advantages for themselves. Nevertheless, the vocabulary of democratic pluralism was always available and used sometimes successfully to challenge the practice. No contrary vocabulary that favored majoritarian democracy or privileged the claims of one group over another has had power enough to uproot or unseat it.

The Israeli War for Independence and Majoritarian Democracy

Compared to the American revolutionary war, which lasted from 1776 to 1783, the Israeli war for independence was a short–lived affair. It began on 14 May 1948 and ended with an armistice at the end of the year. This, of course, provides no measure of the intensity of the conflict, which in fact was a "total war," though regionally confined. Moreover, armed hostilities between Jews and Arabs in Palestine had a long history and, unfortunately, a long future too. Although an armistice agreement was reached among the warring parties in the spring of 1949, no final peace agreement between Israel and its neighbors was signed at the time and the legitimacy of the state remained a violently contested issue. Nevertheless, in January 1949, national elections were held and, on 10 March 1949, the provisional revolutionary government officially turned over the reigns of power to the first constitutional government. Israel was a state and its sovereignty was recognized by nations throughout the world, if not by its Arab neighbors.

For the Israelis, the War for Independence was ideologically driven. The struggle embodied their strong normative commitment to establish a Jewish state in Palestine. The ambition to do so was authorized by a United Nations resolution passed on 29 November 1947. The resolution called for an end to the British Mandate over Palestine, in place since the end of World War I, and planned to partition Palestine into two parts, one allotted to the Jews and the other to the Arabs (with Jerusalem in an "international zone"). But the ambition was not externally imposed. It was a product of the Zionist movement. The Zionist movement began in Europe in the nineteenth century. As Shlomo Avineri (1981) argued, it arose in response to the growing spirit of nationalism there, a spirit that intensified outbursts of anti–Semitism. It responded also to the spread of liberalism that had led to the "emancipation" of Jews but, paradoxically, posed difficult questions about how to define and maintain Jewish identity. The movement was heterogeneous. The central currents within it were liberal and socialist, with strong connections to the labor movement, and also secular, while nonetheless placing a high value on Jewish culture and the potential for development among the Jewish people. Other currents were politically right wing, still secular but traditional and conservative. And, despite the wariness of the Jewish religious establishment, currents of religious faith were present as well. Underneath this heterogeneity, however, was a core commitment to establish a Jewish nation. As stated clearly in the program adopted at the first Zionist Congress, held in Basel, Switzerland in 1897, the primary aim was "to establish for the Jewish people a publicly recognized, legally secure home in Palestine" (Quoted in Goodman 1922: 855). This aim made Zionism a revolutionary movement and, from the viewpoint of Arabs in the region, a subversive movement.

Unlike the American revolutionary war, the Israeli war for independence was not moderated by the belligerents' hopes to win the support and allegiance of the civilian population.[10] It was a "total war" in the sense that it was waged not just between armies, but against civilians as well. It was a war between whole peoples, Arab and Jew. Evidence of this is not difficult to find. During the Arab revolt in 1936 Arab guerrilla forces launched attacks against Jewish settlements, inviting retaliatory attacks against Arabs by the Jewish right wing IZL (National Military Organization). Once it became clear that the UN partition of Palestine would take place, most Palestinian Arabs flatly refused to live under Israeli rule, preferring the alternative of flight into exile. During the war, Arabs shelled Jerusalem destroying thousands of homes and buildings and inflicting 1,200 civilian casualties. They also blocked the supply line to Tel Aviv, laying siege to Jerusalem and threatening mass starvation had not the siege been broken by heroic efforts to reopen the Jerusalem highway. On the Israeli side, in a deed repudiated by the Haganah and punished by the provisional government, IZL forces retaliated against an Arab attack in April 1948 by massacring and then mutilating the bodies of 200 Arab men, women and children. By June 1948 efforts to encourage Palestinian Arabs to remain in their homes were abandoned. When Israeli troops captured Arab villages, they destroyed the Arabs' homes and ousted them from the village, turning them into refugees. Moreover, David Ben–Gurion, leader of the provisional government, argued that Israel should not encourage the return of Palestinian Arabs after the war.

The war was unlike the American revolutionary war (and more like the situation French revolutionaries confronted in the 1790s) in another respect. It was an international conflict. The Israeli Declaration of Independence, proclaimed on 14 May 1948, was taken as an invitation by Israel's Arab neighbors to invade the new state, with the aim of abolishing it before it was established. Israel was hard–pressed to respond. The Arabs were well armed and enjoyed the tacit support of Britain. They attacked at once on three fronts, Lebanon and Syria from the north, the Arab Legion and Iraqi forces from the east, and Egypt from the south. Britain had prevented the Israelis from mobilizing arms or openly raising and positioning armed forces in the months before the expiration of their rule. The Arab forces were variously effective. But, under these circumstances, they were able in short order to threaten Tel Aviv, lay siege to Jerusalem (capturing the Old City) and control the Negev. This is not the place to recount the battles of the war, fought in waves between truces imposed by the United Nations for the rest of the year. It is enough to say that Israel's provisional government was able to mobilize and equip the population for war and to lead its military forces effectively to counter the invasion. It pushed Arab forces back to their original positions and established military control over a larger territory than it was allotted under the United Nations' partition plan. The price of military success was high, quite beyond the dollars spent. The Jewish population in Palestine at the beginning of the war was under 600,000. At the beginning of the conflict, at most 30,000 were mobilized to fight. Within six weeks, another 30,000 were mobilized and by the end of the war 100,000 served in the Israeli Defense Force. Of these, 6,000 were killed and 30,000 were wounded. This represents an extraordinarily high level of military participation, exceeding the high point of American mobilization for World War II, and an equally extraordinary high rate of casualties. Obviously, Israel was a "nation in arms" in the fullest sense; its army was a "people's army." Its existence was owed to the effort and commitment of its citizen soldiers. These soldiers and their military and political leaders embodied the ideals of the revolutionary movement and bore its charisma. They brought the aspiration for a Jewish state into reality.

But what kind of state would it be? How would the charisma concentrated in the revolutionary movement be allocated and routinized in the constitution of the state? The core commitment of the Zionist movement, its heterogeneous ideologies orbiting around the central current of social democracy, and experience of building national institutions in Palestine under the British Mandate provided the essential answers to these questions. The central imperative was that Israel be a state for the Jewish people; it was the normative commitment that defined the Zionist movement. This commitment was underscored since the turn of the century by the experience of the Holocaust and by the dire straits of so many Jewish refugees who needed a home but were encamped in Europe as "displaced persons" following the world war. It was strengthened further when Arab countries responded to Israel's independence by expelling the Jews who had been settled there. Legal expression of this commitment is found in the Law of Return, passed by the first Knesset, and the Nationality Law of 1952, which together gave all Jews in the world a legal right to immigrate to Israel and conferred automatic citizenship on Jews who declared their intention to settle there (Kook 1995: 322–5).

Also critical was that Israel be constituted as a liberal democratic state. As Erik Cohen (1989: 149) observed: "Revolutionary Zionism . . . sought to establish not only a national state but an enlightened nation–state, one based on principles of democracy and equality for all citizens irrespective of race or religion." There was ample precedent for adopting democratic processes in the new state. The Jewish community in Palestine (the Yishuv) had highly developed national institutions to govern over much of Jewish life, including a democratically elected assembly that observed principles of universal suffrage, "one man one vote," and proportional (instead of territorial) representation. The latter choice, carried on in the new state, favored the development of ideological parties and resulted in indecisive election outcomes, requiring the formation of coalition governments. Nevertheless, it also reflected the political reality that before statehood, groups could choose to secede from the assembly rather than obey its authority (as some did). It encouraged development of what Dan Horowitz and Moshe Lissak (1978) have called the "rules of the game," to regulate conflict by compromise rather than by attempting to gain exclusive control. A tradition of abiding by these rules was helpful to establishing a stable democracy. But this was within the Jewish community. There remained an inevitable and continuing tension between these two constitutional principles, the first, exclusive, favoring the Jews, and the second, inclusive, based on liberal norms of civil society favoring universal equality.

The immediate issue, however, was to establish the constitutional authority of the state, to establish its legitimate right to rule. The Declaration of Independence provided for a provisional government to rule immediately. When possible, the provisional government would be replaced by a permanent government to be established (it was thought) in accord with a newly written constitution. In the midst of war, the provisional government had to act decisively to consolidate its control over Israel. That required provisions for domestic law and administration (currency, tax collection, postal system, etc.) to fill the vacuum created by British withdrawal. More important was gaining control over the armed forces. This would require strong efforts to concentrate the revolutionary charisma that in the Yishuv was dispersed more widely, as Horowitz and Lissak (1978: 39) remark, than institutional authority. Under the British Mandate, the Yishuv developed an array of self–defense forces without centralized command. The Palmach, an elite commando unit organized to fight the Germans during World War II, was within the Haganah, but supported by and centered within the kibbutz movement. The Haganah was formed in response to the Arab revolt in 1936, combining control over most of the self–defense forces under the control of the Jewish Agency and the National Council. Yet the right wing Revisionist Party maintained a separate military of its own, the IZL. These decentralized forces were all that was available at the start of the war for independence. The task of the provisional government was to establish one military, the Israeli Defense Forces (IDF), over which it was the sole authority. Ben–Gurion acted swiftly to do this, even to the point of using force against the IZL and risking a mutiny among the Palmach. But he was successful in his aim, thus ensuring that when the war for independence was won, it was one army under one state embodying the charismatic authority to rule over Israel. While a written constitution was not adopted, the provisional government successfully transformed its rule into a permanent and strong central government. It was a democracy and it recognized the legitimacy of disparate ideologies and religious beliefs.

But it was a majoritarian democracy in that the Jewish people—and more particularly the Ashkenazi Jews (from Europe) who carried the Zionist dream to Palestine—were the dominant group in Israeli society. It was their vision of Israel and their rule that prevailed. This was evident foremost in the structure of Israeli citizenship. We have noted already that Jews worldwide had privileged access to immigrate into and become citizens of Israel. But it was not the case that all Jews were treated equally. Oriental Jews, expelled from Arab lands and immigrating to Israel, were not treated as equals in the new state (Margalit 1998). More critically, the Arabs who remained in Israel were granted Israeli citizenship, including the right to vote, but they did not enjoy the same degree of citizenship as Jews (Kook 1995; Peled 1992). Not fully trusted to be loyal, the Israeli Arabs were not mobilized for military service as other citizens were. Their exclusion may be explained in terms of security requirements. Yet it imposed a serious stigma. Under continuing threat of invasion, Israel (until recently perhaps) has remained a nation in arms (Cohen 1995). Service in the IDF has been an important qualification for certain welfare benefits and jobs; to be a citizen soldier enhances one's civilian status; but these are advantages denied to Arab citizens (Horowitz and Lissak 1989: 206). Nor have Arab citizens been free (as Jewish citizens are) to purchase or in some instances even to lease property, an obvious economic disadvantage (Kook 1995). More demeaning, Arab citizens were deprived of equal civil rights from independence until 1966 as they were forced to live under military rule. This discrimination is not evidence that Israel was not democratic. But it shows that the dominant group in Israel did not believe that all its citizens would accept its legitimacy. Its politics were majoritarian and aimed to privilege the dominant group in just the ways that Tocqueville feared. Its strength as a state was limited to the degree that coercion (not consensus) was required to ensure compliance with its rule. This limitation remains a challenge to Israel's domestic peace even as prospects for peace with its neighboring states improve.

A Cautionary Conclusion

The point of this argument is not to suggest that the revolutionary war in the United States established democracy more successfully than revolutionary war in Israel. The evidence here warrants no such historical judgment. The challenges faced by these revolutionary movements were unique. It is not obvious that there were open paths better than the ones followed that leaders could have been chosen. While analytical comparisons are justified for the purposes of building theory, they do not overshadow the substantial achievement of both countries to establish stable and enduring democratic states against long odds.

The point rather is a theoretical one, to remind us that democracies are not all alike, that the prospects for peace within democratic societies vary depending on the kind of democracy they are. We can state the main propositions simply. The more majoritarian the democracy, the more democratic charisma is concentrated in the hands of a dominant group, the more likely it is that the state's right to rule will be challenged by subordinate or marginal groups. The more pluralist the democracy, the more widely democratic charisma is dispersed among all citizens, the more accepted and secure is the state's right to rule. Whether majoritarian or pluralist democracies are established depends, we have seen, to a large degree on the substance of the revolutionary ideology and the intensity of the conflict for independence that that ideology supports. In either case, however, it is likely that democratic states will be more peaceful and just than alternative forms of government will be.

Yet, as we look forward to a time when democracies are more widespread, we should not be blinded by the promise of the long–hoped for democratic peace that seems within our grasp. There are variations in the quality and justness of peace as consequential for the well–being of society as are variations in the quality and the justness of war. That fact is not changed by the triumph of democracy.

Notes

*   This paper is based on ideas originally presented at a conference sponsored by Hebrew University in Jerusalem to honor Professor Moshe Lissak. I am indebted to Professor Lissak and to the other Israeli conference participants for their thoughtful observations about the complexities of Israeli society. I am especially grateful to Professors S. N. Eisenstadt and Eyal Ben–Ari who encouraged me to undertake this comparative project and again to Professor Ben–Ari for giving me time to complete it.

1.   Kant took pains to distinguish a republican constitution from a democratic one. "Democracy," as a form of sovereignty, was appropriately contrasted with autocracy and aristocracy. The term "republican" referred in contrast to the form of the regime. Specifically, a republican regime separated the executive from the legislative powers. Wherever this separation was absent, the government was "despotic." Kant thought—perhaps having in mind the tumult of revolutionary France—that democracy necessarily took the form of despotism. The despotism he worried about was what we more often call the tyranny of the majority. He was right to worry for reasons discussed below. But he is on shaky ground empirically to suppose that democracies may not assume republican forms. Indeed, it is difficult to conceive of modern democratic states that are not representative republics.

2.   A state without legitimacy can hardly be strong. As Holsti (1996: 84) observed: "A totalitarian regime, for example, may rank high on institutionalization and extractive capacity, but that does not make it a strong state. Its rule is ultimately based on fear, force, and coercion rather than on consent or voluntary compliance."

3.   The French Revolution was of course a critical event in moving France toward a democratic state, as was for England its revolution of the seventeenth century. Both provided a store of precedents, symbols and myths without which it is impossible to explain their present democratic forms. But in neither case did the revolutions end with the establishment of new and enduring democratic republics. That was an achievement of future generations.

4.   There are studies that consider how mobilization for war affects the range and distribution of citizens' rights (Turner 1986: 69–70; Barbalet 1988: 37–40). My question is different. I ask how the charisma that citizen soldiers accumulate during revolutionary wars affects the constitution of a democratic

order once the war is ended. To my knowledge, this question has not been raised before.

5. Thomas Hickey was the first to be hanged as a traitor to the revolution on 28 June 1776, even before the independence was formally declared. Hickey was, however, a member of George Washington's personal guard. The Continental Congress moved "cautiously against disaffection among the civilian population," taking their property but only reluctantly and after careful "due process" taking their lives (Kettner 1978: 177–83).

6. Enslaved African Americans, for instance, fought on both sides of the conflict. Whether fighting for the British or the Colonials, blacks hoped to gain social respect and freedom from their service. Unfortunately, that was not the usual outcome (Quarles 1973). (To see how the revolution affected the status of women, see Norton 1980).

7. The argument in this paragraph paraphrases key portions of the argument in Kettner (1978: chap. 7).

8. Neimeyer (1996) has summarized the evidence of resistance by enlisted soldiers in the Continental Army to treatment by Congress or their officers which they thought violated their contract or their rights to contract as "free men" and "citizens" over the terms and conditions of their service.

9. This issue is most extensively treated by Alexander Hamilton in The Federalist No. 29, reprinted in Cooke (1982). For original documents on both sides of the debate, see Bailyn (1993).

10. The facts about the war reported in this paragraph and the next are taken from Sachar (1996: 315–53).

References

Avineri, Shlomo 1981. *The Making of Modern Zionism: The Intellectual Origins of the Jewish State*. New York: Basic Books.

Bailyn, Bernard (ed.) 1993. *The Debate on the Constitution*, 2 vols. New York: Library of America.

Barbalet, J. M. 1988. *Citizenship: Rights, Struggle and Class Inequality*. Minneapolis: University of Minnesota Press.

Cohen, Erik 1989. "The Changing Legitimations of the State of Israel." In *Israel: State and Society, 1848–1988*, edited by Peter Y. Medding, 148–65. New York: Oxford University Press.

Cohen, Stuart 1995. "The Israeli Defense Forces (IDF): From a 'People's Army' to a 'Professional Military'—Causes and Implications." *Armed Forces and Society*, 21, 2: 237–54.

Cooke, Jacob E. (ed.) 1982. *The Federalist*. Middleton, CT: Wesleyan University Press.

Diamond, Larry and Marc F. Plattner (eds.) 1993. *The Global Resurgence of Democracy*. Baltimore, MD: Johns Hopkins University Press.

Dixon, William J. 1994. "Democracy and the Peaceful Settlement of International Conflict." *American Political Science Review*, 88, 1: 14–32.

Gaddis, John Lewis 1997. *We Now Know: Rethinking Cold War History*. New York: Oxford University Press.

Gellner, Ernest 1994. *Conditions of Liberty*. New York: Penguin.

Goldstone, Jack A. 1991. *Revolution and Rebellion in the Early Modern World*. Berkeley and Los Angeles: University of California Press.

Goodman, Paul 1922. "Zionism." In *Encyclopaedia of Religion and Ethics*, 855–858. New York: Charles Scribner's Sons.

Holsti, Kalevi J. 1996. *The State, War, and the State of War*. Cambridge: Cambridge University Press.

Horowitz, Dan and Moshe Lissak 1989. *Trouble in Utopia: The Overburdened Polity of Israel*. Albany: State University of New York Press.

Horowitz, Dan and Moshe Lissak 1978. *Origins of the Israeli Polity*. Chicago: University of Chicago Press.

Jameson, J. Franklin 1926. *The American Revolution Considered as a Social Movement*. Princeton, NJ: Princeton University Press.

Janowitz, Morris 1983. *The Reconstruction of Patriotism*. Chicago: University of Chicago Press.

Kant, Immanuel 1983. *Perpetual Peace and Other Essays*. Indianapolis, IN: Hackett.

Kettner, James H. 1978. *The Development of American Citizenship, 1608–1870*. Chapel Hill: University of North Carolina Press.

Kook, Rebecca 1995. "Dilemmas of Ethnic Minorities in Democracies: The Effect of Peace on the Palestinians in Israel." *Politics and Society* 23, 3: 309–36.

Margalit, Avishai 1998. "The Other Israel." *New York Review of Books* (May 28): 30–35.

McDonald, Forrest 1985. *Norvus Ordo Seclorum: The Intellectual Origins of the Constitution*. Lawrence: University Press of Kansas.

Millet, Allan and Peter Maslowski 1984. *For the Common Defense: A Military History of the United States of America*. New York: Free Press.

Neimeyer, Charles 1996. *America Goes to War: A Social History of the Continental Army*. New York: New York University Press.

Norton, Mary Beth 1980. *Liberty's Daughters: The Revolutionary Experience of American Women, 1750–1800*. Ithaca, NY: Cornell University Press.

Oldfield, Adrian 1990. *Citizenship and Community: Civic Republicanism and the Modern World*. London: Routledge.

Peled, Yoav 1992. "Ethnic Democracy and the Legal Construction of Citizenship: Arab Citizens of the Jewish State." *American Political Science Review*, 86, 2: 432–43.

Perlmutter, Amos 1977. *The Military and Politics in Modern Times*. New Haven, CT: Yale University Press.

Quarles, Bejamin 1973. *The Negro in the American Revolution*. New York: Norton.

Russett, Bruce 1993. *Grasping the Democratic Peace: Principles for a Post–Cold War World*. Princeton, NJ: Princeton University Press.

Sachar, Howard M. 1996. *A History of Israel: From the Rise of Zion to Our Time*. 2nd ed. New York: Alfred A. Knopf.

Shils, Edward 1975. *Center and Periphery*. Chicago: University of Chicago Press.

Shy, John 1990. *A People Numerous and Armed*. Ann Arbor: University of Michigan Press.

Storing, Herbert 1981. *What the Anti–Federalists Were For*. Chicago: University of Chicago Press.

Thompson, William R. 1994. "The Future of Transitional Warfare." In *The Military in New Times*, edited by James Burk, 63–91. Boulder, CO: Westview.

Tocqueville, Alexis de 1945. *Democracy in America*, trans. Henry Reeve. New York: Alfred A. Knopf.

Turner, Bryan S. 1986. *Citizenship and Capitalism*. London: Allen & Unwin.

Weber, Max 1978. *Economy and Society*. Berkeley and Los Angeles: University of California Press.

Wendt, Alexander 1994. "Collective Identity Formation and the International State." *American Political Science Review*, 88, 2: 384–96.

Wood, Gordon S. 1992. *The Radicalism of the American Revolution*. New York: Alfred A. Knopf.

Part 2

The Politics of Civil-Military Relations

3

Civil-Military Relations in Israel in Crisis

Yoram Peri

Civil-military relations in Israel in the late 1990s have reached a critical state, unprecedented since the formative period of Israeli society in the 1920s and 1930s. These relations are in the midst of a process of rapid change; and since it is a polysemic crisis that stems from many sources, it is likely to continue and even deteriorate.

The process in which the Israeli society and military find themselves can be analyzed from various paradigmatic viewpoints, according to the intellectual taste and the inclinations of the researcher. It can be seen as a process of normalization, whereby a society that was involved in a prolonged war, a mobilized society, or a fighting society, becomes a civil society (Barzilai 1996). The process may be seen as one of democratization (Nueberger 1997). The change in Israeli society may also be analyzed as part of a global process of demilitarization, described in the literature as the formation of a "postwar society" (Shaw 1991; Moskos and Burk 1994). In this process war and the military occupy a smaller place in society, the social investment in them is reduced, the weight of the military and its influence decline in relation to the civilian society, and above all, the military ethos is weakened (Ashkenazy 1994).[1]

While agreeing with this description, I believe that in the case of Israel another aspect is involved. I would argue that the structural changes that Israeli society is undergoing are characteristic of a decolonization process, marked by the painful process of suppressing a revolting population and extrication from the occupied territories. Although various manifestations of this crisis have come to the surface in the past decade, Israeli society denies it, and some military officers who dared to expose certain elements of the crisis were silenced and even censured.[2]

Thus, the Chief of Staff aroused a storm when he first referred to this topic in public, at the first memorial ceremony for Prime Minister and Defense Minister Yitzhak Rabin held by the Ministry of Defense on October 8, 1996. Among other things he said, "... the IDF is losing its status [in society]... The sharp criticism that came from the wish to improve, out of caring, has given way to alienation. The direct contact between the IDF and Israeli society, which was so vital to the military, part of its uniqueness and a source of strength, this contact has become a burden, not always appropriate, somewhat stifling... How far we have come from the days when the IDF uniform was a source of pride, an honor."

However, anyone following civil-military relations in Israel should not have been surprised by these words (Peri 1990). Following the Chief of Staff's remarks the political level above him was also forced to recognize this situation: "The most worrisome thing mentioned by the Chief of Staff is the polarization between the people and the military," said Cabinet Secretary Danny Naveh (in an interview on TV Channel 2, October 28, 1996). In this paper I will attempt to elucidate the various components of the crisis, then I will analyze the causes of the crisis and evaluate its implications for Israeli society.[3]

The crisis in civil-military relations encompasses three areas. First of all there is an internal crisis in the IDF, it might be called a crisis of purpose, or rather, an identity crisis. It is partly the result of failure to defeat the Intifada. Second, there is a crisis in relations between the senior military staff and the political echelon above it, mainly due to the colonial condition and the deep schism it created in Israeli society. This is dangerous, because a harmonious relationship between the military and the political level is a precondition for a stable democratic regime, while its disruption opens the way for praetorianism (Perlmutter 1977). Finally, there is a crisis in relations between the military and civilian society, expressed mainly in accelerated demographic changes and the collapse of the (security) meta-narrative of Israeli society. A crisis of this kind could be critical in the case of a "citizens' army" or a "nation in arms," which is the Israeli case (Horowitz and Lissak 1989).

The Military Does Not Represent the Hegemonic Ethos

The special status that the military possessed in Israeli society for over fifty years stemmed not only from its strategic function, but because security was a constitutive ethos of Israeli society. It served as a meta-value, an organizing principle that set the hierarchy of other values in society. The first signs of the undermining of the security ethos began at the end of the 1970s with the development of a reflective and critical approach that exposed the actual existence of this ethos, claiming that "security" is not a given, the result of objective reality, but a social construct, and that the security ethos served the interests of certain social groups (Bar-Tal and Jacobson 1994). For example, Israeli feminist thinking attributed the lack of feminist awareness in Israel to the hegemony of a phallocentric world of concepts that characterizes the culture of a "society at war." Recognition of the limits of power was an even more radical change in the security ethos, although it was internalized only by a section of the political class. It was this recognition that led the government to choose a political solution to the Arab-Israeli conflict (Peri 1996a; Inbar and Sadler 1995).

Though the security ethos did not totally collapse in the 1990s, it has lost its hegemonic status. In a postmodern or late modern situation the old security ethos has become entangled with other types of ethos, individualistic, democratic, or civil ethos, as it is called by different researchers.[4] The military, which was until now the supreme expression of the old, collectivist ethos that included the willingness to sacrifice oneself for the general good; the soldier who inherited the image of the pioneer; military service as an example of civil virtue and good citizenship; all these now face a rival cluster of values and symbols. The role model for young Israelis is the high tech entrepreneur, lawyer or media celebrity. The younger generation's changed attitude to the military, says Colonel Gadi Amir, former head of the behavioral science department in the IDF's Manpower Branch, stems from "general attitudes of the youth towards a range of subjects, not particularly in the military area. There is a shift from the view that the individual should serve the establishment and the ideology to the view that the role of the ideology and the establishment is to serve the individual" (in an interview; see also *Yediot Aharonot*, April 4, 1997).

During the primaries in 1996, Member of Knesset (MK) Haim Ramon, seeking to diminish rivals with a glorious military record said, "The military is simply a security firm, nothing more." The same idea was expressed by the journalist Amir Oren, when he wrote, "The Israeli soldier of the end of the century is an armed policeman, not a national hero (*Ha'aretz*, August 15, 1997).

Thus, we see a change in the attitude towards basic concepts of the previous Israeli culture, such as the hero and heroism, and even towards the victims of the holocaust. The model of heroism was confronted by the model of "the new man" (expressed, for example, in the view that "it is all right for soldiers to cry," or that falling into captivity is preferable to a heroic death, *Sof Shavua, Ma'ariv*, November 5, 1993). The heroes of the security culture of the past (such as Major-General Shmuel Gonen of the book *Exposed in the Turret*) were treated to a much more critical scrutiny (in Shmuel Hasfari's play "Gorodish" at the Cameri Theater in 1994). The attitude to the fallen has changed, and along with it the ways of perpetuating their memory. The resigned acceptance of their loss gave way to a different pattern, of protest and defiance (Haim Guri, *Davar*, September 5,1994).

This change has received emphatic expression in the arts: in painting and sculpture (for example, the exhibition in February 1997— "You are a cannon: images of manhood," mocking the Israeli macho); television broadcasts (like the documentary, "Have you ever shot anybody?" broadcast in April 1996); in the cinema (*Ne'eman*, *Ha'aretz*, April 23, 1996) and in the theater (the play "Fog," on the part played by Ehud Barak, former Chief of Staff, in the "Tse'elim Affair").[5] The change is expressed most clearly in the press, which used to mark national holidays such as Rosh Hashanah (New Year) and Independence Day with special supplements in which the IDF and security were portrayed according to the old security ethos with tales of courage and heroism. In recent years the media present a very different picture: military failures, stories of the bitter fate of prisoners of war, victims of battle fatigue, and so forth (e.g., *Ha'aretz* and Y*ediot Aharanot*, October 10, 1997).

The IDF Does Not Deliver the Goods

What intensifies the crisis in relations between the IDF and civilian society is the fact that for almost twenty years, and particularly in the 1990s, the IDF has been perceived as a body that does not deliver the goods. The important point is not whether this view is "objectively correct," but the very fact that it exists. The argument was first voiced during the war in Lebanon in 1982. From the late 1980s settlers in the territories and supporters of the right blamed the military for "not eliminating the Intifada." Towards 1998 the IDF was harshly criticized for failing to destroy the Hizbulla in south Lebanon.

Not surprisingly, these criticisms elicited an angry response from the IDF high command, which attributed the criticism of the public to the weakened endurance in Israeli society, the mistaken wish for swift achievements. These themes have been repeated in recent years by all the Chiefs of Staff: Shomron, Barak and Shahak (Ofer Shelach, *Ma'ariv*, October 17, 1997). The high-ranking officers responded sharply to the politicians' criticism, blaming them for lack of understanding of security and military issues, and for political and ideological considerations. For example, Major-General Dan Shomron, the Chief of Staff during the Intifada, argued that the demand to root out the Intifada would require the IDF to act in a way that was not acceptable according to the normative codes of civilian society. What they could do, therefore, was suppress it, contain it, but not eliminate it utterly. In the end, the solution to the Intifada was political.

The fact that the Chief of Staff's analysis was right and that a democratic state at the end of the twentieth century is not capable of defeating counter-revolutionary warfare in occupied territories without violating its own democratic character is not relevant to the fact that some sections of Israeli society went on accusing the IDF and its leaders of either professional incapacity or something much worse—making concessions out of political considerations. Similar divisions arose around the war in Lebanon since 1982.

Complaints about the IDF gained momentum with the many accidents, failures and defects in the course of 1997. In this period, the IDF elite units were hit to an unprecedented degree: the air force (in the helicopter disaster that claimed seventy-three victims), the Navy commando unit 13 (in south Lebanon, with twelve casualties), and others, like the Golani and the paratroops' commando units. The General Security Services' failure to prevent terrorist attacks (not to speak of the trauma of failing to prevent the assassination of Prime

Minister Yitzhak Rabin) and the Mossad (who failed miserably in an attempt to assassinate the head of the Hammas political department in Jordan, and later in the discovery of an anebt (who supplied misinformation) both added to the deep disappointment. The large number of casualties—almost 150 dead in one year—was a perceptible and painful balance sheet of ongoing failure.[6]

The series of failures led to some soul searching on the part of the officers, who were used to a tradition of glorious successes. The failures generated a sense of internal crisis. In fact, this was an identity crisis that had begun with the Intifada and became more acute with the beginning of the peace process. The intensity of the internal crisis was mostly hidden from the eyes of the general public (but not from a small group of sharp-eyed outside observers, like analyst Amir Oren in *Ha'aretz*), and only partially revealed from time to time, for example, when the chief military attorney disclosed on December 17, 1996, that the number of charges of violence and brutality filed that year in the IDF had risen by 76 percent, and the total number of charges filed had risen by 15 percent.

Classified internal reports reveal an even worse state of affairs. Such is the picture that emerges from one report summarizing the negative effects of the war against the uprisers in the Intifada: a seriously lowered state of operational preparedness for a future full scale war, a decrease in the level of functioning in combat, in the quality of professional training and in morale. The soldiers experienced feelings of futility and inability to control the situation; increased stress, fear, frustration and anger, expressed in violence, helplessness and apathy. The relationships between commanders and their subordinates deteriorated. The commanders were involved in conflicts such as loyalty to the military system versus the need to support their soldiers, while the soldiers (close to 80 percent) felt that the IDF norms were compromised by their service in the territories. Against this background there was growing support (some 20 percent of the soldiers in compulsory service) for objection to serving in the territories.

A pathological expression of the crisis is the proliferation of court cases against IDF soldiers and officers, not for criminal offenses but for purely professional offenses. Although the Intifada had ended, and the wave of court cases for "deviant acts" committed during it has subsided, the normative base for the operation of special elite units in the territories has suffered a severe blow, as we learn from the cases being conducted over several years against officers of the *"Mistarvim"* (special hit units). The debate that splits the IDF, on the legality of these units' activities, recalls tough periods undergone by colonial armies in the past, such as the British during the revolt in Kenya, or the French army during the revolt in Algeria. History teaches us that it is hard for an army to retain its social high standing in such complex political circumstances.

The IDF's operative failures gave rise to many complaints of conservatism, stagnancy and failure to adjust to the new geopolitical situation, accompanied by suggestions for changes in its structure, strategic thinking and modes of operation. The IDF was criticized for being unprepared for battle (e.g., Ofer Shelach, *Ma'ariv*, December 6, 1996). There were expressions of fear that it would be taken by surprise once again (special supplement of *Yediot Aharonot*, October 17, 1997); suggestions for changing its structure by establishing a land arm, shifting the emphasis from conventional warfare to sub-conventional warfare and setting up a strategic command for unconventional warfare (Major-General Uri Simhoni [Res.], *Ma'ariv*, October 17, 1997). There was also criticism of the strategy of a stationary war of attrition in south Lebanon and suggestions to switch to aerial or other warfare (e.g., Colonel Uri Dromi [Res.], *Ha'aretz*, October 1997). Proposals were also made to cut military service by half (Major-General Shlomo Gazit [Res.], *Ha'aretz*, November 7, 1996), and even to turn the IDF into an all-voluntary force.

The Military Loses its Autonomy

In the IDF's diffuse relationship with civilian society, the relative power of the military establishment and its influence on civilian institutions were salient. This influence was also expressed in the pattern that Moshe Lissak described so well—"the IDF's role expansion." The fragmented boundaries that existed between the IDF and civilian society enabled the military to penetrate the civil sphere and enabled civil spheres, especially the value system, to penetrate the IDF, creating a society that was "military, but not militaristic."[7]

In recent years a reverse process has occurred in the IDF, a process of "contracting roles" (Cohen 1995, 1997). Civilian roles and even military roles have been transferred from the IDF to civilian systems, and this process is expected to continue into the twenty-first century. This role contraction, especially if it entails damage to the salaries and service conditions of the professional officers and a reduced security budget (and the security budget's share in the national budget in fact dropped from over 20 percent ten years ago to approximately 16 percent in 1997), is liable to strengthen corporative processes, change the boundaries between civil society and the military from fragmented into integral boundaries, using Luckham's terminology, and offset the IDF's character as a citizen's army.

Together with the role contraction, the IDF is gradually losing its autonomous status, a status it received from David Ben Gurion, the person who shaped the pattern of relations between the military and society (Peri 1983). In the past decade the civil system has penetrated deeper into the military. The judicial system is one of the outstanding examples of this. Judicial involvement in the operation of the IDF grew to a great extent because of the complex nature of war against the Palestinian uprising, especially in issues concerning the legality of orders. The number of officers and soldiers who were tried in this context is extensive, reaching the hundreds.

Another sign of the erosion of the IDF's operative autonomy is the demand to transfer the authority to investigate training accidents from the IDF to an external factor. A real crisis was created when, in June 1995, the Attorney General, under pressure from bereaved parents and against the opinion of the chief military attorney, decided to press criminal charges against an officer who appeared to have been negligent in the course of military action in Lebanon. The decision to accept judicial consideration of commanders' operational decisions during military action led to an organized protest by senior officers in a style unprecedented in Israel. Scores of officers organized to support and defend the accused officer, and a parliamentary and public lobby was set up (including the President of the State, himself a former general) claiming that placing an officer on criminal trial for an operational decision on the battlefield would lead to a situation whereby officers in the future would be incapable of using professional judgment in the uncertain conditions of the battlefield.[8]

The involvement of soldiers' parents is another visible expression of the weakening of military autonomy. When it began in the 1980s it was welcomed by the IDF. Perhaps this was an attempt by the military to draw legitimacy from the civilian system in view of the depletion of the traditional sources of legitimacy. But the telephone calls from soldiers' parents to the homes of their sons' commanders soon stopped being a subject of jokes and became a serious operative problem. Between the years 1982-1994 the number of complaints sent by parents to the Soldiers' Complaints' Commissioner doubled, and in 1993 the commissioner, Major-General (Res.) Aharon Doron, wrote:

> In the IDF it increasingly appears that it is not just a question of parental involvement but of their intervention. This subject is often brought up in talks with commanders, most of whom speak bitterly of it. The feeling is that the parents' interest goes beyond the limits of positive involvement and becomes a burden on the commanders.

A salient example is the issue of servicemen wounded in action. The public sensitivity to soldiers' lives increased the IDF's effort to reduce the extent of training accidents, however, a great deal of thought was also expended on reducing injuries in battle. Thus, in the war in Lebanon, and even more in the war against terror, methods of action were adopted whose primary aim was to reduce as far as possible the injury to the fighters' lives.

In analyzing the 1982 war in Lebanon there were already some, like Edward Luttwak, who argued that the meticulous care for the fighters' lives severely limited the IDF's fighting ability and reduced its probability of success on the battlefield. In the battle tactics that were adopted in the 1990s—both in micro actions against individual terrorist squads and in large operations in Lebanon—the heavy fire power, particularly by the air force, was designed to cover the avoidance of direct contact of fighters with the enemy, all in order to reduce the soldiers' exposure to injury.

Without taking a professional stand on the question as to whether over-sensitivity for the fighters' lives has reached a dysfunctional point where it hinders the IDF's ability to achieve the aims set for it (as does Schiff, *Ha'aretz*, February 8, 1995), it is suffice to mention that the response to civilian pressure, rather than professional considerations, is what caused the IDF to adopt certain methods of action, which would probably not have been adopted without such pressure. One cannot avoid comparing the policy of the 1990s to an opposite decision that was taken in the early 1950s. In 1953, after IDF units had retreated a few times in the course of retaliatory actions without managing to achieve their goal because of injuries they sustained, the Chief of Staff Moshe Dayan ordered every unit go on performing its appointed task unless it sustained losses of 50 percent of its forces. This decision was one of the causes of improvement of the operational ability of the IDF elite units during those years.[9]

The Media: The "Culture of Criticism"
Penetrates the Sphere of Security

The IDF of the 1990s is facing invasive pressures from various social institutions. The State Comptroller's high level of involvement in security and military matters, or a more critical approach of the academic world, are some examples. The media are a key element in this matter. Although the Israeli media long ago adopted the style of critical, attacking or investigative journalism, in the sphere of security they continued as a mobilized press (Barzilai 1996). In the last decade a change has occurred in this sphere, too, with the war in Lebanon in the early 1980s and the Intifada at the end of that decade serving as a catalyst for the transition from deferential journalism, through adversarial journalism to the confrontational model of journalism.

The reasons for this change are described at length elsewhere (Peri 1999). This is a cultural pattern characteristic of contemporary democracies and is called "the culture of criticism." This culture is fostered by the professions, particularly academic and scientific professions and artists. It is encouraged by the media and contributes to the decline of public trust and cynicism towards institutions of government in general (Cappella and Jamieson 1997). Pressure exerted by the press reduced the power of the military censor and weakened the IDF's mechanisms of control over the media (accreditation, spokesmanship, etc.). Beyond this, self-censorship was lifted and the view was adopted that the role of the media is first of all to criticize public institutions and establishments of all kinds, including the military establishment.

The level of this criticism was revealed in a series of content analyses of newspaper publications in recent years. One analysis was done on all the articles and feature articles on security that were published in the years 1994-1995 in Israel's two major weeklies, *Yediot Aharonot's Shiva Yamim* and *Ma'ariv's Sof Shavua* (their combined circulation in those days was about one million, which meant that one of these two papers was to be found in almost every household in Israel where newspapers were read). In an initial breakdown relating to the attitude of each text—critical of the military and the security establishment, neutral, or supportive of the military—it was found that almost two thirds of the articles were negative, critical of the military, and only one third were positive or neutral. This rate is the reverse of what existed in the 1960s or 1970s (in the 1950s it was almost impossible to find a negative reference to the military).

A more detailed analysis was conducted of all the feature articles that were published in the first half of 1996 in Israel's four dailies: a quality newspaper, *Ha'aretz*; a financial magazine, *Globes*; and two popular newspapers, *Yediot Aharonot* and *Ma'ariv*. The study examined only feature articles, not news items, and analyzed the main characteristics of the IDF as they emerged from these texts. Four specific criteria were examined: the IDF's effectiveness, its relations with the civilian population, its concern for the individual, and its prestige in society. The picture that emerges from the findings reflects a deep crisis of confidence in the military. The IDF is described as ineffective and wasteful, unprofessional, failing to achieve its aims, not behaving according to the accepted norms of civilian society, indifferent towards the soldiers, sexist, and losing prestige. *Globes* criticizes mainly financial and organizational aspects, *Ha'aretz* focuses on issues of management and command, while *Yediot Aharonot* concentrates on the human aspect of concern for the individual, but the common denominator of all four papers is clear.

The culture of criticism was most palpable during those special holidays such as the New Year and Independence Day, when the special festive issues and supplements that usually contained interviews with the Chief of Staff and generals, and songs of praise for the IDF, were now full of penetrating and painful criticism (one example, the supplement, *Iton Yerushalayim* of May 9, 1997, "When Gorodish fled from the battlefield," on the battle of the Harel brigade at the Rafiah junction in the War of Independence in 1948).

A Split in the Relations Between the Political and the Military Level

The identity crisis that the IDF suffers from, a crisis of its self-image and role-perception, is not the only one. A second crisis is the crisis in its relationship with the political echelon above it. Although the war in Lebanon had already caused a crisis in relations between the military and the political level, it was a crisis around the status and functioning of Defense Minister Arik Sharon, and it ended when Sharon was removed from his post. All in all, it, like the Intifada, did not generate a crisis in political-military relations, despite the fact that there were often deep differences on strategic issues, such as when Chief of Staff Dan Shomron supported the policy of inaction during the Gulf War, in opposition to Defense Minister Arens, who thought that the IDF should directly attack the missile launchers in Iraq. The basic agreement that had prevailed between the military level and the political level since the establishment of the state, and the IDF's deep professional appreciation of the political level (personified in the figure of "Mr. Security," Yitzhak Rabin) eased the potential tension between the two levels.

The situation changed with the beginning of peace process following the 1992 elections. The pattern of military-political partnership that had existed in Israel, and Prime Minister Rabin's tendency to use military staff rather than civilian officials, led to the deep involvement of officers in the diplomatic negotiations. This created several problems. There were cases when the officers disagreed with the positions of the political level (Chief of Staff Barak's opposition to the Oslo agreement, or the Chief of Central Command's criticism of the security arrangements on the eve of signing the Oslo B agreement) and did not refrain from airing their views in public. Even sharper was the parliamentary criticism of those who opposed the agreement, accusing the IDF officers of serving political attitudes, involvement in diplomatic activity, and losing their military judgment.

After the general elections in 1996 and with the rise of the Netanyahu government, an open split was created between the top echelons of the IDF and the political level, of an intensity and style previously unknown in the IDF. "For the first time the IDF is learning the political level above it in terms of 'know the enemy' rather than 'know thyself,'" wrote Amir Oren (*Ha'aretz*, August 15, 1997). Time after time the Prime Minister, the Ministers and senior Knesset members such as the chairman of the Foreign Affairs and Security Committee, expressed lack of confidence in the Chief of Staff and senior officers. Again and again the senior officers were accused of "collaborating with the opposition," of being "enslaved to the opposition," of "adopting political positions opposed to government policy." Failures and blunders were followed by exposure of defects in the work routine in the seam between the political and military level, in the reporting procedures, in the absence of orderly consultation or regular cooperation.

An unhealthy situation exists between the new Defense Minister (Maj.-General Ret.) Yitzhak Mordechai and the Chief of Staff (Lieut.-General) Amnon Lipkin-Shahak, on the background of a constitutional deficiency allowing a situation wherein an officer subordinate to the Chief of Staff retired from the army and overnight became the minister superior to the CoS. Added to this is the tension that developed between the two in the context of the failures in Lebanon (*Ma'ariv*, November 1997). But whereas the tension between the Defense Minister and the Chief of Staff is professional and personal, the tension between the Prime Minister and the senior military command (as well as the senior command of the other security branches, the GSS and the Mossad) stems from his accusation that the military adopts a political approach.

The military felt hurt by the attacks from the Prime Minister and the political leadership. They felt that the politicians accused them in order to absolve themselves of responsibility for mistaken decisions.[10] The senior command claimed that its attitudes were not the result of political considerations and certainly not party ones, but the outcome of purely professional perceptions. They were also worried by the government's attempt to cut the security budget just at a time when the likelihood of war was growing due to the retreat from the peace process. The injury to the military's status in the eyes of the general public and the faults in the work arrangements in the seam between the administration and the senior command (such as the absence of regular work meetings between the Chief of Staff and the Prime Minister, leaked accusations and so forth) add negative undercurrents to the bleak picture.

The friction was not just a question of personal "chemistry" and the presumed influence of political leanings, but basic differences on a matter of principle. The feeling of the military high command was that the political echelon assigned it tasks without understanding the professional constraints that prevented the fulfillment of this policy. A discourse among Chiefs of Staff, held by the newspaper

Ma'ariv in a special festive Rosh Hashanah edition in 1997, revealed this division in its full depth, when Chief of Staff Amnon Lipkin-Shahak and Minister Raphael Eitan exchanged sharp words on the subject of the war in Lebanon.

Said the Minister: "These things [that the problem of Lebanon has no military solution but only a political one] may be interpreted as pressure by the military on the political level, as if the military threw the problem to the political level and said 'Solve it... do whatever you want, we have no solution.' The military must not say such things." To this the Chief of Staff replied: "I think the statement that the military throws problems to the political level is not correct. The military, on its own initiative or by request, presents to the political level the reality as it sees it... It is the duty of the military to examine and recommend to the political level what it thinks is the right thing to do with regard to the use of force."

General Shahak did not hide his attitude that "In the war in Lebanon you will not reach a situation where the other side will stop. Since these are the facts it is much better for the military to say so to the political level. Look what happened when we reached Beirut—it [the war in 1982] did not stop and we came back from there. And the Intifada—could have gone on another eight years. I ask why it stopped, and I think it would not have stopped [if not for the political agreement with the PLO] but would simply have added to the list of graves on our side and theirs, and perhaps even got worse. In guerrilla warfare and Intifada you have to understand that the military will stand as long as the political level demands, but the role of the political level is to take the bull by the horns and deal with the political side." After this Chief of Staff Shahak did not hesitate to argue with Minister Eitan about his view of the Arabs and spoke in favor of the possibility of reaching a political settlement with them (*Ma'ariv*, October 1, 1997).

The wish to prevent a future situation in which, after a catastrophe, the military would be accused, led the IDF top brass to deviate from the principles binding public officials to keep arguments behind closed doors. Public wrangling, exchange of accusations in the press and gossip in corridors of power exposed a deep crisis of confidence and an open split between the military and the political level above it. Never had an IDF general attacked the Prime Minister in such sharp words as did Major-General Oren Shachor in 1997, when he was forced to resign from military service and spoke "in the name of the generals who cannot speak while Netanyahu shoots them in the back... There is a state of paranoia and witch hunt [of the Prime Minister against the IDF leaders]" (*Yediot Aharonot*, August 15, 1997). Even if the estimated possibility of a military coup, prevalent in the media and political discourse at the end of 1996, was exaggerated, the very fact that it was voiced shows how new this situation is. This requires further explanation.

The gap between the Likud administration and the military top echelons stems from the difference in their *modus operandi*. While many of the politicians act out of ideological considerations (this was most perceptible among the religious components in Netanyahu's government, but also among some of the politicians with a nationalistic approach), the IDF has traditionally encouraged pragmatic and practical thinking. This tradition of evaluation of the situation is what brought the IDF senior command to support the peace process, although with caution and reservations.[11]

Such professional considerations were what brought the IDF top brass several times in the course of 1997 to reject suggestions for aggressive action that were raised by the political level, especially by the Prime Minister. In contrast with the past, when the military was the force that pushed for military operations in response to actions of Arab armies or against Palestinian organizations, under Netanyahu's government the IDF leaders often found themselves in a restraining and moderating role, blocking "Netanyahu's adventurous initiatives" that were often motivated by internal considerations (Oren, *Ha'aretz*, October 21, 1997).

The different basis—the politicians' ideology versus the officers' pragmatism—was what obstructed the existence of a common language between the two levels at such a sensitive moment for Israeli society. It is interesting that just at this time the military was more sensitive than the politicians to the meaning of its relations with civilian society. The politicians, who were motivated by deep historical and political awareness, underestimated the danger to the IDF from the weakening of its social and moral base. In contrast, the military, worries it would not be able to act to the best of its ability in a divided society, supported political compromise, even a high price, that would decrease the probability of military confrontation.

The crisis in the relations between the military and the political level that was revealed at the end of 1996 reached its climax during the 1999 elections, when tens of retired generals joined opposition parties, and formed new ones, with one aim: to topple down Netanyahu's government, which they did. In the future, the tension between the military and the political level is liable to grow into a real crisis if the military thinks it is being asked to act against its professional principles and perform tasks of which it is not capable—either because they will not be acceptable to large sections of society, or because their high cost in human life will fatally damage its position in society.[12]

The IDF—From a Citizens' Army to a
Professional Force of Part of the Nation

Apart from the cultural developments in Israeli society, recent years have seen demographic changes whose effect on civil-military relations is significant. From the mid-1980s to the mid-1990s the population of men aged 18-21 rose by more than 25 percent, and the potential of military reservists (aged 22 to 51) rose by almost 59 percent. The potential draftees continued to grow by almost 20 percent until the year 2000. The rate of demographic growth since the late 1980s, largely because of the addition of some 800,000 new immigrants from the former Soviet Union, led to a situation whereby the IDF does not need all the potential manpower. Therefore, the IDF began to examine various possibilities for changing its recruitment policy, meaning a profound change in the structure of the IDF and the way it operates: voluntary enlistment; selective recruitment according to certain criteria; differential compulsory service (i.e., different draftees will serve for periods of varying length according to IDF needs); shortening the period of compulsory service; or changing the balance between conscripts and regular forces and the reserve forces (e.g., lowering the upper age limit for reserve service, drastic shortening of the length of reserve service).

Each of these possibilities has far reaching implications not only in professional military terms but also from social and civil viewpoints.[13] In fact, most of the alternatives for "selective reduction" compromise the basic principle of a nation in arms, entailing retreat from the model of a citizen's army and turning the IDF into a professional military. The consequences of this for civil-military relations are clearly revolutionary. These issues were discussed extensively in the IDF in the early 1980s, and at that time it was the Manpower Branch that estimated that such changes were unavoidable (Major-General Gideon Shefer, head of the Manpower Branch, *Ha'aretz*, November 11, 1996), while most of the IDF top brass still adhered to the traditional view that the IDF has to fulfill national and social missions, as in the past.

Understanding that this was an issue of great national significance, the IDF submitted it for discussion by the Knesset Committee on Security and Foreign Affairs and by the Cabinet.[14] These began to discuss the issue in 1994 but avoided taking radical decisions, and a year later the ball was back in the court of the military. Here, too, no decision was made. Another commission that was appointed to examine the matter, the (Maj-General Ret.) Yaron Commission, wrote in its summary of the report presented to the Chief of Staff in November 1996, "It is not right to continue the ongoing debate within the military as to whether its role is to deal with national missions and at the same time pay dues to society—such as soldiers working as teachers or in immigrant absorption... The IDF must decide on its position and understand that this is vital. The IDF's duty to the nation is to build up its power and not to pay dues to society." Perhaps the fact that this Commission was chaired by a former head of the Manpower Branch weighted its position on the side of the "professional faction."

But in practice the IDF chose to deal with the problem by muddling through, rather than making a decision in principle. In the mid-1990s it increased to tens of thousands the number of servicemen/women it "lent out" to other civilian bodies, such as the civil service, *Magen David Adom* (the equivalent of the Red Cross) and the Society for the Protection of Nature. Identifying the core of the problem as growing antagonism of the reservists towards the military, the Chief of Staff decided in May 1995 to make considerable concessions in the reserve service. The period of active reserve duty for combat soldiers was shortened and the age limit for reserve service in combat units was lowered, and the number of "reserve days" was cut by approximately 50 percent compared with the mid-1980s. In the following years further improvements were made in reserves' service.

The change in policy would not have happened if the IDF had not been subjected at the same time to heavy civilian pressure. A forum arose, composed of hundreds of reserve officers who felt trapped between the military and civilian society. They demanded that the IDF and the administration act to ameliorate the situation, particularly by correcting the distortion in the service load, granting various rewards, material and symbolic, to reserve officers, and changing attitudes towards them both in civil society and inside the IDF. In the Knesset a parliamentary lobby has operated for several years, and even succeeded in introducing an amendment to the law pertaining to reserve service, significantly improving the reservists' service conditions.[15]

The second component of the new manpower policy was designed to deal with the problem of surplus manpower. Here the decision was for "active adoption of elements of selective recruitment and differential service," that is, making the principle of compulsory service more flexible by reducing the number of recruits, extending the differential range of service and increasing the number who receive early release. The IDF responded to the requests of religious parties to increase the number of yeshiva (religious academies) students who were exempted from military service, and their number rose from a few thousand after 1977 to over 20,000 in the 1990s (from 2.5 percent to more than 7 percent of each year group). New immigrants also received far-reaching exemption, from complete exemption for immigrants who arrived after the age of 29 to significant reduction or total cancellation of the six months period of service for those eligible for the draft, as well as exemption from reserve service.

The basic entrance requirement for recruits was also raised, and the IDF more easily dispensed with the service of those who had low psychological profiles ("section 21") or had difficulty in adjusting to military life. This definition is objective and professional, but one does not need to be a student of Foucault to see its clear social background. In order to avoid dealing in the IDF with young people who said explicitly that they were unwilling to serve, and in the light of the fact that there was no need for full mobilization, it was easier in practice to do without their service.

Thus the IDF, which started off in the early 1950s as a citizen's army, ends up in the 1990s as a military that forgoes the draft of some quarter of all the men who are eligible for military service, 5 percent of them Israelis living abroad, 7 percent yeshiva students, about 3 percent exempt for medical reasons, and the remainder unsuitable in various ways. Whereas in the past this kind of exemption was a cause of social stigma, it is no longer so today. One indication of this is the fact that the Civil Service Commission decided to stop the practice of examining the IDF records of candidates for the civil service. The meaning of this is that "section 21" will not be a barrier to acceptance for work in the civil service and less of a barrier than before in the private sector.

The greater flexibility of the military is expressed not only in the recruitment policy but also in the actual service. Whereas in the past a lowered medical profile was the main reason for early discharge, in the mid-1990s "unsuitability and reduced functioning" became the main cause of early discharge of more than 8.3 percent of all those doing compulsory service. According to the data of the Manpower Branch, some 20 percent of all conscripts do not complete their term and are discharged before the end of the three years because of unsuitability. In practice about one third of each year group do not complete their full military service as determined by law, namely three years. According to the IDF forecasts, sometime in the next decade only half of the young Israeli man and women aged 18 will serve in the military (Amir Oren, *Ha'aretz*, November 28, 1997). This is a very new version of a "citizens' army."[16]

The Change in Motivation Reflects the Change of Attitude Towards the Military

Since the growth of modern nationalism and the nation-state, military service has been the hallmark of citizenship (Giddens 1985), noticeably in Israel, which was born in war (Peri 1996b). The change in Israeli society's attitude to military service is the most salient expression of the change that has taken place in the security ethos and in the status of the IDF in society. For this reason the decline in motivation to serve in the military has become a popular subject in the public and media discourse in recent years.

Recognition of the need to serve in the IDF is still quite high among young people who are eligible for the draft. In a survey conducted in 1994 by the IDF's behavioral sciences department, 50 percent of the subjects replied that they would volunteer for the full three years if service in the IDF was voluntary, 44 percent replied that they would volunteer for a shorter period and only 6 percent said they would not volunteer at all. This rate has been fairly stable since the mid-1980s (during the war in Lebanon the percentage that opposed the draft was about twice as high). Nor has there been a fundamental change in the potential draftees' motivation to serve in elite units (the number applying to serve in these units in 1996 was still five times higher than needed), or to volunteer for officers' courses (the slight drop does not affect the IDF's ability to recruit each year its required cadre of officers). Nevertheless, there have been significant changes in the motivation for service.

First, there has been a drop in the potential draftees' motivation to serve in combat units, and particularly in unglamorous field units. According to a report by the head of the Manpower Branch, in 1989 64 percent thought that for Israeli youth "service in combat units is a duty." In 1996 only 44 percent gave this answer (*Yediot Aharonot*, October 23, 1996). Incidentally, the shortage of manpower in field units is what led to the surplus of unexploited manpower in administrative units, a surplus that aroused public anger and the claim that the IDF recruits and holds on to far more people than it needs.

Second, there is a decrease in officers' motivation to go on serving in the IDF at the end of their compulsory service, and in the case of those who sign on for an initial period in the regular army—after completing their duties as company commanders, in their mid-twenties. Another change, connected with the nature of the motives for serving, will be discussed below. However, the real crisis in civil-military relations in Israel is not to be found among the potential draftees nor among the regular army officers, but in the reserve forces. This is where the real change has taken place, since the early 1990s, both in motivation and behavior. And it is worth remembering, the IDF, especially the army, is basically dependent on these reserve forces.

For several years the IDF has had a deficit in the reserves. Although the growth in the number of recruits each year should lead to a yearly increase in the number of reservists available, in fact there is an even larger flow of dropouts from reserve service before the age of exemption (45 in combat units, 51 in non-combat units), for reasons of health, psychological problems, other reasons or just plain dodging. (*Ha'aretz*, September 9, 1996). One of the results of this is that the burden of reserve service falls unequally and arouses the anger of those who carry it. MK Raanan Cohen wrote in a letter to Prime Minister Yitzhak Rabin on August 21, 1994, that "30 percent of the reserve soldiers do 80 percent of the reserve days." A study conducted by the IDF revealed that many of the reserves would not report for duty at all if it was up to them: 50 percent of the reservists up to the rank of captain replied that if they had the opportunity they would not report for reserve service (*Ha'aretz*, September 12, 1996). A study conducted by the IDF in 1974 found that only 20 percent replied in this way (*Yediot Aharonot*, October 17, 1997).

With regard to military operations the problem has become even more acute recently. The State Comptroller, who investigated the matter, reported in May 1996 not only that "there is a growing trend of unequal distribution of the load of reserve service and thus the load of service on combat units is heavier than on other units," but there is also "increasing difficulty in recruiting reservists for active service in many of the units. The difficulty is greatest in rear units." And the military correspondent of *Yediot Aharonot* wrote, "The commanders who spoke to the State Comptroller's review team were the ones who pointed to the decreasing motivation, especially for operational duties in Judea, Samaria and the Gaza Strip. The factor that carries most weight is the sense of unfairness in division of the load and the social legitimization granted to evasion of reserve service" (*Yediot Aharonot*, May 1, 1996).

Evasion of reserve service has become a wide-spread phenomenon in the 1990s. According to an estimate published in October 1997, out of every eleven reservists only two actually serve. Even if some of the other nine do not dodge service but are simply not summoned, there is still a very wide gap (*Yediot Aharonot*, October 17, 1997).

The IDF overcame the lack of reserve manpower by the same technique it used in the past to deal with conscientious objectors. Not wishing for a confrontation with the dodgers, it preferred to solve the problem by informal arrangements in the spirit of the strategy used by the U.S. Military with regard to homosexuals—"don't ask, don't tell." The commanders of the reserve units summoned many more soldiers than are actually required and ended up with the required number. The State Comptroller revealed the severity of the problem in her report for the year 1995. Between the years 1992-1995 the number of deserters rose by 54 percent. In order to fulfill their tasks the commanders of grade A units call up a reserve of 150 percent, and Grade B unit commanders summon up to 500 percent.

The reservists' lowered motivation and evasion of active service have far-reaching implications even for military decisions on the strategic level. For this reason, the IDF senior command decided not to send reservists to the war in south Lebanon, using only compulsory and regular units (*Ha'aretz*, October 1, 1997). For the same reason a decision was taken in 1997 that signified a radical change in the IDF's work conception. It was decided to set up regular companies for security service in the territories (*Ha'aretz*, July 17, 1997).[17] But more important than this was the recognition by Prime Minister and Defense Minister Yitzhak Rabin of the change in the reservists' attitude to military service. This is what brought him—although he did not dare to admit it publicly—to adopt a historic decision and choose political compromise rather than following rigid policies which increased the probability of future war. In the cultural climate that had developed in Israel fifty years after the establishment of the state, there was a real danger that not all the reservists would willingly report for a war on which there was no national consensus.[18]

The Change in the Social Basis of the IDF

The IDF became a weighty factor in Israel's stratification system since its inception. Military service became the entrance ticket into Israeli society. The extent of their contribution to the war effort and to security at large determined the position of individuals and social groups along the center-periphery axis (see Lissak and Knei-Paz 1996; Kimmerling 1993).[19] The collapse of the old security ethos created a situation whereby the IDF lost its status as a central stratifying factor. Although military service is still converted into civilian status, and a military career for people coming from relatively low social groups constitutes a ladder for social mobility, the picture is completely different from what it was in the past.

The postmodernist character of Israeli society, and the parallel existence of different types of ethos, led to a decline in the significance of military service. Among central groups in society it is legitimate not to have served in the IDF ("Profile 21 is not a disgrace any more," Ma'ariv, August 3, 1996) and this is manifested in the political sphere. Many of the members of the parliamentary coalition formed in 1996 notably lacked a military background (religious, ultra-orthodox, new immigrants). The political influence of rabbis (from Rabbi Shach to the Kabbalist Kaduri), to whom the military and security are completely foreign, is felt increasingly.

Motivation to serve and willingness to fight are decisive factors in the power of a citizen's army, and hence also in the IDF (Gal 1986). The differential change in the attitudes towards the military among different groups is both an effect and a cause of profound and wide-ranging social changes. Examinations of changes in motivation show that while motivation has declined steeply among certain social groups, among other groups various types of motivation have risen. In studies from the year 1988 researchers at the Institute for Military Studies in Zichron Ya'akov found a decline in the motivation to serve among youth from non-religious urban high schools, or kibbutzim, along with a rise in motivation among young people from religious backgrounds who study in vocational schools. Similar findings have been obtained in later studies (Meisels et al. 1995). The drop in motivation has occurred among those social strata who were previously at the social center, and noticeably in the kibbutz movement, which was over-represented in the IDF, especially in the officer ranks and in elite units. In contrast, there has been a rise in motivation among social groups that were considered "peripheral," from low socio-economic strata, Mizrahi residents of development townships, and particularly religious youth.

The religious variable is the most important one. Many studies in the 1980s and 1990s have found large gaps between the motivation and willingness to serve of religious youth attending schools of the national-religious stream and non-religious youth in regular national schools. In one study, by Yaakov Katz of Bar-Ilan University, the religious students expressed higher willingness than the non-religious to join the IDF after finishing their studies (86 percent as against 76 percent), to serve in the IDF for the full three years (81 percent vs. 68 percent), to serve in combat units (49 percent vs. 34 percent), to serve in elite units such as pilots, special reconnaissance units or boat squads (14 percent vs. 7 percent) and as officers in the regular forces (18 percent among the religious compared with 5 percent among the non-religious) (*Ha'aretz*, September 12, 1996).

However, it is not the degree of willingness to serve among members of various groups that is the most significant change but their motives for serving. In the past the wish to serve in the IDF stemmed from the old collectivist value-system, unaccompanied by classic values such as militarism, authoritarianism, alarmism, a pessimistic view of human nature, nationalism and urge for power (Peri 1996b). Today, other values provide the basis for motivation, the major factors being religious beliefs, a nationalist approach, hatred of Arabs and the desire for revenge on them. These tendencies are clearly visible among those with high motivation, groups with low socioeconomic status for whom the IDF serves as a route of social and economic mobility (Meisels and Gal 1989). If this process continues, and all the signs indicate that it will, it is liable to lead to a radical change in the value-system of the Israeli officer class, which, unlike other professional armies, was not characterized in the past by a "military mind." If the present trend continues, this set of beliefs will become more dominant.

Personal motivation for military service, together with a planned strategy of political groups in the religious sector, has led to a significant growth in the representation of religious groups in the IDF, especially in elite roles. For several years eight yeshiva high schools have trained young people for officer roles and service in elite units. These young people constitute a significant percentage of the company commander rank in combat units, and the trend is rising. Based on estimates (the IDF stubbornly refuses to supply official data), it can be predicted that at the beginning of the twenty-first century this population structure will affect the composition of battalion and brigade commanders, and in the following decade also the general staff (see also Nahum Barnea, *Yediot Aharonot*, August 30, 1996).

Towards the end of 1996 the proportion of religious soldiers in the IDF was 15 percent, but 30 percent in voluntary, elite, units. In certain officers' courses the rate of religious soldiers reaches 40 percent (Zvi Gilat, *Yediot Aharonot*, July 14, 1995). If the air force pilots' course in the past served as a prism for Israel's stratification structure and reflected disproportionately the kibbutz movement and the yuppies (affluent Israelis of western ethnic origin from the center of the country, graduates of elitist academic high schools), here is the profile of the trainees who completed the course in July 1997: for the first time in the history of the air force the number of religious men equals the number from the kibbutz movement—11 percent for each of these two groups.

Conclusion

There are two causes of the crisis in relations between the IDF and Israeli society. The first is the incongruity between the old institutional structure of the IDF as a citizen's army, and the new social reality that exists in Israel fifty years after independence. The second cause is related to the political and security circumstances under which Israel exists, primarily the continuation of the exegetic conflict. Despite the specific nature of the Israeli case, the IDF went through a process similar to many other armed forces in other societies. Moskos defines this development as a transformation from late-modern to the post-modern era. One aspect of this transformation, argued Moskos, is a structural transformation of the military forces.[20]

According to this model, in the modern era, from the beginning of the century until the end of the Second World War, the armed forces were mass, or conscript armies. In the late-modern period (1945-1990) they became big professional armies, and today (since 1990) there are small professional armies. The advocacy for decreasing the size of the IDF lies behind the proposals to change its structure, make it more technologically sophisticated, abolish the three years compulsory service and to build it on professionals serving for longer periods of time. The air force or the navy are the models for such a reform.

Resistance to a change has been quite strong until now, because of deep convictions that the IDF should remain a citizen's army and continue to fulfill social and civilian functions. Therefore the probabilities for the implementation of such a reform are not very high, particularly as long as there is no fundamental change in Israel's foreign relations with its neighbors. If the new model is adopted, however, even when it is gradually implemented, it will bring about a fundamental change in civil-military relations. The IDF will stop being a citizens' army, the boundaries between the armed forces and civilian society will change, civil and military cultures will be differentiated, and civil supervision of the military will also be transformed. As permeable boundaries are precondition for prevention of praetorianism, the crucial issue under the new circumstances is whether alienation will develop between the military and civilian sectors.

In the situation that has arisen we may predict two possible directions of future development, each potentially dangerous. One prediction is that growing convergence between the military and the nationalistic political elite will aggravate the tensions between the military and broad sections of society. The fact that the IDF will not be recruited from society at large will alienate it from those broad sections, eroding its status as a focus of solidarity, a unifying, nation-building factor. The strengthening of its political-ideological character, together with growing professionalization and corporatism, will further accentuate its military mind and its alienation.

According to the other scenario, the military will continue to reflect the same social and professional backgrounds as in the past, while the nationalist camp will become dominant again in civil society. Then, too, the gap between civil and military sectors may grow and the relations between the armed forces and their civilian superiors will deteriorate to a dangerous level, particularly if the military officers feel that the civilian authorities order the implementation of policies that contradicts their professional beliefs, in a similar way to what happened under Netanyahu's government.

At the beginning of the new millennium the relations between the military and the political echelons, due to the existence of personal affinity between the political and the military elites, and because both Barak's government and the supreme command shared the same approach to the peace process. However, the structural causes of the crisis in civil-military relations have remained unchanged, thus the crisis is far from being over.

Thus, entering the post-war era will not necessarily ease this tension. Furthermore, the crisis might escalate due to the contradictions between the new security needs and military interests on one hand and the late modern orientations of large segments of civil society on the other hand. The crisis in civil-military relations is thus far from being over, even when peace is implemented.

Notes

1. The opposite of de-militarization is not militarism but militarization. For a discussion of the difference between the two and its relevance to Israeli society, see Peri 1996b.
2. Two outstanding examples are the Chief of Staff's suppression of a comprehensive study conducted by the IDF's history department two years after the outbreak of the Intifada, and a series of studies by the behavioral sciences department, exposing the crisis, as a result of which the commander of the unit was forced to end his military service.
3. This paper is based, among other things, on talks and interviews with IDF officers and internal classified military documents. To overcome the difficulty involved in using this material, the only references to the subject here will be quoted from newspapers. However, they are also supported by these documents and interviews.
4. Social researchers of Israel use many concepts in attempting to characterize the new era. See, for example, Arian 1996; Ezrahi 1997; Kimmerling, *Ha'aretz*, July 7, 1997.
5. An expression of this cultural change can be seen in veteran press and war photographer Micha Baram's retrospective photo exhibition in 1994. For more on political anti-war statements by Israeli artists, see Meir Shnitzer, *Ma'ariv*, September 24, 1993.
6. See the Rosh Hashanah issue of *Al Hasharon*, October 1, 1997, with a "blacklist" of sixteen events, training accidents, injured in military operations, exceptionally severe criminal acts, suicides, corruption, murder, and even prisoners' revolts—all of which happened in the passing year.
7. For a bibliography (not updated) of Lissak's studies on civil-military relations in Israel, see Lissak 1984, especially the 1977 article on the IDF's "role expansion."
8. Another expression of the depth of the judicial involvement and the attempt to "civilianize" the military in this sense was made in July 1997, when the Supreme Court proposed amending the military law to equalize the period of detention in the military with the period of detention practiced in the civilian system, 48 hours instead of 8 days.
9. Moshe Dayan relates to this in his autobiography (1976: 113).
10. For example, following the bloody fighting caused by the opening of the Western Wall Tunnel in Jerusalem on December 16, 1996. *Ha'aretz* "A senior security source rejects the Prime Minister's evidence on the matter of the tunnel." See also Ze'ev Schiff's biting description of the friction between the military and political levels. "Do the security people have any influence?" *Ha'aretz*, December 19, 1997.
11. When Chief of Staff Amnon Lipkin-Shahak thought that Prime Minister Shimon Peres was conceding too much in negotiations with Syrian President Assad, beyond what the late Prime Minister Yitzhak Rabin was prepared to concede, he expressed his opposition forcefully. This in turn led to a protest by Peres himself.
12. The example of the "flower revolution" in Portugal in 1974 comes to mind. There the military in opposition to the civil administration sought to put an end to the colonial rule in Mozambique and staged a military coup in the metropolis. Of course, the political regime in Portugal was very different than that in Israel.

13. The report of the Shafir Commission thoroughly examined the advantages and disadvantages of each of these options. Shortening the compulsory service, for example, will shrink the period of professional training, leaving the trainees with less experience. Reducing the number of days of active reserve service is liable to lower the level of preparedness for war. Selective recruitment and differential service break the principle of equality and open the way for social unrest.

14. Interview with Major-General Uzi Dayan, former head of the IDF Planning Branch, 1995.

15. The "Reserve Battalion and Regimental Commanders' Forum" even sought professional advice and consulted a sociologist, Professor Eyal Ben-Ari. I wish to thank him for the information he gave me on the subject. On November 26, 1996 the representatives of the forum met for the first time with the Knesset Committee on Foreign Affairs and Security and described to them the wretched situation of the "disintegrating reserve system." Interview with the former chairman of the Committee on Foreign Affairs and Security and Deputy Defense Minister, Major-General (Res.) Uri Orr. MK Cohen's initiative is described in his report of September 1997.

16. Interview with the former head of the Manpower Branch, Major-General Yoram Yair, 1995. See also the head of the Manpower Planning Branch, Brigadier-General Israel Einhorn, *Yediot Aharonot*, January 25, 1996. In face of the reduced length of service the IDF circumvented the law determining three years and made acceptance to voluntary units conditional on extension of the period of service through participation in pre-military courses or commitment to sign on for a period in the regular army after completing compulsory service. Thus a situation is created whereby 13 percent of the men serve for over 36 months, balancing the 4 percent who serve less than 36 months.

17. The wish to avoid setting up an "occupying force," an important principle in the IDF's activities in the territories since 1967, was based on considerations connected with civil-military relations, even if this was counter-productive in terms of military advantage.

18. Yitzhak Rabin admitted to this attitude in internal discussions and also in conversations with me. See Rabin 1996.

19. A list of Dan Horowitz's publications on the subject of civil-military relations in Israel can be found in the book in his memory edited by Lissak and Knei-Paz (1996). On the subject of stratification in the military, see his joint article with Kimmerling (1974).

References

Arian, Asher 1996. *Security Threatened.* Cambridge and New York: Cambridge University Press.

Ashkenazy, Daniela 1994. *The Military in the Service of Society.* Westport, CT and London: Greenwood Press.

Bar-Tal, Daniel and Dan Jacobson 1994. *Security Beliefs Among Israelis: A Psychological Analysis.* Tel-Aviv: Tel Aviv University. (Hebrew).

Barzilai, Gad 1996. "State, Society and National Security: Mass Communication and War." In *Israel Towards the year 2000*, edited by Moshe Lissak and Baruch Knei-Paz, 176-95. Jerusalem: Magness Press. (Hebrew).

Burk, James 1992. "The Decline of the Mass Armed Forces and Compulsory Conscription." *Defence Analysis*, 8: 45-59.

Cappela, Joseph N. and Kathleen Hall Jamieson 1997. *Spiral of Cynicism*. New York and Oxford: Oxford University Press.

Cohen, Stuart 1995. "The Israeli Defence Forces: From a 'People's Army' to a 'Professional Military', Causes and Implications." *Armed Forces and Society*, 21, 2: 237-54.

Cohen, Stuart 1997. "Towards a New Portrait of a (New) Israeli Soldier." *Israel Affairs*, 3: 77-117.

Dayan, Moshe 1976. *Avnei Derech* (Milestones). Tel-Aviv: Idanim. (Hebrew).

Ezrahi, Yaron 1997. *Rubber Bullets*. New York: Farrar, Straus and Giroux.

Gal, Reuven 1986. *A Portrait of the Israeli Soldier*. New York, Westport, CT and London: Greenwood Press.

Giddens, Anthony 1985. *The Nation State and Violence*, Vol. 2 of *A Contemporary Critique of Historical Materialism*. Berkeley: University of California Press.

Horowitz, Dan and Moshe Lissak 1989. *Trouble in Utopia: The Overburdened Polity of Israel*. Albany: State University of New York Press.

Inbar, Efraim and Shmuel Sandler 1995. "The Changing Israeli Strategic Equation: Towards a Security Regime." *Review of International Studies*, 21: 41-59.

Kimmerling, Baruch 1993. "Militarism in Israeli Society." *Theory and Criticism*, 4: 123-40. (Hebrew).

Lissak, Moshe (ed.) 1984. *Israeli Society and its Defense Establishment*. London: Frank Cass.

Lissak, Moshe and Baruch Knei-Paz (eds.) 1996. *Israel Towards the Year 2000— Society, Politics and Culture*. Jerusalem: Magnes. (Hebrew).

Meisels, Ofra and Reuven Gal 1989. *Hatred Towards Arabs Among Jewish High-School Students*. Zikhron Yaakov: The Israel Institute of Military Research. (Hebrew).

Meisels Ofra, Reuven Gal and E. Fishof 1995. *General Perceptions and Attitudes of High School Students Regarding the Peace Proccess, Security and Social* Issues. Zikhron Yaakov: The Israel Institute of Military Research. (Hebrew).

Moskos, Charles C. and James Burk 1994. "The Postmodern Military." In *The Military in New Times: Adapting Armed Forces to a Turbulent World*, edited by James Burk, 141-62. Boulder, CO: Westview Press.

Nueberger, Benyamin 1997. "Peace and the Quality of Israeli Democracy." In *Israeli Society and the Challenge of Transition to Co-Existence*, edited by Tamar Herman and Ephraim Yuchtman-Yaar, 122-32. Tel-Aviv: Konard Adenaur Stiftung and the Tami Steinmetz Center For Peace Research, Tel Aviv University.

Peri, Yoram 1999. "The Changed Security Discourse in the Israeli Media." In *Security Concerns: Insight from the Israeli Experience*, edited by Daniel Bar-Tal, Dan Jacobson and Aharon Kleiman, 215-40. Samford, CT: JAI Press.

Peri, Yoram 1996a. "Afterwards - Rabin: From Mr. Security to Nobel Peace Prize Winner." In *The Rabin Memoirs*, Yitzhak Rabin, 239-80. Berkeley: California University Press.

Peri, Yoram 1996b. "The Radical Social Scientists and Israeli Militarism." *Israel Studies*, 1, 2: 230-66.

Peri, Yoram 1990. "The Impact of the Intifada on the IDF." In *The Seventh War, The Effects of the Intifada on the Israeli Society*, edited by Reuven Gal, 122-9. Tel Aviv: Hakibbutz Hameuchad. (Hebrew).

Peri, Yoram 1983. *Between Battles and Ballots: Israeli Military in Politics*. Cambridge: Cambridge University Press.

Perlmutter, Amos 1977. *The Military and Politics in Modern Times*. New Haven, CT and London: Yale University Press.

4

From Military Role-Expansion to Difficulties in Peace-Making: The Israel Defense Forces 50 Years On

Uri Ben-Eliezer

Introduction

One of the arguments that various researchers repeatedly put forward about Israel is that although it is in a permanent situation of war and various tensions and paradoxes are embedded within the relations of the civil and the military, it has succeeded in maintaining a democratic way of life and its army has never threatened the political realm. The argument is usually based on a certain set of assumptions. By disclaiming them and adducing others, this article arrives at different conclusions concerning the place of the military and military solutions within Israel's society and politics.

The most common argument is based on a paradigm that can be termed "civil-military relations." It was developed mainly in United States after the Second World War, and among those associated with it are scholars such as Huntington (1968), Janowitz (1971), Luckham (1971), Finer (1976), Lissak (1976), Perlmutter (1977), and others.[1] Notwithstanding differences and disagreements among them, which are plainly discernible in their writing, common denominators also exist, holding that in every society the process of differentiation that characterizes the modern era has caused a separation between two spheres: the "civil" and the "military," relations between which may

be balanced or unbalanced; that national leaders tend to act rationally when considering questions of peace and war; that the army is
a profession that offers a service to a client and usually takes a non-
political viewpoint; and that if the army is involved at all in politics,
it operates at most as a pressure group seeking resources. Military
coups and militarism, according to this paradigm, are usually the
result of a crisis, a "pathology," that alters the functional, balanced
relations that exist between the civil and the military.

The "civil-military relations" paradigm has been widely critiqued
for its basic assumptions and the inability of applying them outside
the Anglo-Saxon, or Western industrial, world.[2] However, in studies
of the subject undertaken in Israel, the paradigm's basic concepts
still hold pride of place. Little attention is shown over the possibility that it is problematic to "import" an American theory, that was
propounded in an era that believed violent conflicts were being replaced by a "cold war" and deterrence, and the function of armies
had changed—henceforth they would engage in the prevention of
wars, just as preventive medicine solves problems before they arise
(Ben-Eliezer 1997).

The fundamental, underlying assumption that almost without exception characterizes the study of "civil-military relations" paradigm in Israel is the unquestioning acceptance of the idea that a
separation exists between the two spheres of the civil and the military.[3] Yet no attempt is made to ascertain whether civility as such
even exists in Israel; and if so, what its essence and character might
be.[4] At the same time, together with the unexamined agreement on
the existence of two sectors in Israel, the researchers maintain that in
contrast to the "classical" Western model, which in its archetypal form
displays integral boundaries between the two sectors, in the Israeli case
the boundaries are permeable. This situation, the argument goes, is the
key to understanding Israel's success in preserving a democratic way
of life with priority granted to the security realm. Thus, for example, as
Horowitz and Lissak (1989: 195-230) would have it, the permeable
boundaries allow interaction between the military and civilian sectors
at multiple points of contact, which are highly organized into a balanced relations of civilianization of the military, from one side, and
partial militarization of the civilian sector, from the other.

To this hypothesis of a balance between spheres that reduces the
impact of the prolonged conflict on the functioning of social and

political institutions, two supplementary hypotheses were added. One posits two types of time that exist in Israel—a routine time and a time of emergency situations—along with peculiar arrangements that allow orderly transitions between the two times without serious disruptions.[5] The other refers to the reciprocal relations that exist between the military and political elites. These relations, which are characterized in terms such as partnership and "convergence," are presented as crucial to Israeli civilian control. As described, for example, by Maman and Lissak (1996: 71): "The openness of the communication channels between the two elites [the political and the military] ensures continued civilian control of the military elite, and functions as a safeguard against the formation of a military clique that would endanger the democratic nature of Israeli society" (see also Peri 1981; Horowitz and Lissak 1989: 226-7).

Within the permeable boundaries and an increasing number and variety of points of contact between the two spheres, a key term is suggested, that of the role expansion of the military. The classic definition of an army as being limited exclusively to the military-security sphere is problematic, some researchers have explained, because the army in the modern era has assumed new roles which are unrelated to security in the narrow sense of the term. They are manifested in political issues, in economic and managerial projects, and in the media, health, culture, and so forth. In an earlier version of the "civil-military relations" paradigm, the military's penetration of the civilian sphere was presented as a means to modernization and nation-building. Even military coups were sometimes presented as a positive, effective means to "Third World" modernization given the fact that the political and economic elites in these countries were irresponsible or weak (Johnson 1962; Janowitz 1964; Hanning 1967; Bienen 1971; Lissak 1967, 1976).

The Israeli conception of role expansion seemed somewhat misplaced in this context, since there had not been a military coup in Israel and because it was impossible, based on any criteria, to posit the strong, highly organized, and authoritative political elite as weak. The expansion of the army's role, then, was depicted as an instrument of modernization, nation-building and integration (in a society that absorbed new immigrants on a massive scale), in a manner that did not conflict with the will of the political elite but was undertaken in cooperation with it. Moreover, even though the army's political influence in Israel was acknowledged in conjunction with the perception of this role, it was presented as contributing even to Israeli democracy and to its inherent element of equality, because it prevented military corporatism and reduced the differences between army and society. To quote Perlmutter (1969: 70): "The result [of the role expansion] is that the differentiation between soldier and the civilian is seriously weakened... the model of the professional [isolated] soldier is being changed by 'civilianization of the military elite'" (See also, Horowitz and Lissak 1989: 213-21).

Even when the modernization theory lost its commanding presence, and the army's role expansion—certainly if it generated military coups—was depicted as a form of plunder and usurpation, as well as the realization of narrowed interests, as described, for example, by Stepan (1976), the Israeli claim remained unchanged. The result was that the IDF (Israel Defense Forces) was, concomitantly, portrayed as a contributor, through its role expansion, to immigrant absorption and integration, economic development, agriculture, technology, communication, culture, health services, etc. All in all, it was an army that was a school for the nation and a contributor to civic education (Halpern 1962; Lissak 1971; Horowitz and Kimmerling 1974; Horowitz 1975; Azarya and Kimmerling 1980; Azarya 1983 Ashkenazi, 1994).

However, such a description is probably nowhere more problematic than when applied to Israel. In a country which has been and continues to be involved in organized violence and wars on a large scale, it should give one pause to speak of the "civilianization of the military," "civilian components of the military," "the non-military role of the military", or even "a civilianized army" (Perlmuter 1969: 132; Azarya 1983: 102; Horowitz 1982). Eager to prove that Israel is not a garrison state and does not maintain a separatist army that constitutes a threat to the government, most of the researchers who deal with Israel have neglected the simple fact that the principal role of an army, and certainly of the Israeli army, is to engage in organized violence. It does this in preparations for war, in war itself, in conquest, and in occupation. Indeed, this role may go beyond the army to become a more extensive social project. The researchers, however, did not consider the possibility that the structural linkage between the so-called two sectors bears different political implications apart from the functional role of "mitigating" tensions and preventing the possibility that a separatist military clique might foment a military coup. It is in fact the purpose of this paper to articulate the possibility that the army's role expansion and its linkage to society might be connected not with modernization and nation-building, not even with military coups and the army's direct political intervention, but with militarism and war.

The Nation-State and Violence

To make this case, it is necessary first to posit several basic assumptions on the place of organized violence in society and on the "role" of military role expansion. Such assumptions should not ignore the fact of wars and of organized violence or diminish their social and political significance, as the "civil-military relations" paradigm does, nor should the basic assumptions de-emphasize the importance of the state as a means of domination, standing at a crossroads of two systems: the internal and the external, or the national and the international.

The centrality of state and army, the frequency of wars, and the role of organized violence in the modern era are themes central to the writings of scholars such as Giddens (1987), Mann (1988), Dandeker (1990), Barnett (1992), Porter (1994), Tilly (1995), and others. The most telling phrase in this connection is probably that of Charles Tilly (1975: 42): "War made the state and the state made war." One need only glance at Mann's (1993: 358-401) impressive tables on the high proportion of the state's expenditures on war in relation to other outlays to grasp the importance of these factors in modern history.[6]

The state system which took shape in modern Europe—for example, after the Peace of Westphalia, in 1648—legitimized the use of force by states, without providing for a higher body to which states could turn for peace-making. It would not be exaggerated to contend that one reason the state achieved dominance in the modern era was its flirtation with organized violence and war. In part the flirtation was created by the "vicious circle" cited by the neorealists, according to which reinforcing the security of one state by forming a military force undermined the security of other states, so that ultimately all of them became insecure and vulnerable to wars.[7] However, this set of relations was not caused solely by a problematic structure in the international system. In fact, in the nineteenth century, as Giddens (1987: 181-92), Tilly (1995: 125) and Mann (1988: 405-6) note, the emergence of the state as a structure of domination was accompanied by the neutralization of domestic power centers and the "pacification" of the population by channeling internal disputes into external conflicts. In order to survive the conflicts between them, states needed to squeeze the resources of the population and mobilize the people for war. But first it was necessary to persuade the people. Otherwise, why should they risk their lives? Nationalism and militarism were invoked to this end, the more so as states needed not only presence but active participation and excellence by their subjects in the course of wars which themselves changed their character in the nineteenth century: no longer were wars fought to secure dynastic rule, with the use of small armies, abetted by mercenaries, and led by officer-aristocrats who saw fighting as a status symbol. Where before the civilian population had been largely excluded, a major change occurred with the idea of universal conscription, based on national enthusiasm and the glorification of war, encouraged by the state apparatuses.[8]

It is also important to emphasize that even if modern warfare helped democratize political life—according to the well-known thesis that in return for fighting, the population was granted political rights and representative institutions—all nation-states developed a brew of nationalism, patriotism, and militarism as an inducement for the population to take part in the project called war and in the preparations for it. What had come into being was a system of domination that indoctrinated and socialized its subjects more intensively than it expressed their will.[9]

So effective was the cultural component that, as many historians have commented, the declaration of the First World War, for example, was greeted with an outburst of popular enthusiasm which had long been brewing. As Bond (1985: 72) writes: "The diplomats and businessmen may have made desperate last-minute efforts to avoid the plunge into warfare, but they were powerless in the face of the emotions of a generation which had been physically and mentally prepared for this supposedly inevitable clash between the nations" (see also Tuchman 1966).

We arrive, then, at the conception of the army's expanded role. As it will be presented here, this is one of the most effective means a state has to get its population involved in war and the preparations for war, and to obtain legitimization for the use of organized violence for political purposes. This thesis is utterly at odds with the argument of the "civil-military relations" paradigm which, in the case of Israel, as noted, viewed role expansion as a functional mechanism to ensure civilianism. Moreover, as will be seen later, so great is the significance of role expansion as an instrument to transform organized violence into an asset of the entire society, and thus render it legitimate, that even when meaningful changes occurred in civil-military relations in Israel over the past decade, as Israel embarked on a path of peace accompanied by decolonization and demilitarization, the army's role expansion continues to occupy a paramount place in Israeli politics. It constitutes an effective tool by those who are trying to stymie the peace process and to keep the centrality of organized violence in the society and culture.

This thesis, on the connection between role expansion and militarism, and on its implications for the prospects of peace in Israel, will be clarified in the following three sections. The first section will present Israel as a late version of a type of state organization known as a "nation-in-arms." This is characterized by an extensive expansion of the army's role, the blurring of the distinction between army and society, and the perception of the organized violence as a legitimate means for solving political problems facing the state. The second section will describe the changes Israel has been undergoing in the last ten years characterized by demilitarization and de-escalation. The third section will describe the way role expansion, in the face of these changes, becomes an instrument to further a politics, which is intended to prevent the possibility of peace and keep a society in a mental and practical state of war.

Role Expansion of the Military

In the nineteenth century some states connected nation and war through various typical arrangements and a specific world-view, according to which they can be termed "nations-in-arms." Post-revolutionary France and again from 1870 on, Prussia following its defeat by Napoleon throughout the nineteenth century until the First World War, and Japan in the early years of the Meiji Era (1868-1912) are examples of states which constructed the population into a nation, through the army, for the purpose of war.

In France, war preparation had taken the form of compulsory service in 1793, the famous *levee en masse*, which expressed the aim of creating a strong, patriotic, national, mass army ready to repulse a Prussian invasion. It was the gifted Lazare Carnot who was able to put the economy on a war footing in order to arm and equip the troops, and to mobilize everyone, including women and children, to take part in the war effort. Eventually, Napoleon's *Grand Armee* would number over one million soldiers and be engaged in wars for more than twenty years, based on certain arrangements of the nation-in-arms (Sobul 1974: 268-9; Hayes 1931: 433-83).

Prussia even more than France is an historical example of the calculated manufacture of national feeling and the blurring of the distinction between army and society to help win wars. Most striking in this regard were the reforms carried out within the Prussian army following its defeat by Napoleon in 1807. The new army included both regulars, the *Landsturm*, and a reserve force, a militia known as the *Landwehr*, which extended the population's participation in the army to lower classes. The opening of the army's gates to new groups generated deep concern among the traditional elites, such as the Junkers, who feared for the fate of the status privileges which had accrued to them, among other reasons, due to their army careers. However, there was a common goal that overrode social differences and the danger posed by armed lower classes—the will to be victorious in war (Vagts 1959: 116-28).

Gradually, the change in the perception of war and the role of the military became clear and substantive in Prussia-Germany, as expressed, for example, by General Moltke in the introduction to his history of the Franco-Prussian war, written in 1887-88: "The time has passed when small armies of professional soldiers fought campaigns for dynastic reasons, taking a city or a tract of land and then returning to their winter quarters or concluding peace. In present-day wars whole peoples are called to arms, and there is hardly a family that is not involved. The whole financial power... is brought to bear, and no change of year brings the incessant action to an end" (Foster 1987: 223; for more on the Prussian-German case, see Willems 1986: 49-112; Eley 1987).

In Japan, too, the imperialist thrust brought about a new orientation concerning the relationship between army and society. As a Japanese military academy report explained: "A characteristic of modern war is fighting with the total strength of nations. War in earlier times was decided by the side with the strongest military power. In modern war, fighting is on the level of financial war, ideological war, and strategic war, in addition to the military war" (Cook 1978: 265; for more on Japanese militarism, see Smethurst 1974; Ienaga 1993/4).

Many European states in the second half of the nineteenth century, such as Italy from 1887 on, under the domination of Francesco Crispi, who was a typical "civilian militarist," possessed certain nation-in-arms arrangements (Bond 1985: 61). Even in Britain, which was never a nation-in-arms, some elements of role expansion of the military existed in the late nineteenth century, for the sake of military success. This was made through infrastructural organizations and associations such as youth organizations like the Boy Scouts, the church, and even the public school system, which placed military elements at the center of the collective consciousness (Springhall 1971; Summers 1976; Otley 1978).

The classical example of a nation-in-arms state, however, is Republican France beginning in 1870. The laws of 1872, 1873 and 1875 laid down principles that placed the French army on a new foundation, most significantly universal conscription and a reserve system that enhanced the idea that the whole nation was engaged in war preparations and war efforts. The Carnot of this time was Gambetta who, among other acts, re-established the National Guard, which he saw as the embodiment of the nation-in-arms (Vagts 1959: 215-6). In their attempt to create a huge, potent army, the French imitated Prussia. Thus, a report by a committee on the service law of 1913 rejected the example of the Swiss militia, recommended by politicians with democratic inclinations, and favored the rigid German model.[10] At the time, especially during the period of the "national revival" (*La Reville National*), cooperation between "civil" organizations and the military was widespread in France, deliberately aiming at creating what was termed *rapprochement armee-nation* (E. Weber 1968).

The major role expansion of the army in these places blurred a possible separation between army and society. As such it was an effective mechanism for disseminating throughout the entire society the readiness and enthusiasm of war, and for turning the whole society into militaristic one. As defined here, militarism is a cultural phenomenon that presents war and military solutions as a central, desirable, effective, legitimate and necessary solutions to political problems that vitiate relations between states and nations.[11] In Prussia-Germany, for example, an entire network of "civilian" organizations and associations operated energetically to further a militaristic orientation and to legitimize it throughout the whole population: student associations, societies for the navy (*Flotternverein*), the church, associations of discharged offices, and perhaps most important, the youth organizations (Bond 1985: 40-71). Role expansion in the various youth organizations was manifested through the fact that several well-known "ex-soldiers," including retired generals, established and directed these organizations precisely for the purpose of countering the pacifist youth movements and of spreading war propaganda. More than once, these organizations received financial aid and other support from the army (Bond 1985: 73-6; Van Der Goltz 1913).

It is doubtful whether under conditions of such an invasive role expansion by the army it is possible even to talk about civilianism and a peaceful state of mind. In societies where the non-liberal, collectivist element, which was central in their organization and mobilization, was accompanied by the obsession that develops in Germany in the face of the *Einkreisung*, the fear of a two-front war (Willems 1986: 153); or in Japan, where the nation-in-arms project was intended to enable Japan the place it "deserved," among the nations (Sunoo 1975: 11-65); or in France, which was gripped by the idea of revenge (*la Revanche*) for the defeat of 1870 and the regaining the lost territories of Alsace and Lorraine (Ralston 1967; Porch 1981; Porter 1994: 149-95).

Israel in its first three decades was a late version of the nation-in-arms model. The enactment of Israel's Military Service Law in August 1949 gave legal validity to the establishment of a strong mass army, national and unified, that transcended all the cleavages that existed with the military forces in the pre-state era. Enhancing the idea of strength was the creation of a four-tier military system: a career army; a regular army (men and women, but Jews only); border settlements; and a reserve army (Knesset [Israel's parliament] Protocols, August 29, 1949). Organizational strength, however, was not enough. Nations-in-arms usually have a few distinctive features, which appears in the Israeli case as well. First, the army, the IDF in the Israeli case, is not only a strong mass army, it is also an army-nation. Like the French army-nation, it is perceived as the symbol of the nation, reflecting what is good—*La France Reel*, as it was known.[12] Role expansion here is manifested in the fact that the army is not built as a professional force, separate from the society, but exists as the army of the whole nation, with the idea of participation at its center. "Everyone" is involved, first as conscripts and afterward in the reserves. Indeed, it was General Yigael Yadin who first described the Israeli citizen as "a soldier on ten months' leave." In Japan the same motif was sounded by General Tanaka Gi'ichi', a founder of the Imperial Military Reserve Associate, who commented in 1911 that "all citizens are soldiers."[13]

The second trait of the nation-in-arms is that the boundaries between society and the armed forces are blurred in various domains. Thus, preparations for war, war itself, and indeed the idea of solving political problems, on the national level, through military means, become a project which involves front and rear, men and women, young and old. In the Israeli context, the IDF busied itself absorbing new immigrants, teaching them Hebrew using specially trained female soldiers. For the "underprivileged" the army has offered programs of education and rehabilitation. In a famous project called "the *Makam*," young people from slum neighborhoods, often with criminal records, get a second chance through military service. The IDF intervenes in the general popular culture by means of military bands or troupes that perform for civilian audiences in the big cities, or through Army Radio, *Galei Zahal*, the country's most popular station, where civilians and soldiers work side by side. The Civil Guard, *Mishmar Ezrachi*, in which arms-bearing civilians patrol their neighborhoods at night, is yet another project that erases boundaries between military and society. There are, as well, various funds, of which the most famous is "*Libi*," which raise money from civilians for army use. And there are special military units, like the *Nahal* (Fighting Pioneer Youth), which mix military service with land settlement and agriculture.[14]

Since these were institutional arrangements, they were naturally accompanied by verbal interpretation, which justified the role expansion by arguing that Israel's security required such enlargement into every sphere of life. A good illustration are these remarks by Ben-Gurion (which he liked to reiterate on any number of occasions): "The scope of our security is wider than that of any other country, and does not depend on our army alone... security means the settlement of the empty areas... the dispersal of the population... the establishment of industries throughout the country... the development of agriculture... security means the conquest of the sea and the air, and the transformation of Israel into an important maritime power... security means economic independence... security means the fostering of research and scientific skill..." (*Knesset Protocols,* November 7, 1955).

More informal mechanisms of military role expansion also exist in Israel. One is the involvement of soldiers' parents in the army. Parents regularly visit military bases on the Sabbath, bringing their sons or daughters home-made food. Basic training for combat units often concludes with a forced march of 80 kilometers or more, following which the soldiers receive their insignia and ranks, or the distinctive beret of their corps. Parents often help their sons on this march, and it is not uncommon to see them urging the exhausted youngsters on toward the end of the ordeal (Katriel 1991; "The Mother, the Commander, and the Soldier," *Davar* [daily], May 15, 1992). In such a way, the dispensing of organized violence in the society becomes a project shared by everyone, if only symbolically.

Another non-formal mechanism of role expansion in Israel is the "parachuting" of retired generals to senior managerial positions in industry or to the top of the political pyramid, a kind of "managerial militarism," that was also common in Prussia-Germany (Willems 1986: 79-80; Stargardt 1994: 84) and in Japan (Humphreys 1975). One of the themes stressed by scholars, who view the link between the military and the civilian in Israel as essential for democracy, is the importance of a "second career" for senior officers, who complete their military service at a young age. This possibility, they say, militates against the possible formation of a military junta possessing corporatist self-interests (Peri 1981; Lissak 1984; Maman and Lissak 1996). However, another perspective is also feasible if we accept the French lesson: the discharge of senior officers at an early age is not intended to avert the emergence of a junta, but to create a firm, young and restless army suitable for war purposes. The fact is that in order to increase the chances of promotion and to enhance military efficiency (irrespective of concern about a junta), the French law of 1873 decreed that no general was to hold the command of an army corps longer than three years (Vagts 1959: 225).

Indeed, an examination of the orientation of Israeli officers who are parachuted into politics shows that they view politics as the continuation of their military service by other means, and not vice versa (Ben-Eliezer 2000). It is a matter of some interest that the Knesset's Foreign Affairs and Defense Committee, the most important committee in the Israeli parliament, consists largely of former generals who were catapulted into politics. When one of them, Major General (Res.) Ori Orr, became chairman of the committee, he did not think—in the spirit of the liberal tradition—of declaring that his task was to contribute to parliamentary supervision of the army. On the contrary, he made it clear that as he saw it his mission was to assist the IDF and protect it from various attacks emanating from the public (Nechama Duek, "He got a Promotion," Yediot Achronot [daily], 22 December, 1995).

In fact, the phenomenon of high officers being parachuted into politics should be seen as one aspect of another prominent feature of all nations-in-arms, having to do with relations between elites. I refer to the close collaboration that exists in nations-in-arms between the military elite and the political elite, based on a shared bellicose perception of reality. Such collaboration generally assumes the character of an informal exchange, in which the army acquires considerable political influence, not so much internally, but in foreign and defense policy, and in return it does not threaten to seize power. In societies that are organized democratically, the army usually does not even question the formal division under which it must execute the orders of the political level. In practice, though, this obedience is based on collaboration between sword and scepter deriving from the priority that is accorded to organized violence as a political solution to national problems. Here, too, role expansion is discernible, since through the exchange, and in return for obeying, the army gains deep involvement, great influence and enormous prestige (Ben-Eliezer 1997).

Another feature of the nation-in-arms is relevant in this connection. The army-nation ostensibly places itself above politics, certainly above party politics. Practically, though, to place oneself above narrowed and factional politics is itself a form of politics, a channel to obtain vast political influence since army's decisions are perceived as solutions that are representative of the entire nation. This was the situation in the 1950s' Israel; the same phenomenon existed in Imperial Japan where, as Nakamura and Tobe (1988) explain, the impact of the army and the expansion of its role were based on the distinction that emerged there between two types of politics. One type, party politics, was absolutely off-limits to the army; but as a result of it, the army intervention and influence was considered essential and perceived as legitimate in matters related to the other type, national politics (Nakamura and Tobe 1988). Thus, the separation into two types of politics was an efficient means that avoided the army's politicization on a party, factional, narrow level, but allowed, at the very same time, the army's political influence on the national and international level.

Likewise, the officers in nations-in-arms have a social role, which is directed to bridge the gap between the officer corps, with its potential corporatism, the rank and file soldiers, and the civilian population. Tracing the roots of the concept, we arrive at the French Revolution and the Jacobin state. In a later version the social role of officers and the army's role expansion appear in France following its defeat by Prussia in 1870. In 1891, when Marshal Hubert Lyautey wrote his famous article *"Du Role Sociale de l'Officier,"* he called for a bond between army and society as a recipe for success in war. He was thinking neither about modernization, nation-building, and progress, nor about the civilianization of the military—otherwise, the idea would have never been taken up so enthusiastically under General Andre's war ministry in 1901-1903 (Challener 1965: 52; Bond 1985: 67).

We find a similar situation in Israel of the 1950s and 1960s: a mobilized society with the army at its center and an officer corps with a social role. The driving factor behind the military role expansion or the social role was neither military coups nor civilianism, but militarism. This was the politics (not the function) of the nation-in-arms. Not only to accustom the population "to live with the conflict," by turning the military effort into the project of "everyone," but rather to get legitimation, through this public involvement, with a politics that was geared to take control of the conflict and then shape it and dictate it. Precisely as Dayan and the IDF did in the early 1950s until the 1956 Sinai Campaign, as the Israeli generals tried to do before and after the 1967 war, and as Chief of Staff Rafael

Eitan together with Defense Minister Sharon attempted to do in the 1982 Lebanon War.[15]

Demilitarization and De-Escalation

The 1980s and the early 1990s witnessed a transformation in Israel in which Israeli leaders forsook the idea of constituting the State of Israel in the entire territories that were conquered after the 1967 War and even started to view reality in terms of a "new Middle East," the development of regional trade and industry which would consign the national conflict and wars to the past (Peres 1993).

These changes were related to internal pressures—the consequence of the Lebanon War that divided Israeli society, the Intifada five years later, which demonstrated a Palestinian obstinate national revival—and to external pressures, such as the consequence of the new role assumed by the United States as the guardian of the world following the collapse of the Soviet Union and the end of the Cold War. The result was a peace agreement signed by Israel with two states, Egypt and Jordan, and the most far-reaching development, the Oslo Agreement between Israel and the PLO, which was approved by the Israeli parliament in 1995. With these transformations, a change has occurred in the relations between army, society, and politics in Israel. The change can be described in terms of a relative decline of the nation-in-arms model: a certain recoil from hegemonic cultural militarism, the beginnings of a demarcation between society and army, and even some indications of a "society against the army" phenomenon.

One such indication was the *rosh katan* ("small-head") syndrome, referring to a "know nothing" or "low profile" attitude that manifested itself among Israeli soldiers during the Intifada. Soldiers and officers did what they were told, no more (Gal 1990). In the mid 1990s young Israelis are less motivated to do army service. In March 1995 the Chief of Staff, Amnon Shahak, warned, in the wake of a survey which found a decline in the motivation of high-school pupils to serve, that Israel could pay a high price for evasion of military service. Shahak said he was concerned about the spread of the phenomenon, which he placed in the context of a growing preference for individualism over collectivism in the Israeli society (*Haaretz* [daily], March 22, 1995). Indeed, in the 1990s Israelis have been far less predisposed to do reserve service than they were in the past. "Nowadays you're considered a kind of nut case if you do reserve duty," the chairman of the Association to Shorten Reserve Service told Knesset members when he was invited to meet with them in the parliament building. The association explained that it sought a "more equitable distribution of the burden," but the group's principal message was made perfectly clear: unwillingness to do reserve duty (Report on their visit to the Israeli Parliament, *Yediot Achronot* [daily], December 21, 1994).

Even more striking than the disinclination of young Israelis to do military service is the public criticism and protest against the army. Criticism of the army had already became a sociological phenomenon during the 1982 Lebanon War, when the IDF proved unable to overcome the Palestinian forces. Matters were compounded by the IDF's indirect responsibility, as determined by a judicial commission of inquiry, for the Lebanese Christians' massacre of Palestinian civilians in Sabra and Shatila, two Beirut refugee camps. In the late 1980s, the Israeli public criticized the IDF for fighting women and children who used stones as weapons. Trials of soldiers for excesses committed during the Intifada drew wide publicity. They were accused of brutality, unnecessary killing, and violation of the civil rights of Palestinians (Barzilai 1996: 171-82; Straschnov 1994: 199-282).

The IDF's senior levels have been heavily censured for training accidents in recent years. Even though the statistics show a downtrend in such accidents, irate parents, including mothers and fathers whose children were killed in noncombat activity, have organized to level fierce criticism. Their main demand—that the army place the investigation of accidents in the hands of an external, neutral body—reflects a rift that has sprung up between army and society (*Yediot Achronot* [daily], April 21, 1993; February 16, 1994).

Another issue was the parents' appeal to court calling upon the military to allow them to inscribe a personal expression of their grief on the headstone of their son in a military cemetery—all the military cemeteries in Israel have uniform inscriptions on the headstones. Following a struggle, they were permitted to add a two-line personal inscription on the headstone, a considerable achievement in a society which is accustomed to view the death of a soldier as a collective phenomenon of an individual who sacrificed his life for the state.[16] Yet another example of criticism of the IDF is the aftermath of the so-called "disaster of the helicopters," when seventy-three soldiers were killed in the collision of two transport helicopters en route to Lebanon in February 1997. Twenty years earlier, fifty-four soldiers (most of them paratroopers) were killed in a training accident when the helicopter they were in smashed into the ground. At that time, when Israel was still a classic nation-in-arms, no criticism from the parents was heard and no demand was raised for a commission of inquiry (*Haaretz* [daily], May 11, 1977). However, in 1997 many of the bereaved parents lashed out at the General Staff and at the political level, some even using the word "treason." "Security," it emerges, is no longer an inviolate fetish whose secrets must never be revealed. On the contrary, it has become a matter of routine for parents to ask the Supreme Court to order the IDF to hand over the results of investigations into accidents; the army's claim of "security secrets," they say, is merely a cover with which to hide blunders (*Haaretz* [daily], March 7, 1997; May 4, 1997; February 5, 1998).

The "society vs. army" phenomenon is also seen in "revolts" by soldiers, especially veteran soldiers, in combat units whose privileges are taken away by a new commander or by an officer who does not want to play by the "rules." The soldiers react by simply going home in an act of collective AWOL. The fact is that such incidents are nothing new in the IDF, but in the past the parents would back the army and usually persuade their sons to return to their base. Nowadays, however, the parents tend to blame the army for mistreating "our children" (who used to be "our soldiers" or "our men"). Parental involvement in the army, which was initially a non-formal mechanism of mediation between army and society, is increasingly becoming an instrument that widens the gap between them, and the army is trying to put an end to the phenomenon or reduce its scope ("The People's Army Became the Mother's Army," Maariv [daily], May 15, 1992; "The IDF's Mistake for Opening the Army Before Parents," Yediot Achronot [daily], May 21, 1995).

Additional examples of the change that is underway in Israel—criticism of the army coupled with its deglamorization—can be seen in the new attitude of the media. For years the Israeli media had considered itself responsible for bolstering the nation-in-arms and protecting the army-nation. In the past decade, however, the press has increasingly become the watchdog of democracy and has not hesitated to gnaw at the sacred cow of army and security. Media criticism of military blunders or of high officers, who in the past were considered above public attack, is now commonplace (Negbi 1995: 37-88; see also Ehud Oshri, "The Censor Test," Haaretz [daily], May 15, 1996). An example will illustrate the phenomenon. In December 1995 the IDF came under sharp criticism from the press over the promotion of a brigadier general named Giora Inbar to division commander. The reason for the public dismay was that earlier in the same year Inbar had been reprimanded twice, once for failing to tell the truth in connection with the death of an officer who may have died as a result of "friendly fire" during an incident in Lebanon, and a second time for poor performance during a series of attacks on Israeli forces in Lebanon (Yediot Achronot [daily], December 21, 1995). In years past, when Israel was entirely a nation-in-arms, criticism of this kind was never voiced, the press exercised "voluntary censorship" with regard to combat officers, who sacrificed themselves for the sake of the nation (and also with regard to those who served in noncombat rear units in the big cities). Today, however, the behavior of senior officers is closely examined and often criticized by the media. The forgiving attitude towards the officers no longer exist (see the case of Moni Chorev, Haaretz [daily], June 24, 1996; Yediot Achronot [daily], July 7, 1996).

The altered relations between society and army were accompanied by an organizational reshuffle in the IDF which became less of an army-nation and more of what some view as a professional army involved in a transition from institution to occupation like any other.[17] This became very apparent during the tenure of Chief of Staff Dan Shomron, and in particular of his successor, Ehud Barak. Barak discharged thousands of regular officers and civilians employed by the army, and slashed rehabilitation projects and other cultural and educational programs which were at the heart of the military's role expansion in blurring the boundary between the "social" and the purely military. The army thus adopted "utilitarian" and economic criteria and techniques as an index to determine the scope and character of its involvement in society; in other words, role expansion began to give way to role contraction (Cohen 1996). During this process, the army itself slaughtered some of the main symbols of the nation-in-arms and the army-nation, notably general conscription and all-inclusive participation. Selective service effectively came into being as women's service was reduced by three months, "unsuitable" conscripts were discharged, annual reserve service was significantly abbreviated, and the like. A case in point is the so-called "Profile 24," which signifies exemption from service in the IDF. In the past, when the nation-in-arms was at its glorious height and the participation motif was absolutely central to the collective identity, early discharge, especially if effected by the decision of an army psychiatrist, was a stigma and could affect not only employment opportunities but even the possibility of getting a driver's license. Yet in 1994 the press reported that thousands of reservists and regulars were being discharged early every year as part of the IDF's new policy. The army even ceased differentiating between Profile 24 for psychiatric reasons and the same status for health reasons (such as training accidents, wounds received in combat duty, etc.) (Tel-Aviv [weekly], July 1, 1994). In the 1990s, about 20 percent of all 18-year-old Israeli Jews do not do army service, and another 10 percent are discharged before completing their full term ("Calls to Shorten Military Service," Yediot Achronot [daily], January 26, 1996; "One Out of Every Five Does not Finish Military Service," Yediot Achronot [daily], Septembre 2, 1976; "Seventy Percents of All Men in Israel Do not Serve on Reserve," Yediot Achronot [daily], December 5, 1996). Statements which were once unthinkable, such as "discharge in order to reduce the budget," have become part of the discourse on mobilization, along with a five-day work week in most units and a cutback on expensive reserve duty as the IDF places the emphasis on high-tech weapons and is undergoing something of a "managerial revolution" to ready itself for the next century (on this trend in general, see Coroalles 1996).

Some of the tendencies mentioned above were blocked by the army itself, realizing their damage to its competence. Other developments matched its own inclinations. The question, however, remains: is the Israeli army adjusting itself to "modern times"? Is it on the brink of becoming a "post-modern army," like the Italian Army, for example, which perceives itself in terms of peace-keeping, one which is part of the global tendency of reducing mass armies and replacing military conflicts with economic ones?[18] Or might it be a mistake to reduce the understanding of the complex relations between army, society, and politics in Israel to these trends? Here we must again resort to the concept of the role expansion of the army, which continues to be one of the instruments in Israel through which a constant effort is made by specific groups to curb the peace process and preserve the militaristic ethos, as we will now see.

The Gordian Knot:
From "Social" to "Political" Role Expansion

Gush Emunim (Bloc of the Faithful) has come a long way from its origins as a protest movement of young religious Zionists who in the 1970s opposed the peace treaty with Egypt, and as a settlement organization trying to appropriate the occupied territories in Judea and Samaria and make them part of Israel, to the present situation, in which the fundamentalist settlers and their supporters are a major factor in Israeli politics. The movement seeks to thwart every peace initiative because of its religious faith in the sanctity of territory, its conception of wars as holy wars, and its perception of army and state through the prism of redemption and messianism.[19]

Some observers have wondered whether the ability of the protest/settler movement to maintain itself independently over so many years attested to the existence of a "weak state," which was unable to reverse the movement's ability to "create facts on the ground" in the territories (on the concept of "a weak state," see Migdal 1988). But that is not the situation. Usually, operations carried out by settlers were the informal continuation of government policy and an indirect means by which Israel achieved domination over the local Palestinian population. Indeed, Israeli governments themselves—all of them, both the right-wing Likud and the left-wing Labor—have engaged in settlement activity throughout this entire period and have developed various means of cooperation with the settlers (Newman 1985; Benvenisti 1988; Ben-Eliezer 1998c).

The IDF itself has a lengthy history of close cooperation with the settlers, which takes place both formally and informally. Throughout the three decades of the occupation, the army has guarded the settlers, escorted them in convoys bound for Israel or for other settlements, "ridden shotgun" for their children on the way to school, and the like. In fact, the army, as it turned out, even provided personal security for Rabbi Moshe Levinger, a controversial figure who resides in the Arab city of Hebron in the West Bank together with a small group of settlers, whom the state itself has indicted on numerous counts of disturbing the peace and attacking Arabs (e.g., *Haaretz* [daily], August 29, 1996).

But the true depth and substance of that cooperation was disclosed in testimony given to the commission of inquiry that investigated the massacre by Baruch Goldstein, a physician from the settlement of Kiryat Arba, of thirty Muslims who were praying in the Cave of Machpelah in Hebron. Goldstein was neither a soldier nor a civilian. He was a janus, a hybrid. Otherwise it would be incomprehensible how a settler—a devoted disciple of Rabbi Meir Kahane, the slain leader of a fascist movement that sought the eviction of all Arabs from the Land of Israel—wearing an IDF uniform with officer insignia and legally carrying an army-issued submachine gun, could enter the Cave at a time when, under a special arrangement, the site was reserved for Muslim worshippers only.

Such janus creatures are a pure product of the state and its military role expansion. The source of the duality lies in decisions made in the 1970s by then Chief of Staff Rafael Eitan to organize settlers in the territories in special reservists' units. The system was called "area defense" (*hagana merhavit*). In fact, such units had always been part of the arrangements of the nation-in-arms, but not in the territories. Eitan, though, took a different approach. By applying this aspect of the nation-in-arms in the disputed territories, and by creating a bond between settlers and soldiers, in fact a Gordian knot, he helped legitimize the settlers' contention that they stand at the center of the nation and that the territories are part of Israel. If the possibility of withdrawal from the territories and evacuating settlements ever looms as a genuine option, it was Eitan's hope that area defense would be one of the key institutional settings that would help torpedo the option.[20]

The role expansion area defense, thus, served not only as a neutral means for keeping security and tranquillity, but as a way to advance a narrow politics. This was seen in French Algeria as well. The "Territorial Units" there reflected military cooperation between settlers and the French Army, along the lines of an army-nation and role expansion. However, what began as a military tactic soon turned into politics. In September 1959, when de Gaulle first spoke publicly about detaching Algeria from France, the settlers turned immediately to the military option, employing French Army weapons, ammunition, and equipment (Horne 1977: ch. 17).

In Israel the situation is even more serious because all the settlers have IDF-issue weapons, even the most extreme among them from the fascist groups "Kahane" and "Kahane Lives," some of whom are even branded by other settlers as a "wild bunch" (e.g. *Yediot Achronot* [daily], February 27, 1994). Moreover, public testimony by a senior army officer given to the commission of inquiry that investigated the massacre perpetrated by Baruch Goldstein disclosed that the IDF had a standing order not to shoot at settlers engaged in "purposeful fire" (i.e., shooting to kill) but first to let them finish shooting and then to try and overcome them by other means (Nachum Barnea, "Who Gave the Order," *Yediot Achronot* [daily], March 11, 1994). The conclusions reached by the judicial commission were also problematic. The panel in effect accepted the case put by the army that the massacre had been an aberration, or as Chief of Staff Ehud Barak put it, it had come "like a bolt out of the blue" (*Yediot Achronot* [daily], June 27, 1994).

The Chief of Staff may have been taken by surprise. But anyone who peruses the writings of the settlers from 1994 to 1996, the period of the Labor Party government and the Oslo accords, will immediately grasp how eager they are to restore the situation in which Israel was a nation-in-arms, living by the sword. True, they write, we have only one IDF, but it must be transformed. The IDF as it stands is not ours. Indeed, the army itself is also to blame: because it has generals who take up the scepter of the politicians, and because it claims that the solution to the Arab-Israeli conflict is political, the army has betrayed its trust (*Nekuda* [settlers' monthly], 178, June 1994; *Nekuda*, 194, February 1996). The conclusion that follows from what is perceived as the army's impotence is clear and concrete, and serves the concept of role expansion: "There is something we can do besides expressing concern," they write, "let our finest go to area defense. Today the immediate threat to the state emanates from the Palestinian enemy... Because of the close acquaintance of the area's [Jewish] residents with the terrain and because of their [high] motivation, the area defense units are excellent and can be a first spearheading and intervention force. Reconnaissance units to guide forces" (Azriel Mintz in *Nekuda*, 194, February 1996).

The pre-state forces experienced cleavages and acute conflicts among the various military organizations. In the first years of the state, the arrangements of the statist-nation-in-arms brought about the establishment of the one and undivided army-nation. Now, fifty years on, parochialism has reappeared as the "social role" of the military becomes a "political role." In other words, the transition was from an army that took part in shaping a dominant culture which glorified the importance of war as a statist political means, to an army in which certain segments use the nation-in-arms mechanisms and norms as a means to further their narrowly bellicose perception of reality. The crucial point is that this new phenomenon relies, among other things, on the military's role expansion. A comparison with the French case in Algeria shows that there, too, the political role concept that emerged among groups within the military was in effect based on the role expansion of the military. In Algeria an entire doctrine, known as *"la theorie de la guerre revolutionaire,"* afforded ostensibly strategic justification for the army to expand into various spheres, a development that vested it with immense political influence and afterward enabled it to rebel several times against Paris based on the claim that the army was the embodiment of the nation (Ben-Eliezer 1998b). The rationalization in Israel is not much different. As one rabbi explained, "The national-religious young man is a religious version of the Zionist vision of the proud and self-confident army... the fulfiller of the Zionist vision. In recent years, beholding the changes that have occurred among the secular Zionist public, they [the religious young people] perceive themselves as the guardians of the flame of classical Zionism. They see themselves as the last of the Zionists fighting a tough rearguard action" (Menachem Fruman, "Cypress and a Skullcap on its Top," *Haaretz* [daily], January 7, 1994).

Another possible state setting based on the socio-political role expansion of the military, which can be used to instill sectorial politics, is the arrangement known as the *"hesder yeshiva,"* in which religious high-school graduates combine yeshiva studies with military training and service. These units are manned almost exclusively by settlers and their supporters. They all serve in the same companies or battalions in both the regular army and afterward in the reserves, and the situation has elements of an army within an army.[21]

One such graduate of a hesder yeshiva, who was serving in an elite military unit, Sergeant Arik Schwartz, was arrested shortly after the Rabin assassination. He was later convicted of supplying explosives (such as machine gun bullets, two hand grenades, plastic explosives, a rifle-fire simulation mechanism, and hundreds of rifle bullets) to the Prime Minister's assassin and his accomplice-brother. Schwartz may have thought that the targets would be Arabs. However, after the assassination and the arrest of the killer Schwartz quickly emptied out a whole arsenal of weapons and explosives which he had hidden in his house and hid them to avoid discovery by the police (Lior El-Hai, "Yigal Amir has Asked Me," Yediot Achronot [daily], March 31, 1996). In any event, the critical point is the ease with which the settlers and their supporters can turn army weapons to narrow political purposes.

Following the Rabin assassination public criticism was leveled at the *hesder yeshiva* system, in which the assassin and his brother had also once been enrolled, for acting as a hothouse for these violent phenomena. The deputy defense minister at the time, Ori Orr, urged the system's dismantlement. However, the National Religious Party threatened a no-confidence motion and Prime Minister Shimon Peres ordered a stop to all discussion of Orr's proposal (*Haaretz* [daily], December 13, 1995; December 14, 1995). The opportunity to do away with the phenomenon of "religion-in-arms" in place of a "nation-in-arms" was not realized. In fact, it is being reinforced by means of additional institutional arrangements that blurred a possible distinction between the "civil" and the "military".

Such arrangements are related to the fact that in the past decade the settlers and their supporters have been given access to a channel of action that enables them to pursue a military career or to excel in elite units. To begin with, there are five pre-army religious preparatory programs (four are in the territories), which encourage youngsters who harbor religio-national or ultra-nationalist convictions to serve in elite combat units and/or to join the career army. These programs produce soldiers who are powerfully motivated to enter combat units and take up a military career. The senior officer corps, it emerges, is unable to remain indifferent to such an outburst of enthusiasm and voluntarism, and has warmly welcomed the phenomenon. Field commanders compete for the program's participants, even lobbying them to join their units. One General, Moshe Yaalon, has actively cultivated the program at the West Bank settlement of Eli since he was the commander of the paratroops brigade (Avishai Becker, "Religious Trek," *Maariv* [daily], March 8, 1996) As for the political implications that drive these religious youngsters, the senior officers are apparently unaware of them, or perhaps are simply not disturbed by them. The youngsters will be "good soldiers," and many Israelis still believe that nothing else matters.

Another state-created breeding ground of factionalism within the army is the military-religious boarding school, which has existed since 1980, when Rafael Eitan was Chief of Staff. The boarding school's emblem is a Torah scroll and a sword, a classic nation-in-arms blurring of the distinction between army and society. The school's internal arrangements are also calculated to eliminate such distinctions. It is run by civilian teachers but is funded by the Defense Ministry. After lunch the youngsters at the school are already under the authority of army personnel. Also worth noting is that the boarding school is run by a well-known rabbi and former Knesset Member, Chaim Druckman, an extreme right-wing settler who is one of the most militant advocates of Israel's right to remain in the occupied territories. The curriculum in this institution includes nationalist and religious indoctrination shot through with fundamentalist ideas of "redemption" and "wars of commandment."

Indeed, the religious settlers, in contrast to their secular counterparts, most of whom moved to the occupied territories for economic reasons, have a militaristic perception of reality. It stems from their belief that the Land of Israel (which, unlike the State of Israel, includes also the territories captured in 1967) belongs solely and exclusively to the Jewish people. Since the land is also populated by "gentiles" and "strangers," who have been there for hundreds of years, there is no other way but to fight them for it. This approach was articulated into religious militarism on various occasions by the spiritual father of the entire religious settlement phenomenon in the territories, Rabbi Zvi Kook, who called on young people to wage a "war of commandment." "From a national point of view," Rabbi Kook wrote, "war is a decree that must be accepted... it belongs to the nation's life-agenda. When there is statehood there is war. War from time to time is a normal thing... The conquest of the Land is a commandment... If it is impossible to carry out the commandment of inheriting the Land without bloodshed, we have no other choice but to act by means of our soldiers and our army" (Aviner 1994).

It should also be noted that it is not by accident that believers in total messianism have sanctified a state together with its institutions and enterprises, even if they are defined as secular. This is a classic messianic manifestation, built on a militant, domineering foundation, that views Israel's wars and its military struggles not only as a pure matter of national security, and not even only as a matter of the settlement and conquest of the land, but as a saliently theological, values-grounded project involving a titanic contest for the uprooting of evil and a striving for universal world reform. Thus even defeat in war, as occurred in the 1973 Yom Kippur War, is viewed by the settlers and their supporters not as proof of the need for peace, compromise, and a departure from the path of force but, on the contrary, as another stage of divine experience on a road filled with ordeals that leads to redemption (Ravitzki 1993: 116-7).

The carriers of this perception of reality are the religious young officers, many of them are graduates of the religious pre-army institutions. In a press interview, one such graduate described the essence of the pre-army programs that operate in the territories: "All of them are part of the right-wing political line... [The participants] see themselves as the spearhead of the IDF, the next [generation of] senior command, [producing] the first religious chief of staff." According to a rabbi in one of these institutions: "The difference between one of our officers and a secular officer is that the latter thinks in terms of self-fulfillment, whereas we are fulfilling a mission" (*Yediot Achronot* [daily], November 17, 1995). Another example of the new perception of reality that is developing in the army can be found in the settlers' journal *Nekuda*. A bereaved father eulogizes his soldier-son, who was killed. He was a young man who enrolled in the religious military boarding school and volunteered for the paratroops, and he symbolizes the new religious Israeli who, according to his father, "did not go the army because he got a call-up order but because he had a sense of mission. He had a mission, the army has a mission, the state has a mission, and therefore he has to serve [and, probably in the father's conception, also to die]" (*Nekuda*, 176, March 1994). In view of the decline of the old model combined with the processes of de-escalation and demilitarization within the society, the ideology of the new religious officers can be termed "neomilitarism." They see their role as being to return Israel to the past, to the glorified days of the nation-in-arms, the army-nation, and the habitude of using military means for political purposes.

It should be taken into account that in order to justify their penetration of the army the settlers in the early 1990s consistently said the IDF was factional, subject to left-wing politicization (led by Rabin and Peres), and was not doing its work well, from their point of view, certainly not with regard to controlling the territories. As they put it, "the army has adopted the slogan of a political solution, like the bent floor of those who cannot dance." And, "In the past army officers would urge the politicians to let them activate the forces of the army so it could fulfill its mission. Today the situation is reversed: they are the ones who are making the politicians faint-hearted" (*Nekuda*, 164, December 1992). The result is a situation in which the settlers accuse the army of politicization, which serves them as a good excuse to politicize the army. "The leaders of the army must know," they write, "that the upshot of this politicization, to put it mildly, is a reaction by the counter-force" (*Nekuda*, 174, January 1994).

Serving officers in the IDF are of course forbidden to express political views publicly, and that role is taken, on both the left and the right, by officers in the reserves. Thus *Nekuda* interviewed a right-wing officer who was jailed for refusing to do reserve duty on the day the Israeli-Palestinian agreement on Gaza and Jericho was made public. Captain Motti Karpel is quoted as saying he is unwilling to cooperate with a government that hands over territory. "The law and democracy are not a supreme, absolute value in themselves," he says, adding that the torch of Zionism has now been passed from the secular to the religious public (*Nekuda*, 171, September 1993). He is hardly alone. In April 1994 about 500 colonels and brigadier generals signed a letter calling on the Rabin government to break off the negotiations with the Palestine Liberation Organization (*Haaretz* [daily], April 4, 1994).

The reservists often give the soldiers in the regular army backing and show the way. But the major backing for the political role conception of the religious soldiers and officers came from their rabbis. At the end of March 1994, in the light of the possibility that the Rabin government would evacuate settlements and hand back territory, prominent rabbis with a dominant influence on the settlers issued a religious ruling ordering soldiers to refuse to obey any order to evacuate Jewish settlers from their homes, citing the Torah, which, they said, forbids the evacuation of any part of the Land of Israel (*Yediot Achronot* [daily], March 31, 1994). The rabbis' commandment to violate IDF orders touched off a huge furor among the Israeli public and caused bewilderment and confusion among many religious youngsters. Some said that they would never disobey army orders in any event. But there were also others. One of them explained, "I will allow myself to be critical... of what the rabbi said. At the same time, I acknowledge that as long as they do not retract what they said... and as long as I do not find Torah sages of the same weight who oppose them, I will weep and do what they say" (Yacob Sadan, "Between Refusal and Tears," *Nekuda*, 177, April 1994).

We see, then, how the role expansion of the military can lead, in conditions of crisis and decolonization occurring in a society that is divided and fragmented, among other issues over a peace process, to the development of a narrowed political role in sections of the army. Ironically and paradoxically, it is the arrangements of the nation-in-arms and of the army-nation, which is not a professional army separate from the society but possesses an emphatic ideological aspect from the outset, which enable certain military groups and ideological colonels to view themselves as the true representatives of the nation, and to work for returning Israel to its glorious militaristic past.

An army, of course, is not a homogeneous entity; it accommodates various views and trends. There are major differences between the senior officer corps and the ideological colonels. Today's senior officers reflect a more or less professional role perception. Some of them are officers with "a knife in their teeth," others are the product of the IDF's diminished status and carry the reputation of being full partners to the Oslo process. In any case, their position is not an easy one in the light of the public criticism to which they and the IDF are being subjected. Still, the senior officer corps is too establishment-oriented to intervene directly in politics. Moreover, the second-career pattern is still widespread in Israel, along with the possibility of being "parachuted" into politics, and these two options constitute safety valves against the emergence of praetorian tendencies among IDF generals. If, however, an Israeli government should ever decide to evacuate settlements and settlers, and if that step leads to a Palestinian state, it cannot be ruled out that for the first time in Israel's history part of the army may rise up in revolt.

As in French Algeria, it will be the settlers who initiate the resistance, and they may well be joined by some units and various colonels. Such a scenario will involve a small number of rebels and a small number of units. In Israel, however, it has already seen that even a handful of people can alter the course of history, certainly if they are backed by a social formation and statist role expansionist arrangements and if they evoke ethno-nationalist and militaristic sentiments from the past.

In any case, the question of whether the prospect exists for a military coup, and what the conditions for that would be, is not at the center of this paper. The intention here is to suggest the manner in which the military role expansion, which in the past was part of the war-making arrangements of the state, serves as an instrument that stimulates and strengthens a neomilitaristic outlook. That outlook is carried by groups, both inside and outside the army, which use the mechanisms of military role expansion in order to transform religious militarism into a ruling military culture in Israel and in order to promote their narrow-based politics. They oppose peaceful solutions to the Arab-Israeli conflict, particularly if they entail ceding land, evacuation of settlements, and the establishment of a Palestinian state.

Epilogue

A year or two after this article was written, and just before it went to the printer, it turned out that even though the right-wing Netanyahu period had ended and Ehud Barak had begun his term of office as Prime Minister, the time of Israeli militarism in its religious version and draws on the perception of the army's broad role, had not yet run its course. True, in the light of the Rabin assassination and the continuation of the peace process, some of the settlers are displaying a more moderate tendency, engaging in self-examination, and even showing manifestations of an individualist religious conception which is less collectivist and less fundamentalist than in the past (*Ha'aretz* [daily], October 13, 1999). Yet the Barak government, too, finds it difficult to evacuate even tiny "hilltop outposts," which were established by stealth at the initiative of only a handful of individuals. This is, usually, the initiative of activists from the second generation of settlers, who are defying what they consider the excessive moderation of the *Yesha* Council, the body that represents the settlers in the territories. The young people want to block the trend that has emerged in the Oslo and Wye agreements by setting up new settlement outposts. Some of them are actually empty, or contain only a few individuals, dilapidated structures, a barking dog, a lone generator, and, of course, IDF soldiers to guard the site. In October 1999, Prime Minister Barak reached a compromise with the settlers on the voluntary evacuation of a small fraction of the outposts, but at the same time his government, in return for this, effectively validated more than thirty of the forty-two outposts that were set up. Thus the government was forced to accept sporadic takeovers by the settlers of land throughout Judea and Samaria. This is the form the peace process has taken and these are the types of obstacles it will face in the future. At one of the outposts, Tzofit, located north of Ramallah, a pre-army academy is operating in co-operation with the army. The reader will undoubtedly not be surprised to learn that under the terms of the agreement between Barak and the settlers, the civilians at that outpost will leave but the pre-army academy will remain (*Yediot Achronot* [daily], October 15, 1999). The "Tzofit agreement" all but exempts us of the need to underline the importance of this article, which intended to address the ongoing and continuing interconnection between the military's role expansion, cultural militarism and difficulties in peace-making in Israel.

Notes

1. Subsuming these scholars under the same category of a "civil-military relations" paradigm, is based on Finer (1978) and Edmonds (1990: 70-112).
2. For criticism on the paradigm's tendency to separate between the civil and the military, see Valenzuela (1985); Berghahn (1981: 67-84). For criticism on the idea that leadership tend to act rationally, see Van Creveld (1991: esp. Ch. 5); Keegan (1993); Tuchman (1984).
3. Among the works that can be categorized under this assumption are: Perlmutter (1969); Peri (1983); Horowitz & Lissak (1989); Yaniv (1993); Ben-Meir (1995); Barzilai (1996). Exceptions are: Erlich (1987); Kimmerling (1993); Levi (1997); Izraeli (1997); Helman (1997); Ben-Eliezer (1998a).
4. Some question that issue. See Schiff (1992); Ben-Eliezer (1993); Peled and Shafir (1996).
5. Kimmerling (1985). Since then Kimmerling has completely changed his perspective on the role of the military in Israel. See Kimmerling (1993).
6. On the break-through in the research on the nation-state and violence, see Shaw (1989).
7. On the neorealist basic assumptions, see Rotberg and Rabb (1989). For criticism on the neorealist perspective, see Wendt, (1992).
8. On the connection between nationalism and war, see Howard (1984); Held (1994: 74-8); Bond (1985); Michener (1993); Forster (1987).
9. On the well-known thesis of the connection between war and democracy, see Andreski (1968: 145-6); Janowitz (1976). On the mobilized and coercive character of states, even democracies, see Porter (1994).

10. Vagts, (1959: 219). For more on the way the French imitated the German army's regulations, see (Mitchell, 1984).
11. On the concept of militarism in general, see Berghahn (1981); Mann (1988); Strgardt (1994); Kimmerling (1993); see also on the difference between cultural militarism and militaristic politics in Ben-Eliezer (1998a: 1-15).
12. On the concept of armee-nation in France, see Bankwitz, (1967).
13. See Cook (1978: 271); Ben-Eliezer (1998a: 193-222). In fact, in all nations-in-arms, including Israel, a gap existed between an ideology of equality and full participation of all citizens and a reality of discrimination and inequality within the military. In the Israeli case, see Levi (1997).
14. Occassionaly, the mechanisms that blurred the distinction between army and society were described, almost always presented as a means for civilianization of the military. On the work with new immigrants and underprivileged, see Bowden, (1976: 67-91); Azarya and Kimmerling (1980); Meisels and Gal (1993); more on the Makam Project, *Skira Chodshit* (Monthly Army Officers), No. 10, 1990; on the military troupes, see Yair Rosenblum (1988); on the IDF Radio, see Mann and Gonn-Gross (1991); on the civil guard, see Kimmerling (1978); on the Nahal, see Bowden (1976: 135-58); Keren (1991).
15. On Dayan and the generals in the 1950s, Morris (1993); Barnett (1992: 155-75); Golani, (1997); Ben-Eliezer (1998a: chap. 11-12); on the period preceded the Six-Day War, Rabin (1979); on the territories and the IDF, Peri (1989); Pedatzur (1996: 194-203); on the Lebanon War, Yariv (1985); Schiff and Ya'ari (1984).
16. On the parents' struggle see *Haaretz*, 29 August, 1996. On this cultural phenomenon in general, see Mosse (1990).
17. Cohen, (1995); Gal and Lev (1996). On the general trend in the world, see Moskos (1986: 377-82).
18. For more details on the claim of a global trend, see the whole volume of Armed Forces and Society 23/3 (1997). As for the Italian army, see Battistelli (1997). See also Burk (1994); Dandeker (1994).
19. On the process of radicalization that the religious Zionism has gone through, see Ravitzki (1993). On Gush Emunim in its early years, see Rubinstein (1982); Lustick (1988).
20. See a debate in the Israeli parliament already in October 1977 on the issue of "private armies" in the territories, which were established through the state's initiatives, Knesset Protocols, October 26, 1977. For another debate, eleven years later, see Knesset Protocols, March 2, 1988.
21. For more information on the Hesder Yeshivot arrangement with the state, see Cohen (1993).

References

Andreski, Stanislav 1968. *Military Organization and Society.* Berkeley: University of California Press.

Ashkenazy, Daniella (ed.) 1994. *The Military in the Service of Society and Democracy.* Westport, CT: Greenwood Press, 1994.

Aviner, Shlomo (ed) 1994. *Army Precepts—Responses on Military Issues.* Jerusalem: Yeshivat Ateret Kohanim. (Hebrew).

Azarya, Victor 1983. "The Israeli Armed Forces." In *The Political Education of Soldiers,* edited by Morris Janowitz and Stephen D. Wesbrook, 99-128. Beverly Hills, CA: Sage.

Azarya, Victor and Baruch Kimmerling 1980. "New Immigrants in the Israeli Armed Forces." *Armed Forces and Society*, 6, 3: 22-41.

Bankwitz, Philip 1967. *Maxime Weygand and the Civil-Military Relations in Modern France*. Cambridge, MA: Harvard University Press.

Barnett, Michael 1992. *Confronting the Costs of War—Military Power, State, and Society in Egypt and Israel*. Princeton, NJ: Princeton University Press.

Barzilai, Gad 1996. *Wars, Internal Conflicts, and Political Order—A Jewish Democracy in the Middle East*. Albany: State University of New York Press.

Battistelli, Fabrizio 1997. "Peacekeeping and the Postmodern Soldier." *Armed Forces and Society*, 23, 3: 467-84.

Benvenisti, Meron 1988. *The Sling and the Club*. Jerusalem: Keter. (Hebrew).

Ben-Eliezer, Uri 2000 "Do the Generals Run Israel? The Transition of High-Ranking Army Officers in Politics and its Implications." In *Fifty Years - a Society in Reflection*, edited by Hanna Herzog et al. Tel-Aviv: Ramot.

Ben-Eliezer, Uri. 1998a. *The Making of Israeli Militarism*. Bloomington: Indiana University Press.

Ben-Eliezer, Uri 1998b. "Is a Military Coups Possible in Israel? Israel and French-Algeria in Comparative Historical-Sociological Perspective." *Theory and Society*, 27, 3: 311-49.

Ben-Eliezer, Uri 1998c. "State Versus Civil Society: A Non-Binary Model of Domination Through the Example of Israel." *Journal of Historical Sociology*, 11, 3: 370-96.

Ben-Eliezer, Uri 1997. "Rethinking the Civil-Military Relations Paradigm: The Inverse Relation Between Militarism and Praetorianism Through the Example of Israel." *Comparative Political Studies*, 30, 3: 356-74.

Ben-Eliezer, Uri 1993. "The Meaning of Political Participation in a NonLiberal Democracy." *Comparative Politics*, 25: 397-412.

Ben-Meir, Yehouda 1995. *Civil-Military Relations in Israel*. New York: Columbia University Press.

Berghahn, Volker 1981. *Militarism: The History of International Debate, 1861-1979*. Cambridge: Cambridge University Press.

Bienen, Henry (ed.) 1971. *The Military and Modernization*. Chicago: Aldine Press.

Bond, Brian 1985. *War and Society in Europe, 1870-1970*. New York: St. Martin Press.

Bowden, Tom 1976. *Army in the Service of the State*. Tel-Aviv: University Publishing Project.

Burk, James 1994. *The Military in New Times—Adapting Armed Forces to a Turbulent World*. Boulder, CO: Westview Press.

Challener, Richard 1965. *The French Theory of the Nation-in-Arms, 1866-1939*. New York: Russel and Russel.

Cohen, Stuart 1996. "The IDF and the Israeli Society: Toward a Military Role Extraction?" In *Israel Towards the Year 2000—Society, Politics and Culture*, edited by Moshe Lissak and Brian Knei-Paz, 215-32. Jerusalem: Magnes. (Hebrew).

Cohen, Stuart 1995. "The Israel Defense Forces (IDF): From a 'People Army' to a 'Professional Military'—Causes and Implications." *Armed Forces and Society*, 21: 246-54.

Cohen, Stuart 1993. "The Hesder Yeshivot in Israel: A Church-State Military Arrangement." *Journal of Church and State*, 35, 1: 1123-31.

Cook, Theodore 1978. "The Japanese Reserve Experience: From Nation-in-Arms to Baseline Defense." In *Supplementary Military Forces*, edited by Louis A. Zurcher and Gwyn Harries-Jenkins. London: Sage.

Coroalles, Anthony 1996. "On War in the Information Age: A Conversation with Carl Von Clausewitz." *Army*, May 1996: 24-34.

Dandeker, Christopher 1994. "New Times for the Military: Some Sociological Remarks on the Changing Role and Structure of the Armed Forces of the Advanced Societies." *British Journal of Sociology*, 45, 4: 637-55.

Dandeker, Christopher 1990. *Surveillance, Power, and Modernity*. Cambridge: Polity Press.

Edmonds, Martin 1990. *Armed Services and Society*. Boulder, CO: Westview Press.

Eley, Geoff 1987. "Army, State and Civil Society: Revisiting the Problem of German Militarism." In *Unification to Nazism*, edited by Geoff Eley, 85-109. Boston: Allen and Unwin.

Erlich, Avishai 1987. "Israel: Conflict, War and Social Change." In *The Sociology of War and Peace*, edited by Colin Creighton and Martin Shaw, 121-42. London: Macmillan Press.

Finer, Samuel. 1978. "The Statesmanship of Arms." *Times Literary Supplement*, February 17, 1978.

Finer, Samuel 1976. *The Man on a Horseback: The Role of the Military in Politics*. Harmondsworth: Penguin Books.

Forster, Stig 1987. "Facing 'People War': Moltke the Elder and Germany's Military Options after 1871." *Journal of Strategic Studies*, 10, 2: 209-30.

Gal, Reuven (ed.) 1990. *The Seventh War—The Influence of the Intifada on Israeli Society*. Tel-Aviv: Hakibutz Hameuchad. (Hebrew).

Gal, Reuven and Raphi Lev 1996. "The Military Role—Between Occupation and Institution." *Maarachot*, 347: 44-6. (Hebrew).

Giddens, Anthony 1987. *The Nation-State and Violence*. Berkeley: University of California Press.

Golani, Motti 1997. *There Will be War Next Summer—The Road to the Sinai War, 1955-1956*. Tel-Aviv: Ministry of Defense. (Hebrew).

Halpern, Ben 1976. "The Role of the Military in Israel." In *The Role of the Military in Underdeveloped Countries*, edited by Abraham F. Lowenthal, 317-57. New York: Holmes and Meier Publishers.

Hanning, Hugh 1967. *The Peaceful Uses of Military Forces*. New York: Praeger.

Hayes, Carlton 1931. *The Historical Evolution of Modern Nationalism*. New York: Russel and Russel.

Held, David 1994. "The Development of the Modern State." In *Modernity: An Introduction to Modern Societies*, edited by Stuart Hall et al. London: Blackwell.

Helman, Sara 1997. "Militarism and the Construction of Community." *Journal of Political and Military Sociology*, 25: 305-32.

Horne, Alistair 1977. *A Savage War of Peace, Algeria 1954-1962*. London: McMillan.

Horowitz, Dan 1975. "The Israeli Defense Forces: A Civilized Military in a Partially Militarized Society." In *Soldiers, Peasants and Bureaucrats*, edited by Roman Kolkowicz and Andrei Korbonski, 77-106. London: Allen Lane.

Horowitz, Dan and Baruch Kimmerling 1974. "Some Social Implications of Military Service and the Reserve System in Israel." *Archives European de Sociologie* 15: 262-76.

Horowitz Dan and Moshe Lissak 1989. *Trouble in Utopia—The Overburdened Polity of Israel*. Albany: State University of New York Press.

Howard, Michael 1984. *The Causes of Wars*. London: Unwin.

Humphreys, L.A. 1975. "The Japanese Military Tradition." In *The Modern Japanese Military System*, edited by James Buck. Beverly Hills, CA: Sage.

Huntington, Samuel 1968. *Political Order in Changing Societies*. New Haven, CT: Yale University Press.

Ienaga, Saburo 1993/4. "The Glorification of War in Japanese Education." *International Security*, 18, 3: 113-33.

Izraeli, Dafna 1997. "Gendering Military Service in the Israeli Defense Forces." *Israel Social Science Research*, 12, 1: 129-66.

Janowitz, Morris 1976. "Military Institutions and Citizenship in Western Societies." *Armed Forces and Society*, 2, 2: 185-204.

Janowitz, Morris 1971. *The Professional Soldier, A Social and Political Portrait*. New York: Free Press.

Janowitz, Morris 1964. *The Military in the Development of New Nations: An Essay in Comparative Analysis*. Chicago: University of Chicago Press.

Johnson, John 1962. *The Role of the Military in Underdeveloped Countries*. Princeton, NJ: Princeton University Press.

Katriel, Tamar 1991. "Picnics in a Military Zone: Rituals of Parenting and the Politics of Consensus." In *Communal Web*, edited by Tamar Katriel, 71-91. New York: State University of New York Press.

Keegan, John 1993. *A History of Warfare*. London: Hutchinson.

Keren, Shlomit 1991. *The Plow and the Sword*. Tel-Aviv: Ministry of Defense. (Hebrew).

Kimmerling, Baruch 1993. "Patterns of Militarism in Israel." *European Journal of Sociology*, 34: 196-223.

Kimmerling, Baruch 1985. *The Interrupted System: Israeli Civilians in War and Routine Times*. New Brunswick, NJ: Transaction Books.

Kimmerling, Baruch 1978. "The Israeli Civil Guard." In *Supplementary Military Forces*, edited by C.A. Zurcher and G. Harris-Jenkins, 107-25. New York: Sage.

Levi, Yagil 1997. *Trial and Error—Israel's Route from War to De-Escalation*. Albany: University of New York Press.

Lissak, Moshe 1984. "Paradoxes of Israeli Civil-Military Relations: An Introduction." In *Israeli Society and Its Defense Establishment*, edited by Moshe Lissak. London: Frank Cass.

Lissak, Moshe 1976. *Military Roles in Modernization—Civil-Military Relations in Thailand and Burma*. Beverly Hills, CA: Sage Publications.

Lissak, Moshe 1971. "The Israel Defense Forces as an Agent of Socialization and Education: a Research in Role-Expansion in a Democratic Society." In *The Perceived Role of the Military*, edited by M.R. van Gils, 325-39. Rotterdam: Rotterdam University Press.

Lissak, Moshe 1967. "Modernization and Role-Expansion of the Military in Developing Countries: A Comparative Analysis." *Comparative Studies in Society and History*, 9, 4: 233-55.

Luckham, A.R. 1971. "A Comparative Typology of Civil-Military Relations." *Government and Opposition*, 6: 9-35.

Lustick, Ian 1988. *For the Land and the Lord—Jewish Fundamentalism in Israel*. New York: Council on Foreign Relations.

Maman, Daniel and Moshe Lissak 1996. "Military-Civilian Elite Networks in Israel: A Case in Boundary Structure." In A *Restless Mind: Essays in Honor of Amos Perlmutter*, edited by Benjamin Frankel, 49-79. London: Frank Cass.

Mann, Michael 1988. *States, War and Capitalism*. New York: Basil Blackwell.

Mann, Michael 1993. *The Sources of Social Power*. V. 2. Cambridge: Cambridge University Press.

Mann, Raphael and Tsippy Gon-Gross 1991. *Galey Zahal—Round the Clock*. Tel-Aviv: Ministry of Defense. (Hebrew).

Meisels, Ofra and Reuven Gal 1993. *The Adaptation of Ex-Makam Soldiers to Civil Life*. Zichron Yaacov: Israeli Institute for Military Studies (Hebrew).

Michener, Roger (ed.) 1993. *Nationality, Patriotism and Nationalism in Liberal Democratic Societies*. St. Paul, MN: Paragon House.

Migdal, Joel 1988. *Strong Societies, Weak States*. Princeton, NJ: Princeton University Press.

Mitchell, Allan 1984. *Victors and Vanquished, The German Influence on Army and French in France After 1870*. Chapel Hill: University of North Carolina Press.

Morris, Benny 1993. *Israel's Borders Wars, 1949-1956*. Oxford: Clarendon Press.

Moskos, Charles 1986. "Institutional/Occupational Trends in Armed Forces: An Update." *Armed Forces and Society*, 12, 3: 377-82.

Mosse, George 1990. *Fallen Soldiers—Reshaping the Memory of the World Wars*. New York: Oxford University Press.

Nakamura, Yshihida and Ryoichi Tobe 1988. "The Imperial Japanese Army and Politics." *Armed Forces and Society*, 14: 511-25.

Negbi, Moshe 1995. *Freedom of Press in Israel—The Legal Aspect*. Jerusalem: The Jerusalem Institute for Israel Studies. (Hebrew).

Newman, David (ed.) 1988. *The Impact of Gush Emunim*. London: Croom Helm.

Otley, C.B. 1978. "Militarism and Militarization in the Public School." *British Journal of Sociology*, 29, 3: 325-39.

Pedatzur, Reuven 1996. *The Triumph of Embarrassment—Israel and the Territories After the Six-Day War*. Tel-Aviv: Bitan. (Hebrew).

Peled, Yoav and Gershon Shafir 1996. "The Roots of Peace Making: The Dynamics of Citizenship in Israel, 1948-1993." *Journal of Middle East Studies*, 28: 391-413.

Peres, Shimon 1993. *The New Middle East*. Tel-Aviv: Stematzki.

Perlmutter, Amos 1977. *The Military and Politics in Modern Times*. New Haven, CT: Yale University Press.

Perlmutter, Amos 1969. *Military and Politics in Israel—Nation-Building and Role Expansion*. New York: Praeger.

Peri, Yoram 1989. "The Impact of Occupation on the Military: The Case of the IDF, 1967-1987." In *The Emergence of a Binational Israel*, edited by Ilan Peleg and Ofira Seliktar, 143-168. Boulder, CO: Westview Press.

Peri, Yoram 1983. *Between Battles and Ballots, Israeli Military in Politics*. Cambridge: Cambridge University Press.

Peri, Yoram 1981. "Political-Military Partnership in Israel." *International Political Science Review*, 2: 303-15.

Porch, Douglas 1981. *The March to the Marne—The French Army, 1871-1914*. Cambridge: Cambridge University Press.

Porter, Bruce 1994. *War and the Rise of the State*. New York: Free Press.

Rabin, Yitzhak 1979. *Service Diary*. Tel-Aviv: Maariv. (Hebrew).

Ralston, David 1967. *The Army of the Republic—The Place of the Military in the Political Evolution of France, 1871-1914*. Cambridge, MA: M.I.T. Press.

Ravitzki, Aviezer 1993. *Messianism, Zionism and Jewish Religious Radicalism*. Tel-Aviv: Am-Oved. (Hebrew).

Rosenblum, Yair 1988. "The Military Troupes: Myths and Reality." *Musika*, 1988. (Hebrew).

Rotberg, Robert and Theodore Rabb (eds.) 1989. *The Origins and Prevention of Major War*. Cambridge: Cambridge University Press.

Rubinstein, Danny 1982. *On the Lord's Side: Gush Emunim.* Tel-Aviv: Hakibuutz Hameuchad. (Hebrew).

Schiff, Rebeecca 1992. "Civil-Military Relations Reconsidered: Israel as an 'Uncivil' State." *Security Studies,* 1: 636-58.

Schiff, Ze'ev and Ehud Ya'ari 1984. *Israel's Lebanon War.* New York: Simon and Schuster.

Shaw, Martin 1989. "War and the Nation-State in Social Theory." In *Social Theory of Modern Societies: A. Giddens and his Critics,* edited by David Held and J. B. Thompson, 129-46. Cambridge: Cambridge University Press.

Smethurst, Richard 1974. *A Social Basis for Prewar Japanese Militarism.* Berkeley: University of California Press.

Sobul, Albert 1974. *The French Revolution, 1787-1799.* New York: Vintage Books.

Springhall, J.O. 1971. "The Boy Scouts, Class and Militarism in Relation to British Youth Movements." *International Review of Social History,* 16: 125-58.

Stepan, Alfred 1976. "The New Professionalism of Internal Warfare and Military Role Expansion." In *Armies and Politics in Latin America,* edited by Abraham F. Lowenthal, 244-60. New York: Holmes and Meier Publishers.

Straschnov, Amnon 1994. *Justice Under Fire.* Tel-Aviv: Yediot Achronot Pub. (Hebrew).

Stargardt, Nicholas 1994. *The German Idea of Militarism.* Cambridge: Cambridge University Press.

Summers, A. 1976. "Militarism in Britain Before the Great War." *History Workshop Journal,* 2: 104-23.

Sunoo, Hakwon 1975. *Japanese Militarism—Past and Present.* Chicago: Nelson-Hall.

Tilly, Charles 1995. *Coercion, Capital, and European States.* Cambridge, MA: Basil Blackwell.

Tilly, Charles 1975. "Reflections on the History of European State-Making." In *The Formation of National States in Western Europe,* edited by Charles Tilly, 3-83. Princeton, NJ: Princeton University Press.

Tuchman, Barbara 1984. *The March of Folly, From Troy to Vietnam.* New York: Alfred A. Knopf.

Tuchman, Barbara 1966. *The Proud Power.* New York: Macmillan.

Vagts, Alfred 1959. *A History of Militarism.* New York: Meridian Books.

Valenzuela, Arturo 1985. "A Note on the Military and Social Science Theory." *Third World Quarterly,* 7: 132-43.

Van Der Goltz, Colmar 1913. *The Nation in Arms.* London: Hugh Rees.

Van Creveld, Martin 1991. *The Transformation of War.* New York: Free Press.

Weber, Eugen 1968. *The National Revival in France, 1905-1914.* Berkeley: University of California Press.

Wendt, Alexander 1992. "Anarchy Is What States Make of it: The Social Construction of Power Politics." *International Organization,* 46: 391-425.

Willems, Emillio 1986. *A Way of Life and Death, Three Centuries of Prussian-German Militarism.* Nashville, TN: Vanderbilt University Press.

Yaniv, Avner (ed.) 1993. *National Security and Democracy in Israel.* Boulder, CO: Lynne Rienner.

Yariv, Ahron (ed.) 1985. *War By Choice.* Tel-Aviv: Hakibbutz Hameuchad. (Hebrew).

5

Dimensions of Tension between Religion and Military Service in Contemporary Israel

Stuart A. Cohen

Introduction

Considerable attention, both academic and popular, has been lavished on what Horowitz and Lissak designated the "religious-secular cleavage" in Israeli public life and—more specifically—on the relations between national religious Jewry and other segments of Israeli society (Horowitz and Lissak 1989). Nevertheless, the available literature for long virtually ignored the possibility that the tension which religion engenders in Israel might spill over into the military domain. In part, that neglect was justified by the overarching consensus on national security affairs which generally pervaded all sectors of Zionist Israeli society, religious and secular alike, and which seemed to ensure the IDF's immunization to the sort of ideological rifts which otherwise characterize so much of Israeli public life. More specifically, it reflected a perception that there was nothing particularly distinctive about the military service patterns of conscripts drawn from what is termed the "national religious" segment of Israel's Jewish population.[1]

True, a substantial proportion of national religious females of conscript age were known to have always claimed exemption from service on religious grounds in accordance with the provisions laid down in the 1953 National Service Law, and thus to follow practices even more widespread in the *haredi* communities (Cohen, Y. 1993). But

173

such was not so in the case of males. Unlike their *haredi* counterparts, very few national religious boys of conscript age applied for
deferment of service on the grounds that "the study of the law is
their profession." Moreover, once enlisted, the vast majority displayed no recognizably distinctive military "profile." Altogether,
national religious troops, reservists as well as conscripts, seemed to
constitute a fully integrated component of the IDF's overall complement. Under such circumstances, there seemed little need to investigate their attitudes towards military service, or to suspect that the
distinctiveness of those attitudes might generate tensions of a wider
nature. Indeed, as recently as the mid-1980s, national-religious troops
were simply ignored as a unique category of analysis by the most
authoritative surveys of the IDF's human complement (Gal 1986).

Such is no longer the case. Once considered irrelevant, discrep-
ancies between "religious" and "secular" troops in the IDF have of
late aroused increasing interest. Especially is this at the conscript
level, where the disparities in their service patterns has become particularly marked. Broadly speaking (obviously exceptions abound),
recruits from secular backgrounds are evincing what a former IDF
Chief of Staff once termed "a preference for the individual over the
collective" (Lipkin-Shahak 1996). Hence, even though their willingness to enlist remains high, the reasons for that propensity have
begun to change. Basically, the traditional incentives of community
service and patriotism are being displaced by a newer drive for self-
fulfillment. Graduates of national religious high-schools, however,
display contrary trends. Specifically ideological motives for service
are pronounced amongst this segment, and indeed are inculcated by
a network of both formal and informal pre-conscript educational
systems (Ezrachi and Gal 1995, Cohen, S. 1997).[2]

The results are easily observed. Once comparatively rare, the sight
of a knitted skullcap (*kippah serugah*; the most obtrusive mark of
male national religious affiliation) on the head of an Israeli soldier
on active front-line duty is now commonplace. Particularly is this
so in those formations to which enlistment is elective and selection
especially rigorous. The rate of national religious recruits to elite
"reconnaissance units" (*sayarot*), for instance, now far exceeds their
proportion in the conscript population (perhaps by a ratio of three to
one). Where available, statistics with respect to the sociological
breakdown of NCO's and junior officers tell a similar tale. At a rough

estimate (Becker 1996), some 30 percent of all IDF fighting servicemen in these ranks now wear a *kippah serugah*; as many as 60
percent of those passing out in the first class of NCO infantry courses
between 1994 and 1995 graduated from the national religious high
school system; the relevant figure in the infantry officers' training
school was 100 percent. Similarly, since 1994 the percentage of national religious graduates of the pilot training program has more
than doubled, from 5 to 11 percent (Harel 1999a).

The possible implications of such data arouse conflicting emotions. In broad terms, the affirmative attitude towards military service displayed by so many national religious troops is welcomed,
since it provides the IDF with a particularly cohesive pool of high-
quality and highly-motivated manpower (Galilee 1998). At the same
time, precisely the same attributes also generate concern, principally on political grounds. Many (perhaps most) national religious
troops, it is argued, possess a deep ideological and religious commitment to the retention of Jewish control over what they refer to as
"the greater land of Israel"—i.e., the West Bank, "liberated" by the
IDF in 1967. Hence, they are likely to be particularly reluctant to
carry out any orders which they might be given to dismantle Jewish
settlements in those regions. Even the suspicion that so large a body
of troops—and junior officers—might thus subordinate their professional military duty to their ideological preferences, it has been
suggested, could confuse the chain of command and thereby spread
dissension throughout the Force as a whole. Should the present trajectory of service patterns continue, the situation might be even more
severe. National religious troops might in time come to dominate
the General Staff, and thus be in a position to impose a religiously-
dictated ideological straight-jacket on the conduct of Israel's entire
security policy (Ben-Eliezer 1998).

So intense has been the degree of interest focused on these particular anxieties, that other possible manifestations of tension between religion and military service have been virtually ignored. This
distortion is regrettable. Not only does it tend to exaggerate the salience of what, after all, still remains an entirely hypothetical scenario; more importantly, it also oversimplifies the wider context of
influences—institutional as well as ideological, and personal as well
as political—which might be generating dilemmas between religion
and military service in Israel. The present article seeks to redress

that imbalance. First, it presents a framework for an analysis of the distinctiveness of national religious service patterns. Thereafter, it outlines some of the areas in which that distinctiveness might generate conflicts of interest.

I

Basic to the argument that follows is the contention that conflicts between religion and military service in the IDF, even when most ideological in form and content, conform to a basic structural paradigm. Principally, this is because of the particular context in which they take place. Unlike other manifestations of religious-secular tension, those which affect relations between national religious troops and the IDF unfold within the setting of a military institution, and hence in an environment considerably more restrictive than is likely to be encountered elsewhere in modern society.

Lewis Coser (1974) long ago provided a useful tool of analysis for this sort of framework when he coined the term "greedy institutions," which subsequent observers have applied to military frameworks. The IDF certainly seems to conform to the typology. Even though its overall ambiance is notoriously informal, and characterized by the absence of a rigid insistence on parade-ground discipline, the IDF nevertheless exerts considerable pressure on personnel of all ranks to adhere to its own version of the military code of conduct, whatever their precise location and form of service (Kasher 1996: 72-91). By the standards of all other institutions in Israeli society, the IDF's demands on the individual serviceman and woman are thus, in Coser's language, certainly "omnivorous." More precisely, Israel's armed forces "seek hegemonic loyalty and attempt to reduce the claims of competing roles and status positions on those they wish to encompass in their boundaries" (Coser 1974: 4).

Confronted with such pressures, all troops (especially if they are raw conscripts) doubtless suffer some difficulties of adjustment. But the experience of exposure to the IDF as a "greedy institution" is likely to be particularly discomforting for troops from an orthodox Jewish background. Basically this is because servicemen and women in this category are still committed to the especially "greedy" framework imposed by the religious requirements of orthodox Judaism. In their case, therefore, enlistment cannot denote acquiescence in the normative precedence of a military form of affiliation over all others. On the contrary, from their point of view, it is the IDF which must accommodate itself to religious requirements. Hence, orthodox troops not only demand the creation of conditions which will enable them to retain their earlier standards of ritual practice; if in need of practical and/or spiritual guidance, they also insist on their continued right of appeal to their non-military rabbinic guides, whom they regard as the only legitimate interpreters of divine law.

As thus portrayed, the potential for conflict between two competing "greedy" frameworks, one military and the other religious, has always been inherent in the IDF. Indeed, some degree of tension between them is known to have been present almost from the moment that the Force came into existence.[3] What has changed in recent years, however, has been both the nature of the issues involved and the intensity of debate to which they have given rise. Whereas tussles between religious and military authorities over issues of concern to the IDF at one time centered on relatively abstruse issues of a ritual nature, of concern to only limited coteries of ardent protagonists, they now concentrate on problems of vital interest to secular and religious Jews alike, and involve far larger numbers of persons, both in and out of uniform.

Shifts of that nature cannot be attributed to any single cause. In part, they reflect the increasing ferocity of religious-secular exchanges across a broad spectrum of issues, ranging from the authority of Israel's Supreme Court to the civic status of new immigrants whose claims to Jewish identity do not always meet the strict requirements of traditional Jewish law. In part, the new tone of debate has also been affected by the increasing political leverage which the religious parties in general have attained as a result of the demise of the Labor Party's parliamentary hegemony and the emergence of a more fractured party system. (Cohen, A. and Susser 1996). At root, however, the principal cause for the shift appears to be essentially cultural in nature. In Israel, religious and secular communities (together with the various sub-divisions within those two broad conglomerates) do not only adhere to different values, increasingly they are also adopting divergent life-styles and attributing disparate meanings to concepts and symbols about which there once existed a broad measure of inter- and intra-communal agreement. As a result, the opportunities for religious-secular communication have been considerably narrowed, as indeed have the channels which once made such communication possible (Liebman 1998; Yuchtman-Yaar 1998: 31-7).

The growing divergences between the meanings which the majority of national-religious and secular conscripts attach to the notion of military service in Israel constitute a particularly striking example of that process. Notwithstanding occasional deviations from the norm in both camps (neither of which are, of course, completely homogenous), the overall thrust of developments seems clear. When placed within the context of other currents of public opinion, the distinctiveness of the concepts of "national security" embraced by the national religious community is becoming increasingly marked. So too is the singularity of their attitude towards service in the IDF. Although necessarily linked, these two facets warrant separate treatment.

The Divergence on Security Opinion

As all analysts of Israeli military affairs repeatedly point out, the country's strategic position has undergone a massive transformation during the past two decades (Cohen, E., Eisenstadt and Bacevich 1998). This process has been bifurcate, and proceeded on two entirely contradictory trajectories. One has been auspicious: Quite apart from concluding peace treaties with both Egypt (in 1979) and Jordan (in 1995), successive Israeli governments have also signed several interim accords with the Palestine Authority (1993-1997) and conducted intermittent negotiations with Syria. The second trajectory, however, has been far more menacing. In recent years, acts of Palestinian terrorism and other forms of "low intensity" subversive activities (especially when fueled by a militant brand of Islamic fundamentalism) have become increasingly lethal and burdensome. At the same time, the danger that comparatively remote foes once considered "beyond the horizon" (Iraq, Iran, Algeria and Sudan) might launch air and/or missile strikes—nuclear as well as conventional—against the Israeli heartland has become increasingly real. In this respect, the Iraqi Scud attacks of 1991 are thought to represent only a foretaste of things to come (Feldman 1998).

Whether or not such transformations in Israel's strategic environment might eventually mandate a revision in her existing operational doctrines remains to be seen (Levite 1989). More immediately apparent are their contributions to the erosion of the societal consensus on the definitions of Israel's ultimate security requirements. Admittedly, the latter is not an entirely new development (Barzilai 1996). Both the Lebanon War (1982-1985) and the *intifada* (1987-1993) generated controversy over the difference between wars of "choice" and "no-choice," as well as an unprecedented incidence of conscientious objection to military service on the part of conscripts and reservists (Linn 1996). The impact of more recent developments, however, has been still more profound. The curious admixture of expectations for peace and renewed fears of violence now nurtures— above all else—a sense that previously axiomatic definitions of security might now be loosing their salience. For one thing, the course of events seems to have undermined confidence in the viability of the traditional distinction between threats to Israel's "basic security" (such as were once presented by the prospect of a large-scale military invasion on the part of neighboring Arab armies, which now seems to have receded) and "current security" concerns (defined as sporadic guerrilla attacks, which have only recently come to be recognized as a serious inherent danger). What is more, technological and diplomatic developments have also demanded other re-assessments. Even such central concepts as "strategic depth," "battlefield decisions," and "territorial advantage"—all of which at one time were generally considered integral, and indisputable, ingredients for Israel's survival—are no longer universally regarded as imperatives (Arian 1995: 91-127).

Far from distancing itself from such debates, the national religious community is now one of their most articulate participants. Especially has this been so since September 1993, when the Rabin government concluded the Oslo agreement with the PLO. Altogether, the prospect that, as a result of Oslo, Israel would have to relinquish control over much (perhaps all) of Judea and Samaria aroused visceral emotions, surpassing all other issues as the single most definitive fault-line between Right and Left in public debate. But within the national religious community, feelings ran particularly high. Only a small minority of this segment of the population greeted the Oslo agreement with enthusiasm. For the most part, reactions were adamantly hostile—in some cases, violently so. This attitude was rooted in the belief that the relinquishment of Jewish control over any portion of the Land of Israel, quite apart from being an act of strategic folly, also constituted a sin. Ever since 1967, mainstream religious Zionist thought has been insistent in its declarations that Israel's repossession of its God-given patrimony—including the newly liberated "territories"—constitutes an obvious act of Divine will. Given that background, no leap of theological imagination was required to object to the Oslo accords in the word employed by a former Chief Rabbi of Israel's Sephardi community: "None of us owns the Land of Israel. Hence, none of us has any right to relinquish any portion of the entire country" (Eliyahu 1993).

The impression of wall-to-wall doctrinal unity which such *ex cathedra* pronouncements seek to convey cannot be totally sustained. A more rigorous survey of the available materials indicates, rather, that rabbinic opinion on security matters can—and does—evince almost as many divisions as are to be found in other sectors of Israeli society (Naor 1993; Cohen, S. 1998). What is more, the need to re-assess previously axiomatic security concepts in the light of the possibility of a peace settlement has become just as keenly felt. Hence, it would be entirely wrong to portray national religious opinion on security matters as entirely monolithic. Nevertheless, what does remain true is that divisions of opinion on security matters within this particular segment of the population (still) seem to be far less sharply defined than is elsewhere the case. By and large, national religious public opinion remains distinctly "hawkish" (Barzilai and Inbar 1996: 66-7). Just as significantly, the opinions on security affairs espoused by the majority of the community's spiritual guides and teachers remain (with only occasional exceptions) broadly consistent with precepts which they have been taught and amplified with increasing elaboration ever since 1967.

In terms of the framework employed in the present essay, this situation has produced at least one salient result. In its guise as the supposed embodiment of a security "consensus," the IDF today exerts far less sway over secular popular emotions than was once the case. Specifically: the extent to which it might reasonably expect to be able to satisfy its inherent institutional "greed" within non-religious society—and without recourse to supplementary incentives—has been severely curtailed. The incantation of a notion of "security" no longer exerts a magical effect, certainly not where military enlistment is concerned. Resort must also be had to other inducements to service, some of which are explicitly material in nature (Stern 1998). For many national religious troops, however, such is not the case. The demands imposed by their concept of security, because both ageless and divinely inspired, continue to be as "greedy" as ever. Indeed, so much is this so that they might in some cases now outpace those being made by the military authorities themselves.

Divergences in Attitudes towards the IDF

The growing distinctiveness of national religious Jewry's approach to the priority of narrowly-defined security concerns intersects with (and is reinforced by) that community's image of the IDF as an institution. Here, too, the contrast with currents of opinion in the secular community is marked.

As the IDF's own commanders themselves acknowledge, the corporate communal status of the military in Israel has in recent years undergone a massive decline. Although the IDF remains the most respected of the country's public institutions, its position at the top of that particular league-table no longer seems as assured as was once the case. Precisely why that might be so remains a matter of debate. In a broad sense, the decline in the aura of public deference towards the IDF constitutes an Israeli expression of a shift in cultural priorities (sometimes categorized by the umbrella-term "postmodernism") which has similarly devalued the status of armed forces throughout the western world. But to this must be added more specifically local causes, many of which reflect public dissatisfaction with the IDF's own operational deficiencies. These first became glaringly apparent in 1973, and seemed to be amplified during both the Lebanon campaign of 1982-1985 and the *intifada* (1987-1993). Subsequently, the Force's reputation was further tarnished by reports of training accidents, abuses of human rights in the occupied territories and of financial corruption at some senior levels of command (Van Creveld 1998: 353-62). As a result, the Force no longer benefits from the virtually talismanic veneration which it enjoyed in the wake of the triumphs of 1967. One indication of the new mood is provided by the greater tolerance for persons whose applications for release from service on psychological grounds was once almost universally condemned (Shedmi 1998); another by what a forum of battalion commanders have termed the "epidemic proportions" of absenteeism from reserve duty (Alon 1996) . Most pronounced of all, however, has been the increase in critical public scrutiny of the IDF—as much on the part of parents of conscripts and the courts as on the part of the press (Ben-Dor 1998).

The national religious community is far from being immune to all such developments. On the contrary, it is precisely within this segment of the population that some of them are most pronounced. Particularly is this so where public censure of the IDF is concerned. Individual leaders of the settlement communities on the West Bank (most of whom are members of the national religious community) have on occasion expressed by far the most vociferous criticisms of local IDF commanders, whom they have often accused of pusillanimity in their dealings with the perpetrators of Palestinian attacks on Jewish lives and property (Schiff 1996). In the wake of the Oslo accords, a minority also advocated a policy of conscientious objection to military service.[4] According to at least one of the community's own spiritual leaders, the majority might be susceptible to other pressures. The continuation of national religious Jewry's high rate of motivation to military service, he has warned, is by no means assured. Members of this community, too, might also be affected by the general anti-ideological tide of "liberal individualism," and therefore adopt attitudes not dissimilar from those apparent in secular segments of society (Amital 1997: 118-9).

That said, it nevertheless remains true that the restraints on active criticism of the IDF remain far stronger within the national religious community than is the case elsewhere. In the terms of the framework employed in this essay, the difference can be formulated in the following terms. Whereas an increasing proportion of secular society now questions the right of the IDF to impose "greedy" demands on the population, most of national religious Jewry remains supportive of such behavior. Once again, the reason for this divergence is to be found in the singular framework of beliefs to which the national religious community adheres. All versions of the religious Zionist credo invest military service with transcendental meaning. Those associated with the teachings propounded by Rabbi Zvi Yehudah Kook (1891-1982), the most influential single personality in the national religious world of letters after 1967, also attribute spiritual properties to the IDF itself. As the agency chosen by Divine Providence to fulfill His promise of Jewish re-possession of the Holy Land, Kook taught, the Force cannot constitute merely a mundane instrument of state power. Rather, it deserves to be considered an agency of Redemption, and hence as a integral part of the process whereby God's rulership over the world is made manifest. In many respects, therefore, the IDF is itself "holy." Certainly, service in its ranks must be considered a religious obligation (*mitzvah*), whose fulfillment—because it carries eschatological overtones of cosmic relevance—constitutes a public duty (Kook 1969).

II

Some observers have posited a possible dichotomy between national religious Jewry's interpretation of national security and its affirmative attitude towards service in the IDF. The latter, they claim, is ultimately conditional on the former. To put matters another way, the obedience of national religious troops to military commands is ultimately dependent on their perception that IDF behavior conforms to the religious Zionist understanding of security. Were that not to be the case, then the troops concerned might refuse to obey military commands or, in extreme circumstances, even rebel (Kasher 1997).

The likelihood that they might do so, it has further been argued, has been made even more realistic by the sense of cohesion fostered among national religious youth by the particularly intense web of institutional networks to which many of them are affiliated. Quite apart from an influential and popular youth movement (*B'nei Akivah*), the range also extends to a country-wide system of gender-segregated and residential national religious high schools (*yeshivot tichoni'ot* for boys, *ulpanot* for girls); over a dozen pre-conscription religious academies (*mechinot kedam tzevai'ot*), whose prototype was established adjacent to the West Bank settlement of Eli in 1984 with the express purpose of providing young men with whatever spiritual and physical "fortification" their forthcoming enlistment in the IDF might require; and some thirty *hesder* (lit: "arrangement") academies, most of whose pupils are enlisted en bloc in their own company-sized formations, so that they might intersperse their periods of military service with study in seminaries of advanced learning (*yeshivot*). The *hesder* framework, it is feared, might especially foster a system of dual control, which compels commanders to share authority with the rabbis to whom the troops—even whilst on active service—frequently refer for guidance. Could the rabbis concerned resist the temptation to exercise (or threaten to exercise) the influence thus placed at their disposal, were they to consider a military order to constitute a transgression of religious imperatives? Once in receipt of such a rabbinic edict, could national religious troops nevertheless be trusted to remain loyal to the conventional military chain of command? (Orr 1995)

The alarmist scenarios generated by such questions certainly cannot be dismissed out of hand. After all, a series of murderous incidents—of which undoubtedly the most prominent was the assassination of Prime Minister Rabin in November 1995 by a national religious reservist who had himself served in a *hesder* unit—provide ample evidence of the depths of feelings which can be aroused amongst religious Zionists by what they consider to be treacherous withdrawals from Israel's God-given inheritance. What is more, there exists ample evidence to warrant suggestions that the assassination might have constituted no more than the tip of a much larger iceberg of sentiment. Just the previous summer, a group of fifteen national religions rabbis (four of whom were associated with *hesder yeshivot*), had publicly called upon troops to disobey any order which they might be given to dismantle Jewish settlements in the Land of Israel. Shortly thereafter, the national council of *B'nei Akivah* adopted a resolution which implicitly supported the right to conscientious objection on the grounds that withdrawal from the Land of Israel contravened Divine law (Cohen, S. 1998).

It is significant that, despite all such indications, the fear of massive disaffection on the part of national religious conscripts on political grounds remains just that: a fear—and hence no more than a cause for hypothetical concern. Thus far, nothing in the behavior of the overwhelming majority of national religious troops substantiates the suspicion that they might refuse orders to implement a withdrawal from areas of the West Bank transferred to autonomous Palestinian control. Neither are there any signs of further rabbinical pressure on the troops to resort to such steps. If anything, quite the opposite is the case. Even at the height of the public furor over the implementation of the Oslo accords (in the summer of 1995), rabbinic opinion with respect to the manifesto referred to above was extremely divided. Only a handful of religious Zionism's spiritual guides were prepared to add their signatures to the document; many roundly condemned its content in public, principally on the ground that it threatened to sow dissension within IDF ranks. Similarly, calls for active conscientious objection to service on political grounds generated very little support within this community, and were in fact disavowed even in settler circles.

Since Prime Minister Rabin's assassination, national religious opinion on these matters has been still more cautious. Understandably apprehensive of seeing yet another genie emerge from the bottle, even rabbis known to be most ardent in their support of the retention of "Greater Israel" have tended to stress the overriding necessity of avoiding any action which might impair the unity of Israel's armed forces (Huberman 1997). Since 1997, especially, successive conventions of the council of *B'nei Akivah* have adopted a similarly muted line. Their resolutions have explicitly rejected the notion of a conflict of interests between the movement's commitment to Israel's security and its affirmation of the obligation to serve in the IDF. Instead, they now imply that the latter in fact constitutes a primary concern, to which all other considerations had necessarily to be subordinate. As far as can be judged, the overwhelming majority of national religious servicemen and women appear to concur with that order of priorities, which (one must assume) are in any case reinforced by their socialization into the military framework and hierarchy during their periods of conscript duty. In this context, it is significant that even the IDF's withdrawal from Nablus and Hebron in 1997 created no particular difficulties. Notwithstanding the fact that the names of these cities (collectively known as "the cities of our forefathers") resonate with historic associations to which religious Zionism has always been particular sensitive, there exists no evidence whatsoever to indicate that groups of national religious servicemen refused to participate in the withdrawal, nor that they ever threatened to do so (Cohen, S. 1998).

III

Whereas the specifically *political* dimensions of the potential confrontation between religion and military service in contemporary Israel thus appear to have been exaggerated (and have certainly not been realized), at least two other facets of possible tensions in this domain, as experienced by individual national religious servicemen, seem to have been overlooked. One is the *intellectual* dilemma rooted in the conflicting "greedy" demands which religious injunctions and military service impose on their time. The other is the *organizational* pressures which a growing body of increasingly articulate national religious personnel exert on the military framework as a whole. It is to a brief survey of these two expressions of tension—both of which are already manifest in some degree—that attention must now be turned.

"Study or Service?"

At a very superficial level, the national religious attitude towards military service in the IDF seems to be altogether straight-forward. Quite simply, and has already been noted, enlistment constitutes a *mitzvah* ("religious obligation") as well as a national duty, the performance of which is hence particularly binding upon all orthodox Jewish citizens of the State. Beneath the surface of that somewhat bland doctrinal declaration, however, there lurk the sort of imponderables over which talmudists have for generations loved to pore. What is the precise halakhic status of this particular *mitzvah*? Is its performance binding perennially, or only at times of perceived national crisis? To what extent, even in the latter case, does it possess priority over other Divine commandments (*mitzvot*), with whose observance military service might conflict?

For over a generation now, religious Zionism's spiritual guides have struggled to provide answers to such questions. Considering the paucity of attention to military affairs of any sort in any of the canonical legal compendia upon which halakhic discourse conventionally rests, that exercise has met with considerable success. The period since 1967, especially, has witnessed the publication of an ever-growing stream of detailed rabbinic compositions in the field generically known as "the laws of war" (*dinei milkhamah*). Encompassing an area of public life uncharted by Jewish traditional teachings for centuries, such works (many of them compiled by rabbis themselves in possession of military experience), are not concerned solely with theological polemic or with theoretical justifications for warmaking. Overwhelmingly, they are designed specifically to enable orthodox servicemen to harmonize the demands of ritual practice with those of military service. Based by an array of stunningly erudite sources, they indeed manage to specify the manner in which that task might be accomplished in a vast spectrum of areas. [5]

Notwithstanding the overall success of this enterprise, there nevertheless remains one significant area of halakhic interest in which the potential for tension between religion and military duty has not entirely been resolved. Specifically, religious Zionist thought has yet to harmonize the (new) call for young orthodox Jews to devote a substantial amount of their time service in the armed forces with the much more ancient teaching that the study of the *Torah* (Divine Law) must take precedence over all other activities. Devotion to the latter principle has always been basic to orthodox Jewish culture. Study for its own sake (*lishmah*), both of the Bible and of the multiple layers of commentaries and supra-commentaries generated by the extended and intensive scrutiny of that text as the word of God and the embodiment of eternal truth, has traditionally been endowed with intrinsic instrumental value. According to some ancient teachings, it even takes precedence over prayer as a true form of Divine worship. By pondering those sources and exploring their limitless nuances, generations of sages and their disciples have stimulated and made concrete an intimate relationship with their Maker. In that way, each is said to have re-enacted the theophany at Mount Sinai, thereby fulfilling Israel's Divinely-inspired purpose. Thus perceived, *Torah*-study (*talmud torah*) is not simply an intellectual experience, designed to collate and increase knowledge. Essentially, it constitutes a sacrament: the means whereby the Jew expresses his piety and achieves communion with his God. That, surely, is the meaning of the teaching (first committed to writing in folio 127 of the talmudic tractate *Shabbat*) that *talmud torah* possesses "no fixed measure"; it is equal to all the other activities, "the fruits of which a man enjoys in this world while the capital remains for him in the world to come."

In their most extreme form, the monopolistic claims of *torah*-study (especially vis-a-vis military service) have found most explicit expression in *haredi* circles. It is now calculated that a vast proportion (probably as much as 80 percent) of *haredi* young males of conscription age presently claim—and receive—extensive deferments from enlistment in the IDF on the grounds that "the [study of the] *Torah* is their profession."[6] Indeed, this particular segment of Israeli society now posits as an article of faith the argument that the energies which its members invest in the scholarly vocation contribute as much (if not more) to Israel's ultimate survival than do the exertions of IDF troops (Friedman 1991). What also needs to be recognized, however, is the degree to which the national religious community is itself susceptible to the force of such arguments. Much though religious Zionism takes issue with many of the basic tenets of *haredi* thought, and especially with the latter's view of the State of Israel as nothing more than another version of Jewry's exile, it too must ultimately acknowledge the axiomatic priority which traditional Jewish teachings attach to *Torah*-study as a full time avocation. In the last analysis, religious Zionism's attempts to rationalize a relaxation of that norm (e.g., Lichtenstein 1981) carry the whiff of an apologia, even when they are most eloquently presented and appeal to the imperatives of state security.

The ramifications of the dilemma thus posed on the service patterns of individual national religious conscripts are becoming increasingly apparent. At their most extreme, they have resulted in the tendency of a small (but increasing) number of national religious men of service age to imitate *haredi* practice, by requesting draft deferments for periods of as long as seven or eight years in order that they might pursue their theological studies without interruption.[7] More ambivalent—but also far more wide-spread—are those frameworks which seek to enable young servicemen to discharge both their scholarly and soldierly obligations within institutional settings especially tailored to that purpose. Unlike his *haredi* counterpart, the national religious youth of conscript age does not simply face a stark choice between study or a three-year stint of compulsory military service. He can now choose between a selection of other options: a *hesder* yeshivah, where he will spend five years, of which just two will be spent in army service; a pre-conscription academy (*mekhinah*), which provides a year-long program of religious studies prior to regular enlistment; the *gachelet* program, which combines four years of service in a *NAHAL* unit religious studies; or the *Shiluv* system, in which two years of study envelope three years of regular military service.

Two facets of that roster are particularly worthy of attention. One is the multiplicity of such frameworks, which seek to cater to a variety of tastes and to satisfy a wide range of aspirations. This facet becomes especially apparent when note is taken of the differences existing not only between the various frameworks but also within them; no two *hesder yeshivot,* for example, are exactly alike in either tone or standard, nor in terms of size and geographical location (Don-Yehiya 1994). The second is their numerical growth. The number of *hesder yeshivot,* for instance, has multiplied from just one in 1964 and twelve in 1980 to over thirty in 1999; and of *mekhinot*—whose growing popularity has in fact come to constitute a threat to several *hesder* academies—from one in 1984 to fifteen in 1999 (Dagan 1999). In terms of registration, the figures are still more impressive. Altogether, the number of national religious conscripts enrolled in one or another of the frameworks listed above now encompasses almost half of each annual cohort of troops drawn from this segment of Israel's Jewish population.

Developments of that nature cannot be attributed solely to the influence of fashion. In a far more fundamental sense, the growing popularity of all such programs also seems to answer to a deeply-felt ideological urge amongst national religious youth. Jeremiahs among this community frequently bemoan the fact that over a quarter of the products of the national religious educational system are estimated to adopt secular life-styles within some five years of their graduation (Sa'ar 1999). A less pessimistic audit, however, must focus on the remaining 70 percent who retain their orthodox identity and thus do not succumb to the intense pressures of secularization to which they are subjected, not least during military service. Indeed, a large proportion of the majority even seem to adopt an even more rigidly orthodox life-style than their parents (Lustick 1988: 166, 223) and thus to be particularly intent on restricting the military's "greedy" intrusions into their pursuit of *torah* scholarship to what they consider to be reasonable bounds. In many cases, moreover, they might also have a more specific interest in doing so. There exists a strong impression that the current religious Zionist emphasis on the importance of *torah*-study constitutes a deliberate response to the corporate challenge presented to the scholastic pride of the national religious community by the exponential growth in the number of *haredi* academies and students. In the words of one commentator:

> We have a duty not to relegate Torah study to those who are not cognizant of God's deliverance [i.e. *haredim*].... In order to promote great *Torah* authorities [of our own], those absorbed in *Torah* must be freed from any other yoke.... If [they] are required to combine their study with military service, even for a very limited time, and thereafter be called to reserve duty... the possibility of pro-

ducing the top quality *Torah* leadership that our nation needs, will be seriously impaired. (Melamed 1991: 72; Steinberger 1997)

Set against that background, the increasing popularity of the idiosyncratic frameworks now available to national religious servicemen lends itself to a somewhat different interpretation than is usually provided by conventional analyses. Far from reflecting an aggressive and outward-looking political ambition, the increase in the rise of affiliation on the part of national religious servicemen to their own institutional frameworks in fact seems to respond to the more inner-directed impulse to comply with the imperative of *torah*-study (Shaviv 1987; Neumann 1999). It is the difficulties inherent in combining that duty with the call to national military duty which explains the multiplicity of programs established in order to meet both requirements. By the same token, the very variety of such programs also serves as a reminder of the depth of the dilemmas thus posed. Seeking an equilibrium between two conflicting "greedy" demands, national religious Jewry in Israel continues to experiment with various solutions—no one of which has yet been found to be entirely satisfactory.

"Insulation or Integration?"

Always central to Israel's character as a "nation in arms" has been an insistence on the sociological heterogeneity of the units of which the IDF is composed. This facet of the military structure was stressed with particular force by David Ben-Gurion (1886-1973), Israel's first Prime Minister and Minister of Defense (he held both positions between 1948 and 1953 and from 1955 until 1963), and the man chiefly responsible for creating the IDF and defining its character. From the very first, Ben-Gurion intended Israel's armed forces to become an instrument of new Jewish "nation building" and a symbolic focus of national sentiment. Above all, he envisioned the IDF as a bonding institution within which Israel's otherwise fractured society could be homogenized and welded into a single whole. This vision did not impair the formation of segregated units for the small Druze minority (Peled 1998); neither did it override Ben-Gurion's support of the arrangement whereby volunteers to the *NAHAL* (Fighter Pioneer Youth) corps enlisted in homogeneous "nuclei" (*garinim*) (Doar 1992). But it did certainly invalidate the establishment of separate military units for specifically religious troops. On this point, Ben-Gurion was adamant: "I fear that the creation of religious units will result in the creation of anti-religious units.... It is preferable, and possible, to educate officers and commanders to understand and respect the religious soldier." Hence: "Our army will be a united army, without 'trends'" (Ostfeld 1994: 441).

Although initially somewhat unhappy with this formula (which leaders of the National Religious party indeed attempted to modify on several occasions in subsequent years), the principle of religious-secular "integration" within the IDF in time became a basic component of the overall religious Zionist ethic. The only particularistic demands which the national religious community made of the IDF was for the establishment of a military chaplaincy (*rabbanut tzeva'it*), empowered both to ensure the supply of the ritual articles which religious troops require on a daily basis, and to ensure that IDF practice adhere to orthodox halakhic standards, especially with respect to sabbath observance and the dietary laws (Michaelson 1982; Meir 1998). In all other respects, however, national religious troops neither sought—nor were granted—any particular concessions. On the contrary, they seem overwhelming to have concurred in the projection of equal military service as the most conclusive proof of religious Zionism's commitment to full participation in the entire national enterprise.

Those conditions no longer apply. Although the IDF undoubtedly remains the most comprehensive of all Israeli institutions—and certainly the most obtrusive meeting-ground between citizens who otherwise live their lives in vastly different cultural milieus—its claims to constitute an integrative bridge between religious and secular troops no longer carries quite the same conviction. Instead, the two communities from which the two categories of troops are drawn seem increasingly to have been drifting apart and thus becoming more segregated. Raised in what are now becoming very disparate environments, and espousing distinctive sets of values, they find it increasingly difficult to communicate when thrown into close proximity by the experience of enlistment. One testimony to that circumstance, especially notable because of the audience to which it was addressed, is provided by a cautionary article which two fresh conscripts published in the bulletin of the national religious youth movement, *B'nei Akivah*. "The IDF," they warned younger members to be aware, "is not at all a religious institution." Only in part is this because conditions in the unit mess do not always meet orthodox dietary standards, especially in isolated front-line postings which are too small to billet a military chaplain. Far more significant are the challenges posed by other tests, most of which are all the more traumatic for being so unexpected:

> Quite apart from experiencing the shock to which every conscript is submitted on entering the military framework, the religious soldier in addition is estranged and struck dumb by the comportment of his secular comrades. Even their everyday speech contains phrases and terms which his own mouth, used to prayer, is unable to utter and which his ears, attuned to words of wisdom, refuse to absorb. (Levy and Furstein 1995: 8-9)

A detailed examination of the precise reasons for that situation lie beyond the scope of the present essay, and can therefore only be briefly outlined here. In part, the change can be attributed to forces at work within the secular community, of which undoubtedly the most important is the decline in attachment to (and knowledge of) the vast cargo of traditional Jewish symbols and practices which in a previous generation constituted integral facets of what has been termed Israel's "civil religion" (Liebman and Don-Yehiya 1983). Equally influential, however, have been processes at work within the insular world of national religious Jewry itself. The products of over two generations of a particularly rich, intensive and essentially segregated educational system, the most recent generations of national religious youth have become far more assertive than their forbears. They have also begun to take the initiative in fields which were once considered to be virtually secular monopolies. One example is provided by the prominence of the role played by the national religious community in post-1967 settlement movement and particularly in *Gush Emunim* (The Bloc of the Faithful), established in 1974 with the express purpose of ensuring Jewish control over the territories acquired during the Six Days' War (Rubinstein 1982). Another—of perhaps greater import in the long term—is the degree to which individual members of the national religious community have penetrated into the very highest reaches of the country's economic, judicial, administrative and communications elite without (again, in contrast to many of their forbears) masking their origins (Sheleg 1994).

Thus far, there exists no evidence whatsoever to support the suspicion that such attributes might also be finding parallel expression within the IDF. Particularly doubtful are the fears that the disproportional representation of national religious servicemen at the rank of junior officer might automatically result in their dominance over the General Staff (Ben-Eliezer 1998). The presence of *kippot serugot* amongst officers of senior rank has certainly grown of late. Moreover, a particularly symbolic milestone was reached—and crossed—in September 1998, when, for the very first time, an officer wearing a *kippah serugah* and who was not the IDF Chief Chaplain (General Ya'akov Amidror) was appointed to the General Staff. In purely statistical terms, nevertheless, senior IDF ranks remain overwhelmingly secular preserves. What is more, there are strong grounds for anticipating that such will remain the case. Primarily, this is because what might be termed "greedy" pressures within the national religious community itself create an upper limit to the military profile of its members. One, to which reference has already been made, is their urge to further their religious education as much as possible.[8] Another is their comparatively early age of marriage and their subsequent attachment to family life. For as long as both trends persist, it could be argued, the overwhelming majority of the national religious will continue to leave military service long before their presence becomes felt at the senior command level. Very few will be prepared to undertake the long-term investment of time, energy and total commitment to professional service which advancement to the very highest echelons of the military hierarchy necessarily demands.

The singularity of national religious military behavior in Israel makes itself felt at a much lower level and in an entirely different form. Its most blatant expression is to be found in the growing tendency amongst this segment of the overall IDF complement to bunch together and thus to form almost homogeneous units. To put matters another way, the push towards "integrated" military service, although still acknowledged to be a fundamental article of religious Zionist faith, is increasingly giving way to an "insular" pull.

The root causes for that tendency are not difficult to gauge. Doubtless, most recruits would prefer to serve with persons of their own kind; and that urge is always likely to have been especially strong amongst national religious servicemen, principally because of their particular need for a communal atmosphere conducive to their observance of orthodox Jewish practice. But what until recently constituted little more than a vague aspiration has now virtually attained the status of a communal imperative. National religious troops are becoming increasingly sensitive to the cultural chasm which seems to separate them from the majority of secular troops. Many, and especially those who enroll in *hesder yeshivot* or in the *mekhinot*, are also expressing increasing determination to minimize whatever threats military service might pose to their ability to conform to the rigid demands of Jewish ritual practice. In consequence, they are expressing a growing preference to serve—as groups—in homogeneous formations, rather than as individuals in the IDF as a whole.

The tendency towards an "insular" form of national religious military service is most blatant amongst *hesder* personnel. This is hardly surprising, since the peculiarity of the *hesder* time-table, which permits a far more truncated spell of conscript duty than is performed by most other recruits, virtually mandates that *hesder* personnel be drafted en bloc and serve together. Fearful that the existence of socially homogeneous units might result in the concentration of casualties within a particular segment of society (such as was indeed experienced by the national religious community when an armored formation principally composed of *hesder* troops was mauled at the battle of Sultan Yakub in June 1982), IDF commanders have since the Lebanon War insisted that individual battalions generally contain no more than one *hesder* company. Within that limit, however, the homogeneity of the units concerned remains marked—in infantry brigades as well as in the armored and engineering corps (to which the vast majority of *hesder* personnel are assigned). In each case, the number of such companies, many of which are composed of students of a single academy, has grown in direct proportion to the expansion of the overall *hesder* complement.

Significantly, however, it is not only *hesder* personnel who tend towards an "insulated" form of military service. Similar, albeit far less institutionalized, patterns are also apparent amongst other national religious recruits. A large proportion of national religious female recruits, for instance, serve in the Education Corps—and more specifically in the "Branch for Torah Culture" which (not unexpectedly) is monopolized by graduates of the national religious educational system. Likewise, a large proportion of the graduates of *mekhinot* gravitate towards the IDF's elite *sayarot* (reconnaissance units), where they now constitute a recognizable, and distinct, category of analysis. A parallel tendency towards "bunching" can be observed in several of the other combat formations within which the numbers of national religious conscripts has become particularly marked. Indeed, it has to a large extent been (unwittingly) facilitated by the reforms which the IDF Manpower Branch in 1995 instituted in the overall draft system, with the express purpose of giving potential conscripts a greater say in the determination of their ultimate locus of service. By permitting—indeed, encouraging—recruits to express their own unit preferences (within certain limits), the new system has also made it easier for many of them to co-ordinate their selections, and thus increase the likelihood that they will serve in specific units en bloc.

Observation suggests that such developments certainly impose a number of organizational constraints on the military framework. (Of these, undoubtedly the most widely-felt, especially amongst junior commanders, is the need to display greater sensitivity to the demands of a sizable and invariably articulate body of religious troops for what in an earlier age were generally considered to be extraordinary, and hence unjustifiable, "concessions.") In the long run, however, the ramifications for Israeli society as a whole must be considered much more significant. After all, the IDF has for long prided itself on its character as a "people's army" which (in the words of one observer),

> has helped to break all barriers between men who lived all their lives in vastly different cultural milieus. Boys from religious families could mix freely with antireligious boys from secularist left-wing kibbutzim, learning to give and take, to disagree while respecting the other's right to his own view, to refuse from excesses of behavior and find a deeper unity of purpose. (Rolbant 1970: 154)

Should the tendencies towards segregation and "insulated" service noted here persist, the validity of all such depictions is bound to be impaired. Instead of being a great "nation builder," service in the IDF might become a "nation divider," thereby reinforcing the drift towards a religious-secular divide which in any case threatens Israel's communal unity.

Conclusion

Many analyses of the potential ramifications of current dilemmas between religion and military service in Israel (it has here been argued) are misdirected. The fear that such dilemmas might result in a widespread incidence of military disobedience, although not altogether unwarranted, seems particularly exaggerated. Underplayed, by contrast, are the wider personal and societal dimensions of tensions between religion and military service. These, quite apart from being even more widespread, might—in the long term—prove to be even more insidious. Rooted in impulses deeply embedded in traditional Jewish culture, they are further fueled by the influence of the educational frameworks in which a large proportion of national religious servicemen and women are raised. To ignore the possible impact of those tensions on the unitary framework of the IDF—and indeed on Israeli society at large—is, it seems, to overlook one of the most important features of contemporary Israeli life.

Notes

1. Although the terminology begs several questions (how "national"? or, for that matter, how "religious"?), it nevertheless remains serviceable. By general consent, the "national religious" community encompasses those sections of the population, Ashkenazim and Sephardim, who invest the establishment of the State of Israel with religious significance as a precursor of Jewry's promised Redemption (Liebman and Don-Yehiya 1984: 57-78). Even as thus defined, "national religious" Israelis, who comprise roughly 12 to 15 percent of Israel's total Jewish population, are far from being an altogether homogeneous bloc. Nevertheless, they do constitute a distinct category of sociological and political analysis. Their outlook and life-styles vary from those of the ultra-Orthodox [haredi] communities (together, some 7 to 8 percent of the total Israeli population), whose attachment to traditional standards of religious observance tends to be more rigid and whose attitude towards modern political Zionism less enthusiastic. But, national religious Jews also differ from the predominantly "secular" majority in several important respects— not the least significant of which is the maintenance of separate educational frameworks.
2. Ever since a "crisis" in motivation to military service was declared in the mid-1990s, regular measurements of the relevant indices have become a national pre-occupation. The latest figures show that a nadir was reached in March 1996, when only 68.2 percent of youngsters drafted expressed a readiness for combat duty. The numbers rose to 77.7 percent in August 1996, but then dropped again to 73 percent in November 1997. By August 1999, they had climbed back to 77.7 percent (Harel 1999b).
3. A Cabinet crisis broke out as early as September 1948 (at the very height of the Israel War of Independence), when two religious cooks on military service refused orders to heat up a meal on the sabbath. See Ostfeld (1994: 748).
4. The most obtrusive example was the proclamation issued by the former IDF Chief Chaplain: "It is forbidden to obey a military order which contradicts the commandment of settling the Land of Israel, which is equivalent to all the commandments of the Torah" (Goren 1993: 1). See also the "Manifesto" (He-

brew) issued by "The Union of Rabbis on Behalf of the People of Israel and the Land of Israel," reprinted in *Ha-Tzofeh* (Hebrew: national religious daily, Tel-Aviv) May 15, 1995: 1.

5. For samples of this extensive literature (all in Hebrew) see: Mishpat Ha-Milkhamah ("The Law of Warfare," 1971) by Shemaryahu Arieli; Ha-Hayil ve-ha-Hosen ("Soldiering and Immunity," 1989) by Joshua Hagar-Lau; Shabbat U-Moed Be-Tzahal ("Sabbath and the Festivals in the IDF," 1990); Pe'ilut Mivtza'it Be-Tzahal Al Pi ha-Halakhah ("Operational Activities in the IDF According to the Halakhah," 1991) by Isaac Jakobovitz; Ha-Tzavah Be-Halakhah ("The Army in the Halakhah," 1992) by Isaac Kaufman; Nachum Eliezer Rabinovitch, Melumdie Milkhamah ("Responsa on Matters Concerning the Army and Security," 1993); and, most impressive of all, Shlomo Goren's Meishiv Milkhamah ("Responding to War," 1983-1996), four fat volumes of instruction and polemic by the IDF's first Chief Rabbi.

    Much of the material contained in these works has been edited and reproduced for the use of individual servicemen in single-volume pocket books. Two of the most popular are: Laws Concerning Army and War: A Guide to Students on the Eve of Conscription (Hebrew: ed. Shlomo Min-Hahar et al; Jerusalem: Haskel, 1971), and Zechariah Ben-Mosheh, Laws Concerning the Army (Hebrew: 2nd. edtn. Sha'alvim, 1988). Even so, the need for supplements to the corpus is continuously felt. For the most recent addition, see Michael Rubin, Sefer Ha-Madim be-Keshet ("Responsa on military matters") (Hebron: Machon Ha-Tzevi, 1999).

6. In 1998 Israel's Supreme Court found that during the previous year 28,550 male deferments had been granted, a growth of 16 percent since 1996. This figure represents almost 8 percent of the entire potential male draft cohort, compared to 6.4 percent in 1995, when the total number of Israeli youngsters of service age was somewhat smaller (*Ressler vs. Minister of Defense*, 715/ 98, www.court.gov.il). Even more marked is the reported growth in the proportion of religious females now exercising their legal right to claim exemption, on the grounds that military service might conflict with their Orthodox life-styles. Of all potential female conscripts, 32 percent claimed exemption in 1996; two-thirds of them on religious grounds (Zamir 1997). On the origins of the various "arrangements" permitting this form of non-service, see Friedman 1990.

7. This phenomenon has become particularly marked amongst students registered in the academies (*yeshivot*) directed by pupils of Rabbi Abraham Isaac Kook (1865-1939), and especially the central Kook academy (*Merkaz Harav Kook*) in Jerusalem (Don-Yehiya 1994). The instance is particularly interesting in view of the fact that Kook, the first Ashkenazi Chief Rabbi of Israel, propounded teachings which—especially as interpreted by his son after 1967— served as a basis for a particularly militant brand of Jewish Zionist fundamentalism (Aran 1991).

8. Personal observation highlights two especially interesting indexes of that trend at an early stage of the military career: (1) A high proportion of servicemen in the *hesder* program reject offers to enroll in the IDF junior officers' course, primarily because they prefer to return as soon as possible to their *yeshivot*; and (2) similarly, few of those who do complete the course evince a willingness to extend their contract terms beyond the mandatory year of professional service. While most enter the mainstream of civilian life, a significant number return to their *yeshivot* and—increasingly—*mekhinot*.

References

Alon, Gideon 1996. "Battalion Commanders Present Memorandum to Minister of Defense." *Ha-Aretz*, November 4. (Hebrew).

Amital, Rabbi Yehudah 1997. "Challenges Facing Israel towards the 21st Century." (Paper delivered at a symposium at Tel-Aviv University, February 1997), *Alon Shevut* 10: 3-9. (Hebrew).

Aran, Gideon 1991. "Jewish Zionist Fundamentalism: The Bloc of the Faithful in Israel (Gush Emunim)." In *Fundamentalisms Observed,* edited by Martin Marty and Scott Appleby, 265-344. Chicago: University of Chicago Press.

Arian, Asher 1995. *Security Threatened: Surveying Israeli Opinion on Peace and War.* Cambridge: Cambridge University Press.

Barzilai, Gad 1996. *War, Internal Conflict and Political Order: A Jewish Democracy in the Middle East* Albany: State University of New York Press.

Barzilai, Gad and Efraim Inbar 1996. "The Use of Force: Israeli Public Opinion on Military Options." *Armed Forces and Society* 25: 66-67.

Becker, Avichai 1996. "The March of the Skullcap." *Ma'ariv*, March 8. (Hebrew).

Ben-Dor, Gabriel 1998. "Civil-Military Relations in Israel in the mid 1990s." In *Independence: The First Fifty Years*, edited by Anita Shapira, 471-86. Jerusalem: Shazar Center. (Hebrew).

Ben-Eliezer, Uri 1998. "Is a Military Coup Possible in Israel? Israel and French-Algeria in Comparative Historical-sociological Perspective." *Theory and Society*, 27: 311-49.

Cohen, Asher and Bernard Susser 1996. "From Accommodation to Decision: Transformations in Israel's Religio-Political Life." *Journal of Church and State*, 38: 817-40.

Cohen, Eliot, Michael Eisenstadt and Andrew Bacevich 1998. *Knives, Tanks and Missiles: Israel's Security Revolution.* Washington, DC: Washington Institute for Near East Policy.

Cohen, Stuart A. 1999. "Military Service in Israel: No Longer a Cohesive Force." *Jewish Journal of Sociology*, 39: 5-23.

Cohen, Stuart A. 1998. "Between the Transcendental and the Temporal: Security and the Religious Jewish Community." In *Security Concerns: Insights from the Israeli Experience*, edited by Daniel Bar-Tal, Dan Jacobson and Aharon Klieman, 371-93. Stamford CT: JAI Press.

Cohen, Stuart A. 1997. "Towards a New Profile of the (New) Israeli Soldier." In *From Rabin to Netanyahu: Israel's Troubled Agenda*, edited by Efraim Karsh, 77-114. London: Frank Cass.

Cohen, Yehezkel 1993. *Female Enlistment and National Service: A Halakhic Enquiry.* Tel-Aviv: Ha-Kibbutz Ha-Dati. (Hebrew).

Coser, Lewis 1974. *Greedy Institutions: Patterns of Undivided Attention.* New York: Free Press.

Dagan, Mattie 1999. "Press Conference on the State of Religious Military Frameworks [by the head of Religious Education in the Ministry of Education]." *Ha-Tzofeh,* June 16. (Hebrew).

Doar, Yair 1992. *Ours is the Sickle and the Sword.* Ramat Efal: Yad Tabenkin. (Hebrew).

Don-Yehiya, Eliezer 1994. "The Nationalist Yeshivot and Political Radicalism in Israel." In *Accounting for Fundamentalisms*, edited by Martin Marty and Scott Appleby, 264-302. Chicago: Chicago University Press.

Eliyahu, Rabbi Mordechai 1993. Interview in *Ha-Tzofeh*, September 14. (Hebrew).

Ezrachi, Yaron and Reuven Gal 1995. *General Perceptions and Attitudes of [Is-raeli] High-School Students Regarding the Peace Process, Security and Social Issues.* Zikhron Ya'akov. Carmel Institute for Social Studies. (Hebrew).

Feldman, Shai 1998. "Israel's Defense Policy: The Dilemmas Ahead." In *Challenges to Global and Middle Eastern Security: Conference Report,* edited by Emily Lauder, 41-43. Tel-Aviv: Jaffee Center for Strategic Studies.

Friedman, Menachem 1991. *The Haredi (Ultra-Orthodox) Society—Sources, Trends and Processes.* Jerusalem: Jerusalem Institute for Israel Studies. (Hebrew).

Friedman, Menachem 1990. "This is the Chronology of the *Status Quo*: Religion and State in Israel." In *The Shift from Yishuv to State, 1947-1949: Continuity and Change,* edited by Vera Pilovsky, 62-64. Haifa: Haifa University Press. (Hebrew).

Gal, Reuven 1997. Interview in *Ha-Aretz,* January 23. (Hebrew).

Gal, Reuven 1986. *A Portrait of the Israeli Soldier.* Westport, CT: Greenwood Press.

Galilee, Lili 1996. Interview with Major-General Gideon Sheffer, Commander, IDF Human Resources Branch, *Ha-Aretz,* December 6. (Hebrew).

Goren, Rabbi Shlomo 1993. "Disobedience to an Order." *Bulletin of the Council of Rabbis of Judea and Samaria* 14: 1. (Hebrew).

Harel, Amos 1999a. "Three Graduates of the Religious Pre-conscription College at Yatir are Among those who will Receive their Pilot's Wings Tomorrow." *Ha-Aretz,* June 28. (Hebrew).

Harel, Amos 1999b. "A Rise in the Motivation of the Sons; a Decline in Support of the Parents." *Ha-Aretz,* July 22. (Hebrew).

Horowitz, Dan and Moshe Lissak 1989. *Trouble in Utopia: The Overburdened Polity of Israel.* Albany: State University of New York Press.

Huberman, Haggai 1997. "The Oslo `B' Map as a Final Settlement." *Ha-Tzofeh,* February 21. (Hebrew).

Kasher, Asa 1997. Interview in *Ha-Aretz,* January 23. (Hebrew).

Kasher, Asa 1996. *Military Ethics.* Tel-Aviv: Ministry of Defense Publications. (Hebrew).

Kook, Rabbi Zvi Yehudah 1969. *To the Paths of Israel.* Jerusalem: Zur-Ot. (Hebrew).

Levite, Ariel 1989. *Offense and Defense in Israeli Military Doctrine.* Boulder, CO: Westview Press.

Levy, Ya'akov and Aaron Furstein 1985. "It's Not Easy to be a Religious Soldier." *Zera'im* 8: 8-9. (Hebrew).

Lichtenstien, Rabbi Aaron 1981. "The Ideology of Hesder." *Tradition,* 19: 199-217.

Liebman, Charles S. 1998. "Secular Judaism and Its Prospects." *Israel Affairs,* 4: 29-48.

Liebman, Charles S. and Eliezer Don-Yehiyah 1984. *Religion and Politics in Israel.* Bloomington: Indiana University Press.

Liebman, Charles S. and Eliezer Don-Yehiyah 1983. *Civil Religion in Israel: Traditional Religion and Political Culture in the Jewish State.* Berkeley: University of California Press.

Linn, Ruth 1996. *Conscience at War: The Israeli Soldier as a Moral Critic.* Albany: State University of New York Press.

Lipkin-Shahak, Amnon 1996. "Memorial address on the first anniversary of the death of Prime Minister Yitzchak Rabin." Official transcript. Tel-Aviv: IDF Spokesman's Office.

Lustik, Ian S. 1988. *For the Land and the Lord*. New York: Council on Foreign Relations.

Meir, Rabbi Yitzchak 1998. *Not By Power Nor By Force*. Tel-Aviv: Ministry of Defense Publications. (Hebrew).

Melamed, Rabbi Zalman 1991. "Producing Torah Leadership." *Crossroads*, 4: 65-72.

Michaelson, Benny 1982. "The IDF Rabbinate." In *The IDF and its Arms*, edited by I. Kfir and Y. Erez, 83-132. Tel-Aviv: Revivim. (Hebrew).

Naor, Aryeh 1993. "The National-Religious ('Credo') Argument against the Israel-PLO Accord: A Worldview Tested by Reality." *State and Religion Yearbook*, 1: 54-88. (Hebrew).

Neumann, Yehoshua 1997. "The obligation of active service in the IDF and Torah study." *Ha-Tzofeh*, February 16. (Hebrew).

Orr, Orri Brig.-General (ret.) 1995. Interviews in *Ha-Aretz*, July 5 and December 11. (Hebrew).

Ostfeld, Zahava 1994. *An Army is Born: Main Stages in the Buildup of the Army under the Leadership of David Ben-Gurion* (2 vols.). Tel-Aviv: Ministry of Defense Publications. (Hebrew).

Peled, Alon 1998. *A Question of Loyalties: Military Manpower in Multi-ethnic States*. Ithaca, NY: Cornell University Press.

Rolbant, Samuel 1970. *The Israeli Soldier: Profile of an Army*. New York: Thomas Yoseloff.

Rubinstein, Danny 1982. *On the Lord's Side: Gush Emunim*. Tel-Aviv: Ha-Kibbutz Ha-Meuchad. (Hebrew).

Sa'ar, Reli 1999. "A Decline in the Religiosity of Students Attending Religious High Schools." *Ha-Aretz*, July 20. (Hebrew).

Schiff, Ze'ev 1996. "Has the IDF been Removed from the National Consensus?" *Dapei Elazar*, 18: 26-33. (Hebrew).

Shaviv, Rabbi Yehudah. 1987. "Conflicting *Mitzvah* Obligations (Halakhic Aspects of the *Hesder*)." *Crossroads*, 1: 187-99.

Shedmi, Hayyim 1998. "A User's Guide to [profile number] 21." *Ha-Ir*, December 4. (Hebrew).

Sheleg, Yair 1994. "The New National Religious Character." *Yom Ha-Shishi*, August 19. (Hebrew).

Steinberger, Rabbi Yeshayahu 1997. "Scholastic Excellence in 'Zionist' *Yeshivot*: Vision and Reality." *Ha-Tzofeh*, January 3. (Hebrew).

Stern, Colonel Ron 1998. "A Revolution in IDF Motivation." *Ma'archot*, 360: 50-55. (Hebrew).

Van Creveld, Martin 1998. *The Sword and the Olive: A Critical History of the Israeli Defense Force*. New York: Public Affairs.

Yuchtman-Ya'ar, Efraim 1998. *Personal, Social and National Attitudes of Israeli Youth in the Jubilee Year: First Draft Report*. Tel-Aviv: Friedrich Ebert Stiftung. (Hebrew).

Zamir, Colonel Avi (Head of IDF Draft Board) 1997. Press conference, reported in *Ha-Aretz*, December 15. (Hebrew).

Part 3

The State and War-Making— Creating Citizens, Soldiers and Men and Women

6

Paradoxes of Women's Service in the Israel Defense Forces*

Dafna N. Izraeli

The extensive literature on the Israeli military only occasionally acknowledges women's presence with a passing reference to the exceptional fact that women are conscripted.[1] Women, however, constitute almost a third of the conscripts and 14 percent of the standing army (Liberman 1995). In 1999, the military conscripted 80 percent of the cohort of 18-year-old men and 62 percent of the cohort of 18-year-old women (Harel 1999c).[2] Studies of the military in other countries have revealed that gender dynamics are a major force in constructing the military, both in the historical creation of military structures and in contemporary policies and practices (see Elshtain 1995; Enloe 1983; Yuval-Davis 1997). This chapter analyzes the gender dynamics that shape contemporary policies and practices of the Israel Defense Forces (IDF).

The approach taken in this article owes much to that developed by Connell (1990) for appraising gender relations and state dynamics (see also Acker 1990; Lorber 1994). The military, like the state, needs to be appraised from the start as having a specific location within gender relations and as having a history shaped by a gender dynamic. Gender is thus both internal and external to the military. It is both a characteristic that individuals bring with them to the military, and a collective phenomenon that is embedded in the institutional makeup of the military. Each empirical military has a definable "gender regime" that is the precipitate of social struggles and is linked to—though not a simple reflection of—the wider gender order of society. A regime, following Foucault (1980), has the capacity to structure the situation of others so as to limit their autonomy and life chances. The gender regime refers to the institutional arrangements that produce inequality in the positions that men and women occupy and in their interactions (Connell 1990).

The main argument of this article is that gender is embedded in the structure and practices and even in the very logic of the military as an instrument of violence and as the institution for safeguarding the lives of "women and children" (Enloe 1990). The gender regime of the Israeli military is based on a gendered division of labor and a gendered structure of power, supported by a gendered ideology that combine to sustain and reinforce the taken-for-granted role of women as "the other" and their proverbial role as "helpmates" to men. By making gender a primary basis for classification, the military intensifies gender distinctions and then uses the distinctions as justifications for women's exclusion from positions of prestige and power. As a structure of power and as one of the important agencies that organizes the power relations of gender in society, the military contributes to gender inequality beyond the boundaries of the IDF.

There is a paradoxical aspect to women's military service in Israel. In a nation that prides itself as being the only country in the world where women, like men, are conscripted—seemingly the mark of gender equality—the military is a major force for the production and reproduction of men's domination in society. The institution designated to symbolize and exemplify women's partnership in the national collective is, in effect, an agent of women's marginalization. Another paradox is that although a significant proportion of women conscripts perform a variety of highly important, responsible, and sensitive roles, their military service is widely perceived as secondary and even as nonessential and expendable by some. This discounting of women's contribution is made possible by the gender segregation of jobs and the attribution of great symbolic value to combat roles, to achieving high military rank, and to reserve service, from all of which women are excluded.

These paradoxes and internal contradictions are the basis for internal and external pressures and occasional disruptions that also provide the potential for resistance and change. The 1995 Israeli Supreme Court decision in the case of Alice Miller v. the Minister of Defense (H.C. 4541/94), which required the air force to open its pilot training courses to women, marks such a change whose effects are still reverberating throughout the gender regime of the military.

The military provides men with advantages in accumulating what Bourdieu (1990) called forms of capital, or valued resources that are at stake in the military field. These advantages in cultural capital (knowledge and skills), social capital (valued relationships), and symbolic capital (the legitimate expectations for deference, prestige, and celebrity) are then used to gain additional and different advantages in civilian life, such as financial and political assets. The capital that women accumulate in the military provides significantly fewer advantages for them in civilian life because of both the nature of the capital attainable and the discounted rate at which it is converted in the civilian arena. This article analyzes these structures and social practices that construct and reinforce gender inequality in the Israeli military, as well as their implications for gender inequality in civilian life.

Significance of the Military for the Production of Gender

The close identification of the military with the state gives the military everywhere a kind of influence and privilege that is rarely enjoyed by other social institutions (Enloe 1983). A characteristic feature of the relationship between the military and civil society in Israel is the broad scope of military involvement in every sector of society and the privileged position of the military in the national ethos. The military-industrial complex is the largest single employer. Until the late 1980s, approximately one-quarter of Israel's labor force received its salary from the defense sector, a proportion that has declined because of cuts in the defense spending. Similarly, approximately one-quarter of the national budget was spent on the military until the early 1990s (Kimmerling 1993) but that proportion declined to approximately 16 percent in 1996. The military has been used as a major vehicle for nation building, and there is hardly an area of civilian life on which the hand of the military has not left its imprint.

The close relationship between the military and civil society is reflected in such popular descriptions of Israel as "a nation in uniform" or of the military as "the people's army" (Horowitz and Lissak 1989). Referring to the heavy reliance of the military on the reserves, these phrases incorporate the myth that every citizen is also a soldier and that the burden of service is shared by all—a myth that has important legitimating functions for the military's centrality in Israeli society (Ben-Eliezer 1995). The myth persists despite the fact that women and a significant proportion of men do not serve in the reserves and that Arabs (with some exceptions) do not serve at all.

Kimmerling (1993), who presented a critical perspective on civilian-military relations in Israel, characterizes these relations as "civilian militarism" in the sense that "militarism penetrates both structurally and culturally into the collective state of mind" (p. 129) and becomes part of the taken-for-granted reality of everyday life. "The essence of civilian militarism is in that military considerations and considerations defined as 'national security' will almost always be privileged over other considerations" (p. 129). Civilian militarism, it should be added, is a gendered concept. Because of the privileged position of military considerations in the collective psyche, the political elite is recruited from the military elite. Since women constitute only 12 percent of the Knesset and with rare exception do not participate in any of the civilian forums in which defense issues are debated, they do not participate in defining what is a consideration of national security. Their exclusion from this discussion further re-enforces the belief that only men can run the military.

The coupling of citizens' rights with participation in the military has existed since the French Revolution and continues the tradition of the Greek polis (Yuval Davis 1997: 96). In Israel military service as "a key citizenship-certification process" (Berkovitch 1997) is equated with service to the Jewish collective and, as such, is constructed as the basis for entitlement to full citizenship in the Jewish state and even for consideration as a normal human being. The importance of military service is reflected in a genre of newspaper stories about young men and women, who, after being exempted from service because of some personal disability, struggle with the military to accept them. The following are two examples:

Tubul, 18, is fighting to get into the army. The problem is a matter of size. . . . At 127 cm she is 13 cm below the IDF's minimum height requirement. "It's my dream to serve in the army," said Tubul. "I just want to be like everyone else" (Collins 1995: 12).

The IDF said no, but M.—a highly intelligent, 20-year-old Jerusalemite with a severe disability—won't take no for an answer. She plans to appeal the army's rejection two weeks ago of her offer to volunteer for military service. . . . "I wanted to serve like them [her classmates]; I don't expect to save the world, but being in the army is the Israeli thing to do, and it affects one's life afterwards," she said (Siegel 1994).

The following incident exemplifies the role of the military as a signifier of citizenship. In an interview on the eve of the formation of his political party, Natan Scharansky, the former Russian dissident, listed the marks of his entitlement to be considered a true Israeli: "I am 100 percent Israeli. I serve in the reserves, pay more than 40 percent income tax, I got into a mess with my contractor, I have two mortgages and my car was stolen. How can one be more Israeli than that?" (Galilee 1995). Scharansky immigrated to Israel as an older adult and consequently was exempt from compulsory military service. Nonetheless, his service in the reserves opens the list of the marks of his entitlement. However, his female counterparts, who do not serve in the reserves, have a lesser claim to being fully Israeli than he, even though it is likely that they, too, pay 40 percent income tax, have two mortgages, and have had their cars stolen.

The rhetoric that links military service with civilian entitlement to recognition is highlighted in the following statement issued by the Israeli Cabinet in commemoration of International Women's Day, March 8, 1992:

> The government takes special notice that the women of Israel, the women of the Jewish people, have always played an important role in all the combat units, in the valor and the building of the nation of Israel and the State of Israel, and, therefore, it is fitting that women play a more important role in the public life of the state.

Given the centrality of the military in Israeli society and the fact that women, like men, are conscripted to compulsory service make the military an important site for the enactment, reproduction, and perhaps the transformation of gender relations.

The Gendered Processes of Incorporation

Gender is among the most important and pervasive signifiers for distinguishing among categories of people in the military. The differential impact of the military on men and women begins long before conscription. In a real sense, it begins at birth. Society sends different messages to boys and girls about who they will be in the military and what significance their service will have for them as a rite of initiation into adult life. For men, it marks the central life event; for women, that definition may be more accurately applied to marriage. As Mazali (1993) observed, little boys who fall and scrape their knees may well be comforted with "it'll pass before the army"; little girls are more likely to be told "it'll pass before the wedding."

From a very young age [society] transmits to the son/male child a complex message. First of all the [military] service is ever present. From the very first moment. Everything about the way he is raised relates to this future stage. And the adult men in his environment—the father, the uncles, and the grandfathers—assure and threaten him at the same time: "the army will make you into a man." They remove the young boy from the circle of adults, arouse in him expectations and curiosity together with a sense of inferiority. The young lad will be able to belong only through his actual military service (Mazali 1993).

Women also play an important role, as mothers, sisters, partners, and friends, in the symbolic rites associated with a male's induction into military service. "The women supply a legitimate agreed-upon vehicle for the expression of fear, emotional relief. If they cry, they intensify the full and difficult truth of the rite and supply it with an important part of its significance. Their tears and cries of protest do not challenge this ritual but are rather integral to it" (Mazali 1993).

The long-term psychological impact on boys and girls of growing up in an environment in which the male from birth is treated as a future soldier who may be called upon to sacrifice his life for his country is worthy of study. Speaking of her own children, Mazali (1993) suggested something of this impact:

> I will begin with Noa [aged five and a half]. I don't think about her conscription from the moment she was born. It is not there all the time, behind my every touch of her. She will not endanger her life in her military service. When I think of her possible/future conscription, I can allow myself a sigh of relief. And the sigh that I *cannot* [emphasis in the original] allow myself from the moment that I gave birth to *boys*, is what these boys grow up with [emphasis added]. This difference is an important part of the way I and my society understand the concepts "daughter" and "son" and "woman" and "man."

Until the mid-1990s, it was the military's practice to treat young women and men who were near conscription age so differently that one might say that they experienced different militaries. Through an elaborate system of gender divisions and distinctions, the military intensified the salience of gender and constructed and reinforced gender differences, which in turn legitimated gender inequalities. Boys and girls who went to the same schools and studied in the same classrooms and wrote the same high school examinations were recruited on the basis of different criteria; went through different predraft preparations, tests, and procedures; and were classified, sorted, and assigned by different organizational units using different criteria. With few exceptions, they underwent separate and different military training and served in different positions and for different lengths of time. They were governed by different allowed behaviors and were subject to different systems of command. There were gender divisions of authority, with men monopolizing the highest positions of power. Men initiated and controlled the divisions, and women rarely participated in the forums in which decisions affecting the lives of women were made.

From the mid-1990s there emerged a distinguishable policy to narrow some of the spheres in which gender was made salient, to recognize that sexual harassment was an issue, and to adopt a more politically correct posture regarding women in the military. Supported by the commanding officer (CO) of the Women's Corps, these shifts were largely guarded responses to increasing public criticism. The military opened a number of combat training courses to women and in 1999 announced its intention to integrate basic training for women and men who were not in combat positions, beginning in the year 2000 (Harel 1999a: 1). It became increasingly politically correct for senior officers to express their approval of opening new roles to women previously closed to them (Harel 1999b).

Gender Distinctions

Compulsory military service for women was introduced at the beginning of statehood. During the Knesset debates on the Security Service Law, which was passed in 1949, all the parties, except for the religious parties, supported the principle of compulsory military service for women. These parties were divided, however, on the extent to which women should be treated differently from men, with the more radical Left parties objecting to differential policies (Berkovitch 1997). They agreed, however, that at a time when the very existence of the newborn state was in the balance, women's service was needed. Women's involvement in the military was not new. They had participated in the various Jewish defense organizations since the early twentieth century and in the Palestinian units of the British army during World War II and had played a significant role in the war of independence (1948-49) (Bloom 1982).

David Ben-Gurion, architect of the State of Israel and its first Prime Minister, was committed to women's participation in the public life of the new Jewish state. He was a powerful spokesperson on behalf of compulsory service for women, but viewed women's service as different from that of men. National security, he argued, could not be defended only by the military narrowly defined. It also required that people settle the border areas, that new immigrants be integrated into the society by first learning the Hebrew language and the ways of the new society, and that the Jewish population in Israel be increased. Such a broad definition of national security included women in their roles as soldiers, as farmers in the settlements along the borders, as teachers in the military, and as wives (of soldiers) and mothers (of soldiers to be) (Berkovitch 1997). As citizens, men and women contributed to the security of the nation through compulsory military service. As biological beings, however, they were assigned different roles. According to Ben-Gurion, women also contributed to the national security by bearing and raising children, and the military must not interfere with this important function, which men are unable fill. In the security discourse, childbearing and child rearing were linked, making the second appear as biologically natural as the first. As Yuval-Davis (1997: 22) observed, "often the pressures on women to have or not to have children relate to them not as individuals, workers and /or wives, but as members of specific national collectivities." Enlarging the Jewish population—which Yuval-Davis (1997) referred to as the "people as power discourse"—was defined as vital for the national interest. Ben-Gurion's broad, almost all-encompassing, vision of what was entailed in achieving national security made it possible to incorporate both women and men into the hallowed project of national security in the name of gender equality, while assigning them different roles in the name of national security. Until the mid-1990s, the contradictions and tensions inherent in the "different but equal" thesis were not generally considered problematic.

Legal Distinctions

The Security Service Law, passed in 1949 and amended over the years, defined eligibility for compulsory military service in universal terms as applying to "a citizen of Israel or permanent resident." Furthermore, although the law made no gender distinctions with regard to the occupations open to men and women, it granted priority to women's family roles over their obligations to military service. Married women were exempt from compulsory military service, but not from reserve service, and pregnant women and mothers were exempt from both compulsory and reserve service (Article 39 of the law).

Where the law made no gender distinctions, such as in type of service, the military institutionalized a rigidly differentiated division of labor (see the next section). Whereas the law treated men and women the same, military regulations and practice treated them differently.[3] For example, whereas the law did not exclude even married women from reserve service, although they were required to serve for fewer years,[4] the number of women called to reserve service is negligible. In the early 1980s, in response to demonstrations by young women who had been called to the reserves against the differential treatment among women,[5] the military reduced the maximum age from 29 to 24. Rather than extend reserve duty to all women, and constrained from using a policy of differential call-up of those women it needed, the military decided in favor of reducing the use of women in the reserves to the minimum. In practice, women are not called to reserve duty.

Where the law treated men and women differently, the military increased the gap. For example, whereas the law specified 30 months' service for men and 24 months' for women, men serve 36 months (the additional 6 months with pay). The mandatory 24 months for women's service was cut to 22 months in 1992, to 21 months in 1993, and to fewer than 21 months in 1994 (Protocol 1994a: 4). About a third of the women actually served fewer than 20 months. These changes in military human resource policy were prompted by the oversupply of recruits because of the significant increase in the size of the 18-year-old cohorts beginning at the end of the 1980s,[6] pressure on the military to demonstrate that it was making budget cuts and becoming more "lean and mean," and strong public pressure to continue universal conscription for men and not to introduce differential service. In this constellation of pressures, given the military's preference for men, women were the weakest and most vulnerable category.[7]

Women's shorter length of service and their exemption from reserve duty were raised repeatedly as the touchstone for their differential treatment. For example, in the air force's defense brief to the Israeli Supreme Court in the case of Alice Miller v. the Minister of Defense, the air force justified its refusal to allow Miller to take qualification exams for flight-training not because "women as women" lacked the ability to become pilots but because their short service made the investment in their training highly uneconomical. The recent decrease in the duration of women's service and the consequent declining returns on the investment made it even less economical. Despite the military's initial insistence to the contrary, preserving the gains made during the previous decade in women's participation in training leading to the more prestigious and skilled jobs will be difficult, to say the least. The eight-month training for women electronic technicians was the first such course to be eliminated.

The argument is circular. The socially constructed gendered practices of assignment restricted the military's flexibility in its assignment of women. This constraint, in turn, limited women's usefulness to the military, which, in the face of budget cuts, became the justification for shortening women's military service. Women's shorter military service then made it cost-inefficient to invest in women's training, which reduced women's usefulness even further. The end result was that women, at least the majority of them, began to appear increasingly dispensable.

Administrative Distinctions

All women belong to the Women's Corps, but few actually serve in it.[8] All but a few are assigned to functional units. There is no men's corps; all the other corps are identified by their respective functions. The Hebrew acronym for the Women's Corps, CHEN, which is spelled the same way as the word charm, underscores this difference. Until 1997, the Women's Corps had formal responsibility for all women soldiers with regard to military training (including officer training), job assignments, discipline, and judicial matters, as well as welfare and well-being, including protection from sexual harassment.

CHEN, established in 1949, was created to assist the IDF in recruiting women and then in managing the personnel and training issues emanating from the perceived differences between women and men. It was intended to be a vehicle for implementing military policy and for providing protection to young women recruits operating in a macho culture—not for providing representation for women in the policy-making centers. CHEN was modeled primarily after the ATS—the women's corps of the British army during World War II in which women served in a separate unit under their own female command. This model is in contrast to the alternative model supplied by the Palmach—the Jewish underground and commando unit created to fight both the Arabs and the British in Palestine—in which men and women served in the same units. The first commanding officers of CHEN were former ATS officers.

In recent years, the protective policies operated through a separate women's unit have been criticized as counter to women's strategic interests. For example, the fact that a CHEN officer had sole jurisdiction over a woman soldier in all judicial matters, even if the soldier's functional unit commander was a woman, differentiated the woman soldier from her male counterpart. If she missed a drill and he missed a drill, he would be disciplined by the unit commander and she by the CHEN officer. This situation, which curbed the functional commander's authority over the woman soldier, also had the consequence of making the commander feel less responsible for her (Bloom 1991). Furthermore, the reluctance of the military generally to apply the same disciplinary measures against women, for example, in cases in which women objected to their assignments, then became a justification for differential assignments which in turn supported the view that women did not really bear their fair share of the workload. By the early 1990s, disciplinary functions in the professional army had been transferred from CHEN to a woman's functional commander. The emerging policy, strongly endorsed by the then CO Brigadier General Yisraela Oron, was to limit the intervention of CHEN to matters that were "relevant to the differences between men and women" or where the women's corps had "a relative advantage, such as in matters of sexual harassment" (Binyamin 1996: 64).

No woman is of a high-enough rank to participate in the meetings of the General Staff on a regular basis.[9] Until 1987, the rank of the CO of the Women's Corps was that of colonel. Only after intensive public pressure from women's organizations and women members of the Knesset was the rank raised to that of brigadier general, one rank below the CO of personnel. In response to the question of whether CHEN could influence the prerequisites for assignments to senior positions, Brigadier General (reserves) Yehudit Ben-Natan, the CO of the Women's Corps during the mid-1990s, replied: (Protocol 1994b: 19):

> Yes, it can sit in the meeting in which decisions are taken, and express its opinion, but how will that influence? There is only one Head of the Women's Corps. She sits in part of the discussions of the General Staff, when it discusses the girls' service in the I.D.F. But it is the established priorities that set the tone.

Women do not participate in determining the priorities. The closure of senior positions to women thus also excludes women's participation in making the policies that institutionalize their secondary status in the military. The CO of the Women's Corps reports to the chief of personnel and may advise the Chief of Staff on matters of specific relevance to women. In fact, until recently, the CO of the Women's Corps was rarely consulted on matters of general relevance to women. For example, she was not party to the decisions to cut women's military service from 24 to 22 and then to 21 months.

Unlike other officers who could call on reservists for technical assistance and support, the CO of the Women's Corps had no (female) reserves to assist her. Toward the mid-1980s, Colonel (later Brigadier General) Amira Dotan established a steering committee of prominent women outside the military to support her struggle to expand the number of job categories open to women in compulsory service and to upgrade the rank of the CO of CHEN. The COs immediately following her, however, were discouraged from such involvement of "outsiders."

The Women's Corps never emerged as a collective voice for promoting women's equality, partly because of the strict limitations on protest of any sort, let alone organized protest within the military, and partly because of CHEN's historical lack of clout or its symbolic representation of women's marginality. Women officers serving in the "men's army" tended to view CHEN as an impediment to women's integration in the military and preferred to disassociate themselves from it. The efforts of the head of the Women's Corps to introduce incremental changes in the direction of greater gender equality were greatly enhanced by the establishment of the Knesset Committee on the Status of Women in 1992. The committee's hearings marked the first time that an agency with authority over the military required it to be accountable for treating women more equally.

Functional Distinctions

Military jobs are generally assigned first by gender and then within gender by aptitudes, competencies, and other considerations.

Table 6.1

The Percentage of Women Officers Among All Officers, by Type of Service: 1998

Service	Conscripts and Professionals	Professionals Only
Total IDF	23.7	14.5
1. Armor/tank	1.3	0.9
2. Artillery	1.9	0.3
3. Engineering	2.5	2.1
4. Ammunition/ordnance	6.8	5.2
5. Field command	12.7	3.6
6. Navy	12.1	8.9
7. Air force	14.1	9.9
8. Military police	18.1	6.9
9. Intelligence	16.9	12.7
10. Maintenance	23.5	15.9
11. Signal	23.2	14.5
12. Medical	32.9	32.1
13. General staff	38.8	27.4
14. Rear command	39.7	25.7
15. Field	43.0	18.7
16. Adjuntance (human resources)	68.0	50.9
17. Education	83.3	59.5

Source: IDF spokesperson.

Table 6.1 presents the proportion of women by type of service for professionals and conscripts (column 1) and for professionals only (column 2). In 1994, women constituted 32 percent of the recruits and 14 percent of the professional military (Liberman 1995). As can be seen, women are highly underrepresented in services with the greatest proportion of combat soldiers (1-8) and overrepresented in services with the greatest proportion of soldiers in white-collar, semi-professional, and administrative positions (12-17). As is the case with all occupational data, aggregated categories veil the true extent of segregation. A more detailed analysis would reveal, as was said earlier, that relatively few jobs are performed interchangeably by men and women.

In every service, the proportion of women among the officers in compulsory service is greater than that in the professional army. This reflects both the greater proportion of women relative to men in compulsory service, as well as the concentration of women offic-ers at the lowest ranks of the command hierarchy, those usually at-tained during compulsory service.

Table 6.2 presents the distribution within each gender among the ranks. At the bottom of the hierarchy, women constitute approxi-mately 32 percent of the lieutenants, a rank achieved during com-pulsory military service. Women are very much underrepresented in the four ranks below Chief of Staff. The appearance of a woman

Table 6.2

Distribution of Officers, by Gender and Rank: 1998

Rank	Women	Men
Brigadier	0.0	0.1
Brigadier general	0.1	1.0
Colonel	1.0	3.0
Lieutenant colonel	4.0	14.0
Major	25.0	31.0
Captain	22.0	25.0
Lieutenant	48.0	26.0
Total	100.0	100.0

Source: IDF spokesperson.

brigadier-general reflects the upgrading of the rank of the CO of the Women's Corps in 1988, referred to previously. There were eight women at the rank of colonel in 1985 and eleven in 1999. Between 1985 and 1999, there was a slight increase in the proportion of women officers up to and including the rank of lieutenant colonel. This resulted largely from the expansion of the rear command following the Gulf War and the military's growing confidence in women's professional competence. The data in Tables 6.1 and 6.2 reveal what Grant and Tancred (1992) called "the dual structure of unequal representation." Women are unequally distributed through the hierarchy of the military as a whole. At the same time, those organizational units in which women are well represented and exert a greater measure of authority are peripheral in the internal organization of the military.

Symbolic Distinctions

Women's exclusion from combat roles has been a pivotal mechanism upon which male dominance in the military and beyond it rests. Whereas the military assigns women to serve in combat units, roles defined as combat and jobs performed in combat areas have been, and with few (but increasing) exceptions still are, closed to women. Exclusion sustains dominance in two ways—hierarchically and symbolically—both of which exclude women from positions of power. By defining combat service as a necessary prerequisite for virtually all the most senior positions and the majority of those below the senior levels and then denying women access to this experience, the military can justify blocking women's movement up the ranks. By excluding women from senior positions, men monopolize the power to define the rules of the game that shape social interaction and opportunity in the military.

The military's resistance to opening more senior positions to women limits competition for those positions. As the hierarchy narrows, opportunities for upward mobility become more limited. Men who are derailed from the combat-career route either because they lost out in the competition or as a result of a drop in fitness, may be "cooled out" by assignment to an alternate career line normally accessed from a different route. This practice permits entry to an alternative career path where the male incumbent frequently receives priority over a more qualified woman in line for promotion. This practice enables the military to repay those (men) who spent years "in the trenches" in contrast to the less worthy (women) who spent them in the offices and to retain valued persons (men) who might otherwise leave the military prematurely for employment in the civil market.

Symbolically, the combat soldier—brave, self-sacrificing, and tough—incarnates hegemonic masculinity in the military. In the security discourse, the combat experience is the unique tempering fire that turns a person (a man) into a real soldier. Only activities that are coded as combat bequeath heroism and glory, and only combat roles, especially those at higher ranks, create entitlements to deference greater than those that can be claimed by equivalent ranks in noncombat roles.[10] In other words, in addition to power of position, moving up the combat-career path provides soldiers with what Bourdieu (1990: 112) termed "symbolic capital," a form of power that is not perceived as power but as legitimate demands for recognition, deference, obedience, or the service of others. It is "denied capital" in that it disguises the underlying interested relations as disinterested pursuits. Symbolic capital is preserved through exclusivity. When it is spread too wide or when it is extended to devalued categories, such as women, it becomes devalued. Excluding women is thus important for preserving the hallowed character, the sanctity, of the combat role. According to Bourdieu (1990), symbolic labor produces power by transforming relations of interest into disinterested meanings and by legitimating arbitrary relations of power as the natural order of things.

The arbitrary nature of the combat prerequisite for virtually all senior positions is not lost on the women officers in the professional military, who perceive a contradiction between the declared ideology and their understanding of military practices in this regard. A 1996 survey on the subject of women's advancement in the IDF found that among professional officers, 82 percent of the men and 75 percent of the women indicated a strong desire for advancement. The two main factors cited as explanations given by the military for the gender difference in rate of advancement were that women did not serve in combat roles (cited by 80 percent of the women and 72 percent of the men) and that the military was concerned that motherhood hinders women's ability to devote all their energies to work (cited by 81 percent of the women and 52 percent of the men.) When asked how justified they thought these explanations were, only 30 percent of the women and 54 percent of the men considered that the first explanation was justified, and only 13 percent of the women and 36 percent of the men considered that the second was justified.

Women officers argue that combat experience should cease to be a requisite for many senior positions that women claim do not require such experience. A policy based on the belief that whoever does not crawl under the barbed-wire fence or does not storm some fortified target cannot be a successful commander is anachronistic and blocks the promotion of capable women. Positions at the rank of colonel and above, which women claim do not require combat experience and, therefore, are unwarrantedly closed to women, include chief educational officer, CO of intelligence, CO of personnel, military spokesperson, chief medical officer, chief mental health officer, chief military prosecutor, and president of the appeals court. The recent appointment of a noncombat officer as CO of personnel supports this argument. The women also believe that if the military command were to perceive it in their interest to do so, they could provide promising women with the equivalent of combat experience in the field.

What Women Do in and for the Military

Women perform three major functions for the military. The first is that they execute all those jobs that the military regularly assigns to women soldiers, among which clerical work and various forms of personnel and educational work are the most obvious and prominent. The second is that they provide a reservoir of labor for regulating the utilization of human resources and replace men in what are classified as "men-replaceable" jobs. The third is that they serve as "women" in "trophy jobs" that contribute to supporting the glorification of the masculine in general and of hegemonic masculinity in particular. Each of these functions is considered briefly in the following sections.

Women's Jobs

Policies concerning women's assignment have been dictated primarily by the changing needs of the military. Two factors were especially important for conducing the military to experiment with new uses of womanpower. The expansion of the military to cope with the new territories under Israeli control following the 1967 war created a need for more combat units. The development and introduction of increasingly sophisticated military technology, which reduced the necessity for engaging in face-to-face combat, made differences in physical strength between men and women less consequential (Yuval-Davis 1997) and increased the need for cognitive and other skills that gave men no special advantage. The changing needs of the military resonated with the demands emanating from the liberal feminist women's organizations to open more job categories to women. The demand was voiced initially by the Commission on the Status of Women appointed in late 1975 in conjunction with the United Nations Year of the Woman. The commission's report (1978) found that in 1976, 210 out of a total of 709 jobs were open to women, but women actually served in only about half of them; approximately 70 percent of the women were in clerical positions. In 1988, women actually served in 234 of the approximately 500 jobs that were then open to them (Bloom 1991: 35). In 1998 they served in 330 of the 551 jobs open to them; 187 were closed to women for reasons of their association with either combat or religious service (IDF Spokesperson). Thus, over two decades, the number of jobs open to women more than doubled, and the number of jobs in which women actually served more than tripled.

Men-Replaceable Jobs

Since the establishment of the IDF, women's participation in the military has been talked about and justified in terms of "freeing men for combat."[11] This phrase is a metaphor for the asymmetry of status ascribed to each gender and for the differential nature of men's and women's incorporation in the military. As Enloe (1983) noted, men are the military; women are in the military. Women free men not only for combat jobs but for technical jobs, jobs that require physical strength and those that are performed under dangerous conditions.

In the 1980s, in response to budget cuts, women were used to replace men in the standing army, as well as in the reserves. Sometimes women replaced male conscripts who, in turn, replaced reservists (Liberman 1995). Women also replaced men as instructors for training men in various field-soldiering and combat skills, including driving tanks and target shooting. Women's representation among instructors, while still small, grew by over 400 percent between 1983 and 1993 and was about 2,000 in 1994 (Liberman 1995). Serving as instructors for men-only units is the closest women get to a combat role and is among the most prestigious jobs open to them.

Instructors' jobs that were previously held by men, however, were usually redefined, made narrower and more specialized so they could be learned in a shorter time, and did not include a combat experience requirement. This redefinition preserved the distinction between what women and what men did—even when it seemed that the divisions were being dropped. For example, men usually became instructors after a period of combat training, while women were sent directly to instructor training. Women instructors, furthermore, were confined to the classroom and generally did not accompany the men during a military drill in the field and certainly never in real-life situations. As the deputy CO of personnel pointed out to the Knesset Committee on the Status of Women: "There is a big difference between being an instructor in the classroom and having to maneuver a tank on the battlefield. I think we are still far away from taking a tank instructor from the Armored Corps training school and putting her now on a tank to make a fighting reconnaissance in South Lebanon" (Protocol 1994a: 6). Women's lack of experience under real-life conditions made them less credible as instructors. What works in the books, the soldiers could say, may not work in practice.

Women who enter men's occupations potentially face the pressures associated with being "the wrong person" in the job. Jobs differ in the extent to which they are gendered, that is, the extent to which being a man or a woman is perceived to be a necessary qualification for their successful accomplishment. Jobs may be gendered in three respects: (1) being of a certain gender is considered an essential prerequisite for the performance of a job; (2) for promotion from the job to a job at the next rank in the same career track; or (3) the context within which the job is performed privileges one gender. All three parameters of genderedness may exist contemporaneously. The more intensely gendered the occupation, the greater the performance pressure on those of the wrong gender attempting to accomplish it. The extent of the perceived genderedness of a particular job, however, is not a constant; rather, it responds to changes in practice.

In the early 1980s, the military introduced the use of women noncomissioned officers and junior officers to serve as instructors for combat units, and toward the end of the 1980s, experimented with using women as platoon commanders of ten to fifteen soldiers for new male recruits in noncombat units. The platoon commander role is gendered in all three respects: a commander (in contrast to an instructor) of men is perceived as a man's job. The commander of men works in a primarily male-dominated setting, and few women get promoted from the rank of platoon commander to company commander.

The following is the author's summary and interpretation of a military report that evaluated the performance of women platoon commanders and highlighted the pressures for women of being in a man's job and women's responses to these pressures.

The platoon commander is in continuous close proximity with her recruits. For approximately one month, she is responsible for training the new recruits in basic soldiering skills, socializing them into the military framework, and guiding the new recruits through this first period of adjustment to the military. The women who serve as commanders of all-male units are scrupulously selected for their high level of expertise in the various skills of soldiering, high level of competence in training ability, personal values and skill in working with young men from different socioeconomic backgrounds, and strong identification with the military. The base in question is situated outside the official borders of Israel, in what is considered "unfriendly" territory. The women commanders sleep in quarters separate from both the men commanders and from the other women on the base who are in service jobs. Symbolically, this arrangement sets them apart from both.

The woman commander is set apart from "the norm" in a number of additional ways. Most important, she is responsible for a group of men and has positional authority over them. What is special is not her being a commander, since there are women commanders over women, but that her subordinates are men who are new and not socialized into military discipline. She is in a different situation from women instructors of combat units. The instructor's authority rests on her expertise in a particular skill or set of skills. She tells men what to do because she knows best. The commander, in the military ethos, is a leader, an object of identification, the incarnation of the military at the grassroots level. Despite the similarity of title and rank, the woman platoon commander is restricted by formal rules and informal norms not applied to men. First, rules of modesty limit her entry into the men's quarters. She is excluded from this private space, which is accessible to men commanders. Second, because of the camp's location in dangerous surroundings and the military's differential policies regarding the movement of men and women in such places, she is not allowed to perform certain duties most associated with soldiering and military prowess. For example, she does not do night guard duty or take charge of the camp at night when the officer is off duty, as her male counterparts do. These gendered practices of everyday life undermine the authenticity of her performance as a commander.

The women responded to these performance pressures in ways similar to those described by Kanter (1977) for token women managers and managers lacking in power: First, they worked extra hard to maintain a reputation for excellence. The evaluation report noted: "To my surprise I had difficulty finding mediocre women commanders to serve as a control group for my study." The more senior officers whom the author of the report interviewed told her they had no mediocre women instructors and had they any, such instructors would promptly be assigned to a job elsewhere." Second, they "did masculinity" (West and Zimmerman 1987), and in the view of the report, they overdid it—imitating the verbal culture—tone of voice, slang, and posture and demeanor of men commanders—"because then they forget you're a woman." In the powerful macho culture of the military, men provided the only legitimate models of leadership behavior and style for women commanders over men. There is a certain irony in that the women felt pressured to avoid "doing femininity" to prove that women were as capable as men, which, according to the report, was one of the motivating factors for women to volunteer for such jobs. In adopting a masculine genre, they also inadvertently reinforced the association of leadership with masculinity. Third, they tended to stick rigidly to the rules that supported their authority. The latter pattern opened them up to criticism for being sticklers about the rules and too technocratic, treating rules as ends rather than as means for achieving wider purposes. Senior officers pointed to this "technocratic" behavior as indicative of women commanders' difficulty connecting their subordinates to the wider purposes of the military and as proof of their lack of leadership. According to the report, they viewed women more as instructors than as commanders.

For most women platoon commanders, promotion within the career track was blocked because men were preferred for positions at the next level. The women were consequently forced to leave the combat track. They usually joined the other women in more traditional jobs. It is not surprising that the report found a significantly higher level of burnout among the women than the men platoon commanders.

Kanter (1977) underlined the catch-22 nature of being a woman in a man's job. As tokens, the women were under pressure to prove that they were worthy of joining the men, no less good at what men did than the men themselves. At the same time, they were expected to remain "true women" and not threaten the men. The tension between the two demands made it difficult for the women to adopt a consistently winning strategy. The following case of the women target-practice instructors provides an example in which achieving excellence was a necessary condition for women's access to prestigious male-dominated positions but that excellence also became the justification for a demand for their removal from these positions.

In the 1980s, the IDF opened the position of target-practice instructors for men to women, and in 1992, the then head of the Infantry and Parachute Corps attempted to close the job to women and return the responsibility to the direct commander. He explained:

> Since the women began as target-practice instructors, there has been a decline in the level of teaching and of target shooting of the men commanders. Instead of turning to their commanders, soldiers are turning to the women target-practice instructors with every question or problem related to target practice because the commanders don't know the answers. Therefore, we decided to revert the job of teaching shooting back to the men commanders. The women target-practice instructors will become instructors in good-training skills and will assist the commanders.

The women instructors expressed their concern that their hard-achieved entry into this and other such high-prestige jobs would be curtailed in the future on the grounds that they were women (Sadeh 1992). From the perspective of the military, there is a logic to the argument of the CO of the Infantry Corps.[12] Given that accurate shooting is at the heart of the infantry soldier's job and that the military discourse requires the commander to lead by personal example, it follows that the immediate officer should be an expert marksman. The women's success was defined as the cause of the male commanders' loss of authority in relation to subordinates—a symbolic castration threat perhaps. Instead of removing the male commanders who failed in their jobs, the women were held responsible for the men's failure and were subsequently to be penalized. Their role would be redefined from that of experts in charge to that of assistants to the men in charge. The proposed change was not implemented, but the demand to remove the women from their jobs reveals the precariousness of the arrangements that tamper with the gendered order of things.

Once women entered a job and performed it to the satisfaction of the military, it tended to be redesignated as a woman's job. Speaking of the "progress" made by women in recent years, the CO of personnel pointed out proudly: "Instructor occupations are closed to boys in many places—including the most attractive places, the most combatlike places—there, only girls are instructors. Only women train our pilots, the combat pilots, on the simulation machines" (Protocol 1994b: 10). What he failed to appreciate is that the redesignation of these jobs into women's occupations, while improving job opportunities for women, also preserved the gendered division of labor upon which the gender structure of power rests.

Trophy Jobs

There were several ways in which women served in the military as women. Women who served with all-male units were perceived as bringing a touch of home to the otherwise cold, military world of boot camp. They became the personification of the collective wife/mother/sister. They functioned as morale boosters, symbolically personifying for the men some of the rewards of home —bestowed by women on men. One feminist journalist commenting on the role of the company clerk considered that it symbolized the gender dynamics of the Israeli military:

The Company Clerk— the miserable dream of too many serious girls, the closest a young woman can get to the field, to the fighters, to the real thing. They march

80 km with the company, bake a cake at home, and never forget anyone's birth-day. The Company Clerk is a symbol of what men and women learn about them-selves and about each other in the military. They [the men] are the warriors, they [the women] are always, always the "helpmate unto him." To be helpmates they are recruited to the military and that is what most of them learn there. (Hareven 1995: 3)

Women are also trophies for heroes. Their beauty is commodified and allocated to the most worthy. A popular adage loosely trans-lated says that "the best men become pilots, the best women are for the pilots." The best women are, in this case, those with movie-star femininity—tall, slim, and beautiful. For some jobs—those close to senior officers—physical appearance is an important qualification. The commander's access to this scarce resource says something about his status. The most beautiful women are his prize, as well as the mark of his military achievement. Their feminine qualities reflect upon his manliness and validate and enhance it.

A journalist who investigated the way female clerical staff are as-signed to senior officers on the basis of their good looks, concluded:

All the models and the beauty queens, the prettiest girls in the IDF, serve as clerical staff in the office of the major generals. The most beautiful serve in the office of the Chief of Staff, or the CO of the Absorption and Classification Base. There is no paragraph in the orders of the Chief of Staff regarding the assignment of pretty women soldiers as clerical staff for senior officers. There is also no such verbal instruction. But, like the *zubor* [slang term for hazing routines performed on new recruits] and like the coffee served to the com-mander, this custom is deeply rooted in the system and is transmitted from generation to generation. (Rosenblum 1995)

The top brass siphons off the most desirable women. Officers fur-ther down the pecking order exert what influence they can to get the prettiest from the remainder assigned to them, as indicated in the following story:

When I was a young officer, we were stationed in Sharm-el-Sheih [south Sinai Peninsula], where all the girls assigned to Sharm-el-Sheih, as well as to the secondary bases in the outlying areas, were sent. We used to sit, all the young officers, and mark each for the person who sorted the girls. Thumbs up meant pretty soldier, leave her in Sharm. Thumbs down—send her to the outlying bases. Later when I was commander of one of the secondary bases, I called that same sorter and told him: "Don't you dare send me any more of those who get the thumbs down." (Rosenblum 1995)

In some cases, the presence of women poses a challenge to mas-culinity that the military sees as beneficial to the training process.

Referring to the outstanding success of women in their new role as platoon sergeants for male soldiers in basic training programs, an army public relations pamphlet explained: "New recruits do not dare complain of muscle aches and pains or drop out of a long distance run when it is being led by a female sergeant" (quoted from Gal 1986). Why they dared not complain is not explained; presumably it is obvious to the reader: No man who considers himself a man would wish to be outdone by a woman. Furthermore, the implication is that men are motivated to prove that they can do better than women or that to demonstrate their superior toughness is an important performance incentive. This military publication, which presumably intended to praise women's accomplishments, subtly reconstituted their taken-for-granted inferiority.

There has been almost no research to date on the impact of the Women's Corps on women. Sasson-Levy's (1997) study of women who served in nontraditional roles noted that "away from home, women have the opportunity to experiment with alternative possible selves from among the multiple femininities within the military. For some women, especially those in traditionally male roles, the military provides an opportunity to free themselves from the constraints of 'Israeli femininity.'" The study's findings revealed, however, that these women soldiers often pay a price for their newfound freedom and empowerment. In identifying with men, they tend to adopt the same disparaging attitude that men have toward other women. Women pay for their newfound empowerment and self-confidence by, as Sasson-Levy indicated, becoming misogynous. If for men, military service resonates with their masculinity and enhances it, for women, especially those in the most prestigious positions, it frequently involves a rejection of femininity and of themselves as women. By associating achievement with the activities that men do, the military intensifies the contradiction between femininity and achievement. More research is required to identify the extent to which the male monopoly of senior positions and the lack of female role models in those positions impact beliefs about women's abilities for leadership roles and their entitlement to them.

The Reproduction of Gender Inequality in Civilian Life

The military provides important opportunities to develop social capital—the social networks so important for access to information and support, as well as to people, places, and jobs in civilian society. Both women and men use such ties to get jobs and other advantages, but the military gives men more opportunities for accumulating social capital than it does women.[13] Men serve for a longer period and often in more varied jobs and locations. The reserves bring together people from many different walks of life who might otherwise not meet one another. Serving together creates social bonds of mutual obligation that bypass status differences in civilian life and often extend beyond the service. For example, in her study of civilian-military relations in Israel, Etzioni-Halevy (1996) found that senior officers meet civilian elites and prepare their second careers while still in the military.

The civilian employer views the military as a valuable training ground for both general attributes and specific skills.[14] In some occupations, the link between the military and the civilian is institutionalized—most noticeably in high-tech industries. In an investigative article, Aloni (1994) quoted a manager who compared the burgeoning growth in start-up companies in Israel in the mid-1990s with what occurred in Silicon Valley in the early 1980s: "In contrast to the entrepreneurs of Palo Alto, Israeli entrepreneurs have technological experience and prefer teams that already worked together in the army or the defense industry." Or as another manager put it: "All our workers served in the same unit and that is how it will be in the near future. In general, the high-tech market in Israel is very tightly networked. People know one another and that is very helpful." Aloni concluded: "The uniqueness of the Israeli high-tech success lies in these army grounded networks and not in 'the Israeli brain' as is commonly thought" (p. 18).

In some occupational fields, such as security-related jobs, a specific type of military experience is a condition for entry. In others, the civilian employers recruit those trained by the military. For example, El-Al, Israel's national airline, recruited its pilots exclusively from the military. According to the El-Al chief executive, the reason for the preference was strictly professional: "One doesn't have to be a man to fly a passenger aircraft but if I can select for El Al the best pilots, those who are better equipped to cope with stressful situations because of their military training, there is no reason that I should forego this advantage. If the air force will train women, I will accept women" (Schochat 1995: 1). In the mid-1990s, this policy was challenged in the Labor Court on the basis of the Equal Opportunity in Employment Law of 1988. After El Al's policy was declared discriminatory, it hired its first woman pilot.

When senior officers enter civilian organizations, they frequently bring with them other officers who were their colleagues or subordinates in the military. Once in the civilian sector, they recruit fellow retirees with whom they weathered emotionally charged experiences, who have proved their worth and their loyalty, and who speak the same language they do. For example, the equivalent of the following headline that appeared in a Tel Aviv newspaper in September 1988 could have appeared following the 1999 election as well: "One more than in the general staff: 22 senior reserve officers fill senior positions in the municipality. Lahat [then mayor of Tel Aviv and senior reserve officer] chooses most of them"(Avnieli 1988: 15). One senior officer hired as the CEO of an important museum complex replaced the existing staff from top to bottom with military retirees from his unit, even bringing in his personal ralashit (the military term for the female office manager of a senior officer).

Men reap greater value from the symbolic capital they accrue from serving in the military than do women. Despite the fact that only a small proportion of men become combat soldiers, the symbolic value endowed by the military reflects on all men in a way that it does not reflect on women. This differential symbolic effect of the military is exemplified in Herzog's (1999) study of women in Israeli local politics that found that women had a harder time than men exchanging their military rank for political advantage in political life. The political parties placed them lower on the party list than men, thus reducing their chances for election. This differential symbolic effect of the military is further exemplified in a study of gender bias in the Israeli courtroom, which found that military service was among the reasons judges gave for leniency in sentencing defendants (Bogosh and Don-Yechiya 1999). The reference to military service as a consideration, however, applied only to male defendants—both those who had actually served and those who would, presumably, serve in the future. That military service was never mentioned in relation to a woman defendant raises the question of whether Israelis think of women's military service as actual soldiering.

The relationship between the military and women's status in Israeli society is circular. A feedback-loop dynamic leads from women's marginalization in the military to women's disadvantage in civilian life and back again. First, the gendered processes by which women and men are incorporated into the military intensify the perceived differences between them and marginalize women. Second, the differential treatment of men and women in the military and women's marginalization produce differential opportunities for mobility, both within the military and in civilian life, that privilege men. Third, the advantages men derive from military service are converted into advantages in civilian life. Military elites slip into roles in civilian elites in which they contribute to the reproduction of gender inequality and to the perpetuation of gendered processes within the military.

Patterns of Resistance

Despite its importance as a site for gender reproduction, until the end of the 1970s the military was not a focus for interest-group formation and mobilization in sexual politics. The brilliant military victory of 1967 was an affirmation of the hegemony of the military and raised it to almost sacred proportions. After the 1973 war, when the Israeli military was caught not adequately prepared, the symbolic wall that had protected it from public criticism weakened (Horowitz and Lissak 1989), creating a social climate that was more receptive to a critical reconsideration of women's status within it. The emergence of a feminist movement in the 1970s created the consciousness that there was a gender problem, and the work of the Prime Minister's Commission on the Status of Women (1978) provided both a mechanism for gathering systematic data and a forum for discussing the issue (Azmon and Izraeli 1993).

During the 1980s, public discourse on women in the military was framed primarily in terms of increasing the range of job categories open to women, especially jobs that were interesting and prestigious, not in terms of promoting equality. There was widespread agreement with the military's interpretation that its mandate of promoting national security exempted it from the constraints of the laws requiring gender equality.[15] This interpretation was, until recently, the taken-for-granted ideology. The idea that the categorical treatment and exclusion of women are necessary consequences of the IDF's commitment to defend the country and, therefore, should not be subject to the same norms of equality governing other sectors of society was expressed by MK Yael Dayan in a Knesset committee hearing on the status of women in the military (Protocol 1994b: 3) as common wisdom:

> Zahal [the IDF] is not obligated to the State of Israel and to the women of Israel to accept equality. I think the issue of achieving equality should not concentrate on Zahal. . . . I don't see this [equality in the military] as equivalent to [equality in] the labor force [and] personal [religious] law and all the other areas [where there is currently inequality]. Zahal is responsible for the security of Israel and for combat.

The Israeli Supreme Court thought otherwise. Its 1995 decision in the case of Alice Miller v. the Minister of Defense challenged the gender regime of the military. The Israel Women's Network, a feminist lobby, in cooperation with the Association for Civil Rights in Israel, petitioned the Supreme Court in the name of Miller following the military's refusal to allow her—a 23-year-old new immigrant to Israel from South Africa who had a degree in aeronautical engineering and a pilot's license—to take the entry examinations that determine qualification for the most prestigious pilot training course.

This case was the first time the Supreme Court intervened in a matter of gender discrimination in the military. By instructing the military to invite Miller to be tested for admission to pilot training and, if successful, to admit her to the course, the Court redefined the grounds for acceptable gender distinctions. It rejected the military's claim that its differential treatment of men and women was merely a ramification of gender distinctions embedded in the law and that, given these a priori legal distinctions, the principle of gender-equal treatment did not apply to the military. The Court also rejected the military's claim that its differential treatment of men and women rested on relevant differences between them and, therefore, constituted a permissible and not invidious distinction. Finally, it rejected the military's argument that the high financial cost of adjusting the conditions of pilot training to women's needs, as well as the difficulties involved in personnel planning caused by women's reproductive and mothering roles, were legitimate reasons for unequal treatment. In the words of Supreme Court Justice Eliyahu Matza:

> Declarations of equality are not enough; because the real test of equality is its realization in practice as a social norm that determines outcomes. This normative obligation applies to the IDF as well. The tremendous influence of the ways of the military on the way we live our lives is well known. The IDF cannot stand outside the process of entrenching the consciousness of the importance of basic laws. It too must contribute its share.

Supreme Court Justice Dalia Dorner acknowledged differences in the law make gender a relevant basis for differentiation, but then placed the obligation for correcting or neutralizing the effects of relevant differences on the military. The differences in the service of men and women as defined in the law, she argued, is a factor that the IDF must take into consideration in its planning but cannot be a cause for permitting discriminatory practice in relation to woman soldiers. In its decision, the Court stated that the military must take into account the different life experiences of women, which include pregnancy and childbirth, and accommodate its policies regarding requirements for the participation of pilots accordingly (Ziv 1999).

The Miller case was also an opportunity to challenge the myth of the higher cost of women pilots. When the air force refused to grant the lawyers for the defense detailed data on the dropout rate for pilots, on grounds of confidentiality, the court registrar, a woman, insisted. Upon examination of the data, the lawyers discovered that the dropout rate was far greater than the Court was led to believe from the air force testimony, thus reducing the anticipated differential cost of training men and women, a fact that influenced the judges' decision in favor of opening the course to women.[16]

The military found Miller not qualified to take the course. Continued military resistance to women's full integration may be gleaned by the fact that after the completion of five cycles of the pilot training course, only one woman had graduated and not as a combat pilot. The military argued that it would not change its policy regarding women in combat roles until instructed to do so by the Knesset. In December 1998, the Defense and Foreign Affairs Committee of the Knesset approved a law proposed by MK Naomi Chazan thatdecared all military jobs open to women. The proposed legislation passed in the fifteenth Knesset, led to the expansion in the number of jobs filled by women, including some combat jobs.

Some Israelis favor eliminating compulsory military service for women. Radical feminists argue that as an instrument of violence, the military is inherently a masculine organization that is oppressive of women and violates women's value of life. Some Liberal feminists, men among them, contend that releasing women from compulsory service would give them a head start relative to men in higher education and thus would compensate them somewhat for the cost that motherhood exacts from their careers. The majority of Israelis, however, oppose such differential treatment, both on principle and in recognition of the symbolic importance of military service for civilian life.

It is unlikely that there will be a significant number of women in the centers of military policy making in the near future, not withstanding the recent declaration of the Chief of Staff that "perhaps one day we will also see a brigadier in the general staff" (Harel 1999b). A greater measure of equality could be achieved, however, if universal compulsory military service were to be replaced by voluntary service and greater professionalization. The culture of professionalism is less intensely gendered than the macho culture of military heroism. Furthermore, the need for employers to compete in an open labor market for educated personnel has historically worked in women's favor (Reskin and Roos 1990). The military would then most likely become a less powerful force in shaping the ideologies and practices of everyday life, and gender inequality within the military will have a less significant impact on gender relations in society. Voluntary service is more likely to be introduced if and when Israel achieves a more stable peace agreement with its neighbors, something to be desired on all accounts.

Notes

*   The author expresses her sincerest gratitude to her colleagues, who generously read this paper at various stages of its development, offered helpful comments, and spared her from embarrassing errors: Delila Amir, Rivka Bar-Yosef, Orly Benjamin, Deborah Bernstein, Naomi Chazan, Hanna Herzog, Neri Horowitz, Baruch Kimmerling, Molly Levine, Judith Lorber, and Yisraela Oron. She thanks the IDF for providing the data reported in Tables 6.1 and 6.2. An earlier version of this paper was published in Israel Social Science Research 12:129-66, 1997.
1.   For a history of women in the military, see Bloom (1982, 1991). On compulsory service for women, see Bar-Yosef and Paden-Eisenstark (1977), Berkovitch (1997), Bloom and Bar-Yosef (1983), Eshkol et al. (1987), and Gal (1986). For a critical appraisal of women's military service, see Hazleton (1977), Rein (1979), Robbins (1994), Sharoni (1992), and Yuval-Davis (1976, 1985). For a critical look at the impact of military experience on female identity, see Sasson-Levy (1997).

2. The figures do not include Arabs who are legally eligible but, except for Druze men, are exempted from military service. Twenty-six percent of the Jewish women received exemptions on the basis of a declaration of religious belief and conscience; 8 percent of the Jewish men were exempted because of devotion to religious study. The exemptions on the basis of religion date back to the establishment of the IDF. For women, they are automatic upon declaration; for men, they require proof of enrollment in a religious school for higher learning.

3. In principle, men and women get equal pay for equal work performed in equal circumstances. Policies concerning the allocation of a variety of fringe benefits, however, have gender effects. For example, professional soldiers serving in combat units receive a bonus that they retain even when they move to a noncombat unit; fewer women than men serve in such units. Similarly, certain jobs that are filled primarily by men receive a "danger" increment. In addition, the military occasionally sponsors housing projects, and whereas access to these low-cost housing units is by raffle, categories, such as combat soldiers, engineers, and wounded in service, are given priority.

4. Men were eligible for reserve duty service until the age of 54 and women until the age of 38 (Article 29 of the law). For physicians and dentists, there was no gender difference in the maximum eligibility age for both recruitment and reserve service.

5. The women, in their early twenties, publicly tore up their reserve identity cards, and one father took the issue to the Supreme Court. It was resolved, however, by the change in military policy.

6. The increase was the result of two demographic factors. The first was the growth in the number of women of reproduction age from 1967 to 1976; these women were offspring of immigrants with large families who immigrated during the 1950s. The number of women of reproduction age peaked again in 1985, so that an especially large cohort of conscription age will be available to the military again in the year 2003 (Sabatello 1992). The second was the large immigration from the former Soviet Union, which reached its height in 1990-91 and had increased the population by more than 10 percent within a decade.

7. The legal distinctions did not present insurmountable barriers to women's service. Notwithstanding these categorical distinctions, the law (Article 17) permitted anyone (with the approval of the Minister of Defense) who was not obligated to serve to volunteer either for compulsory service, even for an extended time beyond that specified in the law, or for reserve service. The commitment to serve for an additional period is binding even if the women marry. An enabling clause in the law, on the basis of which the Minister of Defense introduced regulations in 1951 defining combat and other jobs as closed to women, was omitted in 1987 for reasons unrelated to women in the military. This omission, however, did not affect military policy, which continued to define combat positions as closed to women.

8. Of the several thousand women officers in the military, only some twenty-five serve in CHEN.

9. Chair MK N. Chazan: "To what extent are you currently involved in formulating the [personnel] policy?" CO Yisraela Oron: "I am not involved." Chair M.K. Chazan: "Do they consult with you?" CO Oron: "From time to time on specific subjects. They are not obligated to do so; there is no mechanism that requires that I be consulted" (Protocol 1996).

10. For example, in a survey of political and social attitudes among a representative sample of 15-18-year-old Jewish youths (Van Leer Institute 1987: 18), the respondents were asked to indicate which groups of a list of fourteen were entitled to greater esteem than others. Combat soldier was in first place, selected by 49 percent; followed by IDF officers, selected by 36 percent; and then by noncombat soldier, selected by 6 percent. The remaining eleven groups were nonmilitary. So noncombat men also lose out and, like women, are precluded from accumulating valued symbolic capital and the entitlements such capital entails.

11. The recent COs of CHEN have attempted to alter the terms of the discourse from that of "freeing men for combat" to that of "realizing women's potential," but with limited success.

12. Personal communication from Brigadier General Yisraela Oron, CO of the Women's Corps, August 1996.

13. To say that men are advantaged is not to ignore the problem senior officers face in finding jobs of equivalent status to those they had in the military—especially with the increase in recent years of retiring generals, the limited number of positions at the top of the political and public sector elites, and greater competition from younger university MBAs and their equivalents (see Rami Rosen 1994, 1996). A special unit was established to assist retiring officers to find employment, and increasingly they are enabled to complete academic degrees prior to retirement.

14. In 1995, a Knesset amendment made it illegal to use a person's military profile score as a criterion for hiring in the public service. This amendment was passed to prevent discrimination against individuals who are assigned low scores on the grounds of poor health or other reasons of no relevance to their potential job performance. The need for such a regulation, however, reflects the importance that employers assign to a person's military experience and career.

15. The Equal Opportunity in Work Law of 1988 does not cover the relationship between a soldier and the military, since the law does not recognize it as employee-employer relations (see Ben-Israel 1989: 390).

16. Personal communication from Neta Ziv, lawyer for the defense for the Association of Civil Rights in Israel, May 1999.

References

Acker, Joan 1990. "Hierarchies, Jobs and Bodies: A Theory of Gendered Organizations." *Gender & Society*, 4: 139-58.

Aloni, Roni 1992, February. "The Good Years of the Entrepreneurs." *Ha'ir*, 24: 16-8. (Hebrew).

Avnieli, Anat 1988, September. "One More than in the General Staff." *Ha'ir*, 16:15. (Hebrew).

Azmon, Yael and Dafna N. Izraeli (eds.) 1993. *Women in Israel*. New Brunswick, NJ: Transaction Press.

Bar-Yosef, Rivka and Dorit Paden-Eisenstark 1977. "Role Systems Under Stress: Sex Roles in War." *Social Problems*, 25: 135-45.

Ben-Eliezer, Uri 1995. "A Nation-in-Arms: State, Nation and Militarism in Israel's First Years." *Comparative Studies in Society and History*, 37: 264-85.

Ben-Israel, Ruth 1989. *Labor Law in Israel*. Tel Aviv: Open University. (Hebrew).

Berkovitch, Nitza 1997. "Motherhood as a National Mission: The Construction of

Womanhood in the Legal Discourse in Israel." *Women's Studies International Forum*, 20, 5-6: 605-19.

Binyamin, Shlomit 1996, October. "Redesigning Feminism: Interview with the C.O. of the Women's Corps." *Status: The Monthly Magazine for Managerial Thought*, 64: 64-9. (Hebrew).

Bloom, Anne R. 1991. "Women in the Defense Forces." In *Calling the Equality Bluff*, edited by Barbara Swirski and Marilyn Safir, 128-38. New York: Pergamon Press.

Bloom, Anne R. 1982. "Israel: The Longest War." In *Female Soldiers—Combatants or Non-combatants? Theoretical and Contemporary Perspectives*, edited by Nancy Goldmann, 137-62. Westport, CT: Greenwood Press.

Bloom, Anne R. and Rivka Bar-Yosef 1983. "Israeli Women and Military Experience: A Socialization Experience." In *Women's Worlds*, edited by M. P. Safir, M. Mednick, D. N. Izraeli, and J. Bernard. New York: Praeger.

Bogosh, Bryna and Rochelle Don-Yichiya 1996. *The Gender of Justice: Discrimination against Women in Israeli Courts*. Unpublished research report, Jerusalem Institute for Israeli Studies. (Hebrew).

Bourdieu, Pierre 1990. *The Logic of Practice*. Stanford, CA.: Stanford University Press.

Collins, Liat 1995, March 6. "Woman Fighting for Short People's Right to Enlist." *Jerusalem Post*.

Commission on the Status of Women 1978. *Discussion and Findings*. Jerusalem: Prime Minister's Office. (Hebrew).

Connell, R.W. 1990. "The State Gender and Sexual Politics: Theory and Appraisal." *Theory and Society*, 19: 506-44.

Elshtain, Jean Bethke 1995. *Women and War*. Chicago: University of Chicago Press.

Enloe, Cynthia 1990, September 25. "'Women and Children' Making Sense of the Persian Gulf Crisis." *Village Voice*. (cited in Yuval-Davis 1997).

Enloe, Cynthia 1983. *Does Khaki Become You? The Militarization of Women's Lives*. Boston, MA: South End Press.

Eshkol, Eva, Amia Lieblich, Rivka Bar-Yosef, and Hadas Wiseman 1987. "Some Basic Correlates of Adjustment of Israel Women Soldiers to Their Military Roles." *Israel Social Science Research*, 5: 17-28.

Etzioni-Halevy, Eva 1996. "Civil-Military Relations and Democracy: The Case of the Military-Political Elites' Connection in Israel." *Armed Forces and Society*, 22 :401-17.

Foucault, Michel 1980. *Power/Knowledge: Selected Interviews and Other Writings, 1972-1977*. Edited by Colin Gordon. New York: Pantheon Books.

Galilee, Lili 1995, June 9. "They Are Interpreting My Life for Me." *Ha'aretz*. (Hebrew).

Gal, Reuven 1986. *The Israeli Female Soldier: Myth and Reality*. Zichron Yaakov: Israel Institute for Military Studies.

Grant, Judith and Peta Tancred. 1992. "A Feminist Perspective on State Bureaucracy." In *Gendering Organizational Analysis*, edited by A. J. Mills and P. Tancred, 112-28. Newbury Park, CA: Sage.

Harel, Amos 1999a, April 9. "For the First Time Male and Female Conscripts Will Undergo Officers' Course Together." *Ha'aretz*. (Hebrew).

Harel, Amos 1999b, September 19. "Not in One Year and Not in Two." *Ha'aretz*. (Hebrew).

Harel, Amos 1999c, April 10. "Number of Soldiers Requesting Combat Units Stable at 59%." *Ha'aretz*. (Hebrew).

Hareven, Gail 1995, January 29. *Maariv*. (Hebrew).

Hazleton, Lesley 1977. *Israeli Women: The Reality Behind the Myth*. New York: Simon and Schuster.

Herzog, Hanna 1999. *Gendering Politics: Women in Israel*. Ann Arbor: University of Michigan Press.

Horowitz, Dan and Moshe Lissak 1989. *Trouble in Utopia*. Albany: State University of New York Press.

Kanter, Rosabeth Moss 1977. *Men and Women of the Corporation*. New York: Basic Books.

Kimmerling, Baruch 1993. "Militarism in Israeli Society." *Theory and Criticism: An Israeli Forum*, 4: 123-40. (Hebrew).

Liberman, Ami 1995. "The Utilization of Women Soldiers' Service in the I.D.F. 1983-1993." In *Women and I.D.F. Service: Reality, Wish and Vision*, 13-17. Proceedings of a seminar held at Tel Aviv University, February 21, 1995. The Israel Women's Network. (Hebrew).

Lorber, Judith 1994. *Paradoxes of Gender*. New Haven, CT: Yale University Press.

Mazali, Rela 1993. "As a Traveler from Afar: A Look at Compulsory Military Service in Israel." *Challenge*, 4, 4: 36-8. (Hebrew).

Protocol of the Knesset Standing Committee on the Status of Women 1996, February 27. No. 196. Jerusalem: The Knesset.

Protocol of the Knesset Standing Committee on the Status of Women 1994a, February 8. No. 71. Jerusalem: The Knesset.

Protocol of the Knesset Standing Committee on the Status of Women 1994b, July 19. No. 98. Jerusalem: The Knesset.

Rein, Natalie 1979. *Daughters of Rachel: Women in Israel*. London: Penguin.

Reskin, Barbara F. and Patricia A. Roos 1990. *Job Queues, Gender Queues: Explaining Women's Inroads into Male Occupations*. Philadelphia, PA: Temple University Press.

Robbins, Joyce 1994. "Not Soldiers in the Ordinary Sense: The Israeli Woman Soldier and the Politics of the Non-military Roles in the 1950s." Unpublished master's degree thesis, Tel Aviv University.

Rosen, Rami 1996, September 20. "Try to Accommodate Him." *Ha'aretz*. (Hebrew).

Rosen, Rami 1994, August 5. "A Sudden Feeling of Emptiness." *Ha'aretz*. (Hebrew).

Rosenblum, Kineret 1995, November 24. "How Do You Want Your Secretary, Officer?" *Zman Tel Aviv*. (Hebrew).

Sabatello, Eitan 1992. "The Real Reason for Fertility." (Letter to the editor) *Ha'aretz*. (Hebrew).

Sadeh, Dani 1992, September 13. "Women Under Surveillance." *Yediot Aharonot*. (Hebrew).

Sasson-Levy, Orna 1997. "They Walk Upright and Proud: The Power and the Price of Military Service for Women Soldiers in Men's Jobs." *NOGA—A Feminist Journal*, 32: 21-30. (Hebrew).

Sharoni, Simona 1992. "Every Woman Is an Occupied Territory: The Politics of Militarism and Sexism and the Israeli-Palestinian Conflict." *Journal of Gender Studies*, 1: 447-62.

Schochat, Orit 1995, July 31. "The Last Masculine Occupation." *Ha'aretz*. (Hebrew).

Siegel, Judy 1994, October 21. "Disabled Woman Pressing to Serve in I.D.F." *Jerusalem Post*.

Van Leer Institute 1987, September. "Political and Social Attitudes Among Youth." Research Report. (Hebrew).

West, Candace and Don Zimmerman. 1987. "Doing Gender." *Gender & Society*, 1: 125-51.

Yuval-Davis, Nira 1997. *Gender and Nation*. London: Sage.

Yuval-Davis, Nira 1985. "Front and Rear: The Sexual Division of Labor in the Israeli Army." *Feminist Studies*, 11: 649-75.

Yuval-Davis, Nira 1976. "Israeli Women and Men: Divisions Behind the Unity." Change Report No. 6.

Ziv, Neta 1999. In press. "Civil Rights and Disability Law in Israel and the United States— A Comparative Perspective." *Israel Yearbook of Human Rights*, 28:171-202.

7

Tests of Soldierhood, Trials of Manhood: Military Service and Male Ideals in Israel*

Eyal Ben-Ari with the assistance of
Galeet Dardashti

"I have changed completely. The military service uncovered my real nature. Before I entered the service, I lacked confidence, I was full of inferiority complexes. In the army I chose the hardest course, and in that way I made a man of myself. I made it by applying my will, and it gave me tremendous satisfaction. I grew up and built my esteem. Now I know that if I make enough of an effort I'll be able to achieve any goal I may set myself. That the army is a real test of adulthood. If you complete your service honorably, you're a man"—unnamed soldier, Lieblich 1989: 156.

Introduction

In this paper I examine the relationship between "manhood" and "military service" as it emerges from the accounts found in Amia Lieblich's (1989) significant book, *The Transition to Adulthood During Military Service: The Israeli Case.* By manhood, I refer to the main ideals of approved ways of being a Jewish male in Israeli society. These ideals are not a set of psychological traits which specific individuals may or may not possess, but rather a group of culturally available, recognized and legitimate themes which are more or less identified with certain aspects of being a man in this society (see Gilmore 1990: 1). Military service embraces the variety of role components related to soldiering, and the set of practices and arrangements that structure the experience of individuals within the Israel Defense Forces (IDF).

It is probably true that the warrior is still a key symbol of masculinity in Western societies (Arkin and Dobrofsky 1978; Cameron 1994; Morgan 1994). In Israel, the significance of this symbol has been reinforced by the wider historical context of protracted conflict. As many scholars have noted (Kimmerling 1993; Horowitz and Lissak 1989), this context has granted the military a central place in Israeli society and culture. An important manifestation of this centrality is the still rather widespread assumption that military service for Jewish-Israeli men is a natural, taken-for-granted matter. No less important, as Lieblich (1989) observes, is the prevailing assumption that national service is also a period of transition from youthood to adulthood. As a psychologist, Lieblich was interested in the connection between personality and social structure and the patterns of development into adulthood that military service represents: with the ways in which military service and war have important effects on personality development, values and courses of growth. Using in-depth interviews, she analyzed two themes: the acquisition of coping skills and the expansion of personal boundaries which are defined by the men as "adulthood" (Lieblich and Perlow 1988; 40). The following account is that of an unnamed soldier:

> Perhaps every person has to go through a difficult time between childhood and adulthood, and that's how you become an adult. In my case this was military service. It was terribly hard, but it made me strong (Lieblich 1989: 156).

Interestingly, her views reflect a major cultural assumption of (Jewish) Israelis, for Lieblich chose to focus her analysis exclusively on men. While allowing that women also serve in the IDF, she (1989: ix) nevertheless observed:

> Military service for Israeli men is longer and apparently more demanding. Only men participate directly in combat and continue to be 'soldiers' via compulsory reserve duty throughout the major part of their adult lives and the all too frequent wars. Therefore, the military experience touches men more profoundly, and affects their lives and identity to a greater extent.

While Lieblich was interested in the personal, psychological aspects of military experience, I suggest a shift of focus from the individual to the cultural level. More specifically, I argue that we may benefit by asking about the kind of cultural expectations about manhood that lie at base of the experiences Lieblich investigated. Indeed, when I re-read her material—eight in-depth accounts and a host of shorter passages garnered from soldiers she interviewed—I found numerous hints and intimations related to a set of ideals associated with manhood. Take the following words of a medical orderly talking of the war in Lebanon:

> That's when I saw severely injured men for the first time. I reacted... outstandingly. I was cool. Professional. A Leader in the situation. I acted as if I were a different person there, with a fantastic ability not to panic, to cope under stress. I think that never before or afterwards could I be that man (Lieblich 1989: 128).

Here we find at its most explicit, the gendering of the soldier's role in terms of manhood. Indeed, in an aside situated within a section on self-perceptions of change, Lieblich (1989: 155) states that "maturity" and "manhood" were often used interchangeably by her informants. My contention is that the "gendering" of the soldierly role pertains to military service in general.

A host of scholars have observed that military service forms a rite of passage in a variety of complex societies around the world (Adams 1993; Bourne 1967; Cockerham 1973; Dornbusch 1955; Vidich and Stein 1960), but it has been Aronoff (1989: 132; see also Izraeli 1997; Sion 1997) who has most explicitly applied this insight to the Israeli case:

> The primary rite of passage that initiates one into full membership in the Zionist civil religion is service in Zahal (the Israel Defense Forces). It is the single most important test, particularly for males, for individual and group acceptance in the mainstream of Israeli society.

But Aronoff's stress on understanding military service as "a rite of passage"—a period whose aim is to convey individuals from one social position or status to another—is still too general. It is too general because such a postulate begs questions about the internal structure, dynamics and complexity of this "rite" and the concrete manner by which it is linked to various ideals of manhood. It is in this light that the question that will guide my analysis should be seen: how are the concrete arrangements and practices of military service related to the accomplishment of ideals of manhood in Jewish-Israeli society.[1]

Myths of Warriorhood Personalized: Individual Accounts

In contemporary Israel, notions of warriorhood are still explicated and legitimated through a profusion of stories, sites, and rituals devoted to heroism and sacrifice. Because of the frequent wars that Israel has participated in, the state has had to mobilize young men to fight in its armed forces. As a consequence, since the establishment of the Israeli national entity, it has been necessary to construct what Lomsky-Feder (1994), following Mosse (1990), calls the "myth of participation in war." As many scholars have shown, Israeli society has consistently glorified military service—and especially participation in battles—and developed a set of myths about courage and personal sacrifice (Gal 1986; Sivan 1991), cultural sites commemorating fallen soldiers (Kalderon 1988; Meiron 1992; Bruner and Gorfain 1983; Schwarz et al. 1986; Ben-Ze'ev and Ben-Ari 1996), and rituals marking the importance of war for the very survival of the nation-state (for example, Liebman and Don Yehiya 1983; Handelman and Katz 1995; Handelman and Shamgar-Handelman 1996). Through these various public narratives and rites, ideals of manhood are constantly linked to notions of legitimate national goals such as security and military service.

It is no surprise, then, that the accounts found in Lieblich's book both resonate with wider cultural myths, and personalize them in individual experiences. Take the specific ideal image of a man: the fighter, the combat soldier (*lochem*). David, who joined the paratroopers, notes:

> I knew that I wanted to be a combat soldier, and that I would do well in the army—for myself, not because others expected me to. It is a feeling you grow up with. Maybe it had to do with the fact that my uncle had been killed in the War of Independence. Somehow it all seemed related. (Lieblich 1989: 3)

David's beliefs reverberate both with his uncle's death in war, and with wider notions about the importance of sacrifice in battle. Ido (Lieblich 1989: 71-6) talks of his experience in a comparable manner:

> I started to think about my service when I was about 17. I knew I wanted to serve in an elite infantry unit. I had many reasons for this... I like to walk and hike, and thought that would be a good place for me. I wanted to serve with high quality people and to have an interesting service. Also, as much as it's hard to admit, to serve in an elite unit is prestigious. And, last but not least, there was also some Zionism involved—I wanted to contribute...

Along similar lines, Danny, who eventually ends up in the maintenance corps, tells us:

> I was interested only in combat units and argued about it with my parents, who preferred that I get a less risky job. I told my parents that combat service was the most important and that every young man had to contribute his utmost to the security effort. I didn't mind interrupting my sports career for the sake of full military service. But this was the prevailing atmosphere among all my friends—we wanted to give the maximum. (Lieblich 1989: 61)

In this account, as in the previous ones, notions such as "prestigious," "Zionism," "most important," "contribute his utmost" and "give the maximum" are personalized: they are linked to the specific circumstances of people's lives and ambitions. In a similar vein, Danny explains how the "girls"—concrete individuals—in his high school participated in creating his desire to serve as a combat soldier:

> Many of the girls in my class had boyfriends in the army. Those girls who had boyfriends who used to arrive for leave once in two weeks all dirty and exhausted, had "men" as [boy]friends, not "babies" like us. Only later did I understand that girls really preferred boyfriends who came home every evening. But in high school, the girls certainly pretended that they preferred "fighters." (Lieblich 1989: 61)

In this short passage, Danny evokes three contrasts that delineate the confines of manhood: between "men" and "babies"; between "men" and "women"; and between "men-as-fighters" and "men-as-non-combatants."

Yet the designation of an ideal man as a fighter or warrior that deals with the nation-state's very survival, is only the first tier of the association between soldiering and masculinity. A second tier involves images of how manhood can be accomplished through military service. In other words, I suggest distinguishing between images of what it means to be a man and images of how to become a man.

A Rite of Passage: Trials and Tests

Almost all of the narratives found in Lieblich's volume can be interpreted in a Turnerian fashion (Turner 1974: chap 3) as descriptions of how military service—and especially its beginning stages—serves as a rite of passage: a ritual aimed at conveying individuals from the status of civilian youths to one of fully-fledged soldiers and adults. According to this explanation, soldiers undergo a tripartite process within the rite: they are disconnected from civilian life (through their physical separation from the rest of society), "something" is done to them while in the military, and they then "reemerge" as soldier-men (often through formal occasions marking their new status). But what is of importance are less the ceremonial aspects of this rite, as its internal structure as a series of tests, a series of concrete trials that boys must undergo in order to become men. Gilmore (1990: 11, 223-4) has developed Turner's notions in this direction. His idea is that manhood involves the idea of a "threshold" that boys must pass through testing, and that to pass this test boys must steel themselves, and must be prepared by various sorts of tempering and toughening.

This idea fits the dynamics of military service in many parts of the industrialized world. For example, recruits to both the French Foreign Legion and Marines (who claim to be the most "manly" of the American services) undergo a regime characterized by iron discipline, extreme conformism, exhausting physical training, and mockery and humiliation in order to become "complete" soldiers (Badinter 1992: 76). In the American army the "long gruelling marches are particularly suited to building a man under high stress conditions, where marching is *the means* not the end of this objective" (Arkin and Dubrofsky 1978: 160, emphasis added). Indeed, the metaphor Ido uses to interpret stretcher hikes is revealing in this respect: "You simply see who breaks down and who doesn't... Every one of us broke down at this stage, each in his own way. Some cried a lot, some started to act funny." This image links the idea of military service-as-test to the logic of rites of passage as "breaking" initiates and then "rebuilding" them anew. As Turner (1982: 25) suggests: "initiations humble people before permanently elevating them."

Yet the manner by which men are "broken" and then "re-built" is not a simple matter. What seems to be involved is a subtle interplay of public performance and individual "self-discovery" about abilities to withstand physical and mental hardships. Morgan (1994: 167) suggests that a fruitful way of understanding the relationship between military life and gender is through an examination of the "body practices"—the set of concrete social arrangements—by which the male body is constructed in the armed forces. Let me take up his lead in a number of directions.

In warrior societies around the world, rites of passage usually involve the dramatic enactment of trials on a public stage. These trials give youngsters an opportunity to display (before their community) their courage when faced with pain and mortifications. In the army, soldiers must display qualities of fortitude, tenacity and endurance through practices publicly enacted before their superiors, equals and (later) underlings. It is in this light that the harassment, punishments and humiliation that men suffer during the first stages of military service should be seen. Danny, for example, talks about the prepatory course he attended in aviators' school:

> From the first moment they started to treat us very strictly. They ordered us to run, using various forms of harassment and humiliation... They used very rough language... When a soldier answered a question, the commander would say something like: "Don't throw up in my ear." (Lieblich 1989: 62)

Often, the indignities suffered by young trainees are accompanied by harsh conditions such as extreme cold, lack of sleep or an unending sense of uncertainty. Ido notes:

> For the first week we had no sleep from Sunday to Thursday... Next week, we were given 90 minutes of sleep per night... Frankly, I hardly remember anything from that time—I was in a haze. Sleep deprivation is the most difficult thing... You see things, but you're not there. (Lieblich 1989: 73)

Ido's stress on "being" and "not being" there is an uncanny reminder of a central characteristic of the "liminal" period explicated by Turner (1982: 26) which involves ambiguity and paradox since the initiates are "at once dying from or dead to their former status and life, and being born and growing into new ones." David observes, "We built little tents in the field, and were soaked by the rain at night. We got used to being wet all the time, but we had to keep our rifles dry, and this was a lot of work. People got ill quite often. It was rough" (Lieblich 1989: 5). Clearly, the aim of such practices is to test the youngsters' perseverance and fortitude in the face of various mortifications and harassment. But Eitan's words evoke another dimension to these trials:

> But you have no friends in basic training. You do everything in the unit together, you are punished together. The individual does not stand out and you

> don't form any special relationships. During basic training you may find, at
> most, relationships based on reciprocal help, but it isn't love... We had good
> squad leaders but a very cruel sergeant and sergeant-major. The discipline and
> punishment were out of proportion to the offense. I remember one time I was
> made to cry, but I don't recall the details of the episode. (Lieblich, 1989: 56)

The exposure of individuals' weaknesses in front of superiors and
peers fits with the more general process by which soldiers are "broken" or "disintegrated" and then "rebuilt" or "reintegrated." Moreover, such practices can be seen as a "levelling" process in which
signs of the initiates preliminary status and ties are destroyed and signs
of their liminal state and connections applied so that the initiates become uniform and (socially) anonymous (Turner 1982: 2).

But there is more to the link between body practices and military
service. Connel (1987: 85) suggests that the "social definition of
men as holders of power is translated not only into mental body
images and fantasies, but into muscle tensions, postures, the feel
and texture of the body." Notice the way David talks about walking:

> You admire the performance of your squad leader—he walks, while you run,
> and you still cannot catch up with him. Gradually you discover that he knows
> *how* to walk—and you learn this too. People develop their marching style. After 14
> months of training together, I was able to identify my buddies from a distance by
> their way of walking, you know. (Lieblich 1989: 4, emphasis in original)

As a body practice, walking, a learned and internalized habit carried out with others in the context of trials of endurance, is related
to the process of becoming a man. By following squad (or other)
leaders the soldier learns to carry in his body the appropriate movements and motions related to being a man. Learning to walk is thus
also part of the implicit ways in which the initiates are re-built by
reproducing the walking styles of "real" men/soldiers.

But David's insights reveal yet another dimension to body practices. He remarks that walking also involves self-discovery: "Gradually you learn to love those night marches, and you get to know
yourself really well. Your odor, the sound of your footsteps" (Lieblich
1989: 4). The self-awareness described in this depiction is strictly
physical. But the internal aspects of this body knowledge go beyond
learning to know oneself. In many of the accounts soldiers talked
about how they had gained confidence, discovered personal power,
and realized their ability to withstand hardships through the various
trials and tests they underwent.

Indeed, it is only once David learned that his body could endure tremendous feats and pressures, that he "learned to love those night marches." This theme comes up most explicitly in Ido's account when he says "What you get out of this training is the sense that there is no limit to your abilities. Or, that the limit is where you stop trying, and this shouldn't happen until you actually fall and faint" (Lieblich 1989: 75). Similarly, Gil recollects how during officers' course, marches were tests of performance through which he learnt something about himself: "It was hard. For the first time I felt like a real soldier, not a 'chocolate' one... The hikes were really difficult for me, but I struggled and completed each one" (Lieblich 1989: 88). David links his account of self-discovery through mastery of his body to the common metaphor of new life stages as new chapters in a book:

> I remember getting back from a successful navigation of 60 km on a rainy day. The feeling that I had mastered it, that I could lead myself alone, even continuing for another 10 km if necessary was exhilarating. After the week of navigations I felt that I had overcome my own limits, that I had opened up a new page. (Lieblich 1989: 9)

So strong is the link between physical achievement and the logic of military service as a rite of passage that in many cases bodies are misused, mishandled or abused. For example, Ido describes how he had severe shin splits but had to undergo the long trek marking the end of his training another time.

Yet the situation is more complex because military service is actually made up of a series of tests. For young recruits the initial test is that of basic training, but testing usually goes on each time there is a transition in the recruit's service: for example, going to a specialist course, taking up new tasks and missions, entering a new unit or undertaking a new role. In all of these transitions the interplay (albeit often in attenuated form) of public scrutiny (by superiors, equals and underlings) and self-discovery continues. In terms of ideals of manhood this situation implies—as Gilmore (1990) insightfully observes—a stress on continuous performance. Masculinity is never simply achieved once and for all, it must constantly be accomplished.

Take the example of officers. Even after they have attained their formal rank, these men continue to be tested in terms of endurance, perseverance and physical hardships. David notes that "I really succeeded as an officer. It wasn't easy, mind you. I never slept more than three hours a night" (Lieblich 1989: 13).

Combat: The Ultimate Test

For all of this, my depiction of military service does not stray far from descriptions of rites of passage around the world. What distinguishes military organizations is the unique kind of environment they are supposed to function in, are trained to perform in: combat (Ben-Ari 1998). At the risk of stating the obvious let me emphasize that the focal environment at the level of soldiers is combat and not war in general. What preoccupies soldiers most of all is the localized, violent encounter of two armed organizations (Boene 1990:29). When I went over the accounts found in Lieblich's volume, I began to appreciate that the portrayal of combat harbors the distinctiveness and the strength of the military as a rite of passage to manhood.

In military organizations around the world combat is seen as the core of masculinity (Arkin and Dubrofsky 1978: 16; Morgan 1994: 167; Vidich and Stein 1960). The prime trial of men—the chief ordeal for achieving manhood—is that of mastering the harsh and stressful conditions of battle. It is no surprise then, that similar notions can be found in the IDF. The excerpt which appears in the introduction well encapsulates these ideas, and Alon states of his behavior during battle in Lebanon: "I acted like a different man. The good side is the clarity, precision and awareness that I was able to act under extremely difficult conditions. During this stress situation, I became a leader, better able to cope with the problems than all the officers around me" (Lieblich 1989: 162).

What is the cultural logic at base of such statements? In combat the problem is one of agency: who will be master? Situation or person, circumstances or man. The reasoning here is a simple one: by mastering the situation of combat—through controlling one's emotions, persisting despite extremely difficult conditions and carrying out one's military role—an individual participates in missions assigned to his unit. Yet symbolically speaking, by overcoming the highly stressful circumstances of combat, an individual is also passing the ultimate test of manhood. It is the paramount trial of manhood because the firefight is seen to require of individuals all of the essential characteristics of young men: endurance, self-control, perseverance and composure.[2] Performing well in battle, then, becomes an indicator, a mark that one has successfully become a man.

Even before battle men must steel themselves to accept their role and the risks they are to take. David discloses this point in regard to the Lebanon War:

> We were told to prepare for battle against the Syrian commando forces at dawn. Each one of us went separately to check his weapons and ammunition, to prepare the gear. I remember I prepared myself psychologically as well. I knew that I would be right at the front line, and that my chances of being hit were high. This was... (trails off). (Lieblich 1989: 11)

Similarly, Gil (then already an officer) describes the tremendous pressure and fear he experienced as his battalion awaited its call to battle:

> I started to worry about my friends. I felt a restlessness that I couldn't let my men see. They shouldn't feel my anxiety. I was a very young officer, remember, and I felt that this was another test of my leadership. Inside however, I was quite scared. (Lieblich 1989: 92)

Keeping one's emotions in check even before battle is thus related— via the soldierly role—to the stress of battle. Gad also discusses the war in Lebanon:

> It was frightening to go on those patrols. I knew the odds. I was afraid too, of a sudden attack... We went on a fixed route at a fixed time every day, we were very easy targets for an attack, like in a video game. To help my men control their panic, I used to order them to fire their personal guns all the time we were patrolling the road. This gave them rifle practice, but mostly it helped disperse their fears. (Lieblich 1989: 47)

All three passages illustrate the essentially precarious nature of achievement at base of combat: emotional control, conquering of fear and performance are not givens but must constantly be struggled for. Gad tells us:

> The whole battle took place in the streets of the city. Suddenly I saw burning tires being rolled down the street where we were advancing. Our ammunition truck was standing in the middle of the road, at risk of a huge explosion. I had to make up my mind right away if I was ready for a really brave act. I didn't think much, I ran into the truck, and drove it into an alley, so the tires wouldn't reach it. (Lieblich 1989: 47)

Along these lines, while the ultimate test of manhood is the firefight, the red badge of courage is not the wound. Rather, the emblem of manhood is participation and performance in a very highly pressured situation by displaying "manly" characteristics. Indeed, Lieblich (1989: 128) observes that many "of the men expressed pride and satisfaction in their performance." Several reminded me that they had fought "in the very front line," "in the leading tank," "in the first APC [Armored Personnel Carrier] to enter Lebanon."

Indeed, the "negative" examples, the "failures" documented in Lieblich's volume, often shed a harsh light on these ideals. During the Lebanon war, Yochanan (an instructor in a youth training camp) does not consider himself a "real" soldier since he is not involved in combat, and in his free time he volunteers for a position as near as possible to real soldiering: "I volunteered to help load the planes with bombs and ammunition. What a job! I experienced how real, useful soldiers were spending their time in the army" (Lieblich 1989: 32). Later when invited to a ceremony to receive decorations for participating in the Lebanon war he notes: "Since I had not taken any active role in the war, I decided I would not use the decoration anyway and did not come to receive it" (Lieblich 1989: 33). Yochanan feels satisfied only when he is in a combat zone a few months after the war: "You see, my dream of becoming a medic at the front had really materialized. I even had one chance of treating Israeli paratroopers immediately after a skirmish" (Lieblich 1989: 37).

Manhood and the Simulation of Battle

But what are we to make of the fact that only a minority of men participate in field units and only a smaller minority actually participate in various kinds of firefights (full-scale battles, more limited engagements or other kinds of combat)? How do these individuals put into effect ideals of soldierly manhood? The answer to these questions involves three points. First, the idea that combat is the means for achieving and proving one's manhood may explain the eagerness and enthusiasm of many men to participate in battle. Many of the soldiers interviewed by Lieblich addressed this issue clearly. Alon: "Look it wasn't heroism or anything like it. All I thought of at the moment was that I wanted to join my buddies, and perhaps that I wanted the experience of combat, just for myself" (Lieblich 1989: 20). Gil states:

After two days we were already impatient... The other Nahal battalions were at the front, but we were kept waiting. The combat soldiers in the outposts, men who had trained for this war, were terribly upset. They wanted to participate. So did I. To be with everybody else. (Lieblich 1989: 92)

In the same manner, the deep disappointments some men suffered also attest to the importance attached to participation in battle. Eitan (Lieblich 1989: 58) mentions that "we were transferred to Lebanon, but we were given some marginal task, and actually didn't fight. I felt that my skills were wasted." And Ido recalls,

[T]he feeling that you have all the time... is frustration. Because you practice a lot, and never get the chance to perform even the smallest part of what you have been training for. As a civilian, and a peace-lover, I obviously understand this. But as a soldier who has put so much effort into training, even the most ardent pacifist wants to apply his skills. It's not a political matter. It is a matter of being professional. (Lieblich 1989: 77)

When Gad is injured in the Lebanon war, he says,

The doctors wanted to send me to a hospital in Haifa, but I objected because I wanted to [re]join my unit as soon as possible... I knew that I saw and did things that would never be repeated. I didn't want to miss the experience. (Lieblich 1989: 47)

Then, alluding the myth of the military hero, Gad continues,

One always hears all these stories about the wounded who escaped from hospitals to the front. So, I said, let me be one of those. I put on my helmet on top of this huge bandage, and returned to my battalion... My wound wasn't serious, and my men knew it, so what would they have said: "The commander took advantage of his slight wound and took off to safety." (Lieblich 1989: 128, 48)

Such statements well underline not only how combat is perceived to be the ultimate test of manhood, but how such trials include a constant public exposure to the appraisal of other men. Being with other men in combat is not only related (as we shall presently see) to membership in cohesive units, but connected to the communal affirmation of one's achievement by others.

Second, because only a minority of men participate in firefights, military service provides a host of other situations that are based on a similarity to combat. In other words, the army affords a range of simulations of combat in which individuals can perform, be tested, and prove themselves as men. Along these lines, the series of tests that I outlined before and that are most pronounced during the first stages of service are all based on, or simulations of, the conditions found in combat. These trials include not only battles or skirmishes (although these are the most serious of ordeals), but not less importantly basic training, forced marches, military schooling (gunners, drivers or snipers routes, for instance), various courses (NCO and officer training, for example), and assorted kinds of maneuvers and exercises. Because, albeit in a limited manner, all of these occurrences include elements of physical and emotional difficulties, anxieties and stresses, necessitate self-control and achievement, and involve public inspection and affirmation of individual performance, they are thought to be good occasions for the examination and verification of the characteristics and abilities of men. Put somewhat crudely, even if the opportunity to participate in combat does not arise, if one keeps one's cool and performs well under the harsh conditions afforded by such situations, then one becomes a man.

Along these lines, many of the soldiers view the ability to control their emotions when in acute physical pain as an essential ingredient for mastering difficult circumstances. Eitan:

> Right at the beginning of the tau [an anti-tank missile] course I had an appendectomy, and I was limited in the physical efforts I could exert. It was a tough period. I was in agony, but I knew that if I didn't improve my fitness, I'd be kicked out of the unit. Once when we had to run carrying a stretcher, I couldn't take it... I wanted to scream with pain, but I shut up. (Lieblich 1989: 57)

Similarly, Eitan talks about how parachuting course had "frightening moments" (Lieblich 1989: 56). In the following account, David (also a parachutist) describes being flung between feelings of vulnerability and masculinity:

> There's a beauty to jumping, but it is terribly frightening. You never adjust to it. On the contrary, I think that every jump is more scary, and gradually you get more afraid and you hate it. But when we all gathered at the meeting point on the ground we behaved like big boys again. (Lieblich 1989: 6)

Later, when David is already a commander, he is accidentally shot in the leg by one of his men during training. He states, "I didn't faint, and I acted as if nothing happened. You should never show weakness in front of your men" (Lieblich 1989: 12). Thus, in times of severe anxiety and duress, a "real" soldier/man demonstrates no trace of desperation. Indeed, to successfully pass such tests is the precondition for moving from a state of "little" boys to that of "big" boys.

The third point derives from the previous two. Given its centrality, service in combat units is directly related to status: status is
dependent on proximity to, or distance from, the epitome of the serviceman—the combat soldier. Thus soldiers are ranged along a status continuum ranging from fighters (say the infantry or armour), to
men in support units close to direct combat (say the artillery), soldiers
in auxiliary positions in the field (various maintenance roles in field
units) and all the way to what in the IDF are known as "jobniks," i.e.,
soldiers who have a desk job during their military service (see Lieblich
1989: 69). From our perspective, however, what is of importance is that
men who are not accepted into combat units, and are therefore not
provided with opportunities to prove themselves as fighters, often
express shame and disillusionment. In one account, Gil, a newly
recruited soldier describes his experience at the induction base when
he and group of soldiers are called in by the selection officer:

> You low profilers, [the selection officer] says, you are good for nothing. There
> are three jobs I can offer you, transportation, maintenance or personnel. I came
> out completely heartbroken. What is it, I'm a second-rate soldier?! What a shock.
> (Lieblich 1989: 84)

Gil then describes his first leave after being placed in the maintenance corps. Because the truth is too humiliating, he finds himself
blatantly lying to his friends when they ask him about his placement
in the army: "I could not admit that I had failed in the army" (Lieblich
1989: 85). Gil views his placement as "failure" for it conflicts with
the masculine image that he wishes to portray to himself and to his
friends. Yet when he is later chosen for officer's training he is expected to perform gruelling physical feats. Gil reminiscences about
the prep course leading to officer training,

> I remember the happiness at the end of a march, when the lights of the base
> could finally be seen in the distance. It was as if you had surpassed yourself.
> The motto was to work. Not to flunk. Failing would be such a shame. (Lieblich
> 1989: 88)

This point marks a major shift in Gil's attitude toward army service. Although not given the opportunity to serve in combat roles
during his service, he was expected to complete several courses which
included tests of physical strength and emotional endurance.

Careers and roles in the Israeli army are not simply gender-neu-
tral elements related to military professionalism. Because they are

all marked by the degree to which they encapsulate masculinity, they are central means for designating "proper" male and female domains (see Kratz 1990). Indeed, a striking example of how non-combat roles are gendered female is found in Yochanan's account. At the beginning of his interview he describes how he had been serving in a military youth camp with children and refers to his job as "satisfying." But he then goes on to depict an episode which served as a turning point for him, when he no longer felt that he was contributing anything of significance to the army. He states:

> This became evident to me one Sunday, when I returned late in the day from my weekend at home, and the youngsters had already arrived and were drilled by a female youth leader. I watched the drill exercises and suddenly it became so ridiculous in my eyes, so utterly unimportant. I decided to look for something else to do. (Lieblich 1989: 31)

Once Yochanan saw how easy it was to be replaced by a woman, continuing in that job seems the ultimate blow to his masculinity. His accomplishments at the youth camp no longer gave him pride, and he asked to be moved to another, combat-related assignment.

Administrators: Another Version of Manhood

In a path-breaking paper about gender and military service, Morgan (1994) suggests that we talk about a plurality of masculinities found in the armed forces. Accordingly, alongside the man-as-fighter version of masculinity, another version of manhood can be found in the accounts furnished by Lieblich, that of the "professional organizer," the administrator. While fulfilling the ideals of this version also includes an interplay of public performance and self-discovery, the criteria for success are related to how men apply various administrative expertise and skills. In this respect, military service involves actualizing manly ideals that are closely akin to those found in "ordinary" civilian workplaces: becoming a man means being able to make it in the workplace as a (blue-collar) craftsman and technician or (as a white-collar) manager (see also Arkin and Dubrofsky 1978: 159). While Lieblich provides only a few examples of this kind of masculinity, Gil, who becomes a maintenance officer, is a paradigmatic example. He talks of the kinds of hardships he faced upon taking up his new role:

> How will I manage such responsibility?... "Don't worry," said the Chief Officer... Two weeks later, he came to inspect my battalion and gave me a perfect

evaluation. "This is one of the best organized battalions I have visited," he said in front of all the officers. It made me happy, of course. (Lieblich 1989: 93)

This is when I started to feel that I really understood the military system. Personal problems, power struggles between officers, as well as the mentality of drivers and cooks—and their wives—it all fell into place for me. Often I thought: I'm just 19 or 20, but I see the world as an adult (Lieblich 1989: 93).

A few months later we had an inspection for which I worked like a mule, together with all my men. For the first time I conducted my own inspection of the men, we cleaned and shined the whole place, and our evaluation was 94 (out of 100)—a terrific achievement... Once after the inspection, I overheard the superior officers talking about me. "Where did you get this energetic young officer?" somebody asked. And the Chief Officer of Maintenance bragged: "This is a kid who grows straight up. He does not look towards his home, he is all here. You'll hear about him in the future, you'll see." Of course I was proud. (Lieblich 1989: 94)

In this version of manhood, the mutuality of public recognition and personal pleasure is related to "a job well done." But in contrast to combat roles, jobs carried out by such men all involve passing (formal and informal) tests related to organizing men and materiel. As Barrett (1996: 28) observes in regard to administrators in the American Navy, such men also gain recognition from competing with other men in similar roles or from competing with themselves:

For men like this, it is impossible to ignore the hegemonic masculine ideal of control, autonomy, and authority. They may not fly a jet or command a ship, but they control people and information systems. The search for a stable masculine identity is embedded in a theme of upward mobility. This preoccupation with hierarchical advancement and competition for career progress is a common theme in middle-class men in organizations. (Barrett 1996: 28)

Indeed, for administrators, ideals of masculinity are embedded in a combination of professional rationality (Barrett 1996: 25), diligence, hard work and responsibility. In another part of her book, Lieblich provides short examples which are illuminating in this regard:

I learned to conduct a logical analysis of a situation, and I discovered that I am a careful, hardworking person. (Lieblich 1989: 157)

And

I learned to think: to take a goal, analyze it down to its components, and plan separately how to achieve each component. I learned to plan my steps, not to act impulsively. Make a plan, and include an alternative in case it doesn't work. All these help me tremendously outside the army. (Lieblich 1989: 158-9)

While Barrett (1996: 3) rightly notes that the link between masculinity, violence, and the military is more complex than the image of man-the-warrior might suggest, we must recognize that the situation is not one of a simple diversity of masculinities. Rather, masculinities are linked to each other, hierarchically. In the military, versions of manhood akin to civilian jobs are subordinate to the warrior ideal. The warrior ideal, in other words, is the hegemonic masculinity (Morgan 1994).

Caring

But there is another element related to manhood which—like the experience of soldiers around the world (Bourne 1967; Vidich and Stein 1960)—has a central place in Lieblich's interviews: caring. While caring has received scant notice in the scholarly literature on masculinity in the military, I argue that it entails two complementary ideals. Both involve a blend of concern, emotional expression and support, but while one corresponds to the brother/friend relation and is egalitarian in nature, the other is based on the fatherly tie and is hierarchical in its character. Several men in Lieblich's study describe a feeling of 'brotherhood' in their accounts. Take Ido's words:

You get to know people and you also develop the famous 'soldiers' camaraderie.' Your life depends on each other. After the whole process the team that remained was so close together that I don't know how to describe it. There is no parallel relationship in any other life circumstances. None of us thought of leaving the Unit of our free will. Not because of our earlier reasons for joining, but simply because of our brotherhood. (Lieblich 1989: 74)

And Eitan, a member of an anti-tank unit, says:

Every team gradually becomes a very cohesive group of people... We became such close friends that it was a pleasure to return to the army on Sunday after our leave at home. Our training and tasks were interesting, but the main attraction was the friendship... and the good time we were having together in the field. (Lieblich 1989: 56)

Later,

When it was proposed that I go to the officers' course, I declined, since I felt so good among my friends. At this stage you become very close to others in the unit. You have more time to sit and chat. You know them better and you love them more. This is the famous love among soldiers, which is as much a mystery as love of women. (Lieblich 1989: 57)

While not made explicit in the actual accounts, it could well be proposed that the mutual exposure of individuals' weaknesses (during the initial stages of military service) contributes to soldiers' learning to accept each other as basically equal. Thus, subjection to harassment by leaders is at one and the same time a process of "disintegration" into the basic human material which is then "reintegrated" into the new soldier/man and a process of creating a group of equal individuals. Moreover, the separation from women, which often accompanies such periods, also conforms to this overall logic of becoming a member of a closely knit all-male group. Gil remarks that,

We had difficulties with our girlfriends in town, who were used to us coming home every night. And now, for more than six months we saw them only once every couple of weeks. Gradually, it broke up our relationships... My only regret was that my girlfriend had left me towards the end. (Lieblich 1989: 88-9)

More importantly, after Alon's close friend Zvika is killed in Lebanon, he has trouble relating to his girlfriend. Intimacy is found elsewhere in the supportive male group:

I could not even make love to her, although the whole time we were in Beirut I kept dreaming about making love to her. Only the soldiers... who shared the same experience, could provide some support. (Lieblich 1989: 23)

Caring also appears in hierarchical relationships between commanders or officers and their "men." David examines this quality of relations and his story culminates in combat:

As a communications man, a special relationship develops between you and your platoon leader. It is a kind of romance, you know. After 14 months together, you know each other in and out. I originally had thought that he was a hard man, but I discovered that he was an actor, and very human underneath. When I went on action with him, I saw him as a human being, even afraid. Once he pushed me behind a tree, turned to face me and said: "Take care!" His voice was strange. I knew that he was concerned about me. (Lieblich 1989: 5)

In another passage about the attack on Beirut, he notes: "This time it was different, because my trainees were following me in that street, and I was scared for them. An infantry commander is somewhat like a mother to his soldiers, you know" (Lieblich 1989: 129). Eitan talks about his commander:

> I enjoyed tremendously my work with my superior. He was a good and serious man, and model commander. He would find time to listen to each of the youngest soldiers, to try and understand each individual and his problems... At the same time, he was an outstanding professional soldier. I learned a lot from him... It was especially exciting when he, with all his experience, consulted me about our moves. It gave me a feeling of worth. (Lieblich 1989: 59-60)

Both passages are marked by emphases on sharing and caring. From the perspective of commanders, such sentiments often involve sharing the difficulties "ordinary" men face. Indeed, commanders are often particular that the men see that they are eating last ("after the soldiers"), carry heavy loads, and make the same kinds of physical efforts (like climbing and running) as their men. Gil says "I served as an officer during an extremely difficult winter. I was freezing, but I was the last one to take a snowsuit or a heater for my room. It was important for me to set an example for the other soldiers" (Lieblich 1989: 89).

Yet the hierarchical relationship is ideally based—like the fatherly role—on both concern and discipline. David's account:

> I realized that the essence of leadership was to create a balance between the maintenance of a distance from your men, and, at the same time, making them know that they can trust you and come to you when they really hurt. (Lieblich 1989: 9)

He later adds,

> I met my squad in the field. Gradually I learned to know and to love them. At the same time I developed my own style as a commander... Basically, you care about your men, but you don't show it... you make a severe face, you never smile and you are strict with your men... I used humor with my trainees, but have never laughed in front of them... [During the war] I was worried about men, and the responsibility for their safety was my greatest burden. (Lieblich 1989: 9-11)

In both excerpts concern and discipline are related to the wider role of military commanders (that is, to the accomplishment of missions), but they also touch upon the kinds of caring relations considered exemplary of hierarchical relations between men.

Contexts: (Military) Organization, Nation, and State

Against this background, we must be wary of drawing a direct parallel between military service in the IDF (or in the armed forces of other industrialized societies) and rites of passage in smaller, tribal societies. In both kinds of societies such rites can be seen as moving individuals from one male status to another and as a response "to a need universally felt by the male child: to be recognized as a man; to be one who has broken with the weakness and dependence of childhood" (Badinter 1992: 75; Herdt 1981). But in Jewish-Israeli society the transition is problematic. It is problematic not only because the power of ritual is inevitably diffused and attenuated in the contexts of large-scale and pluralistic societies (where some segements of the population are distanced from central values and myths) (Turner 1982). It is also problematic because in the Israeli context, ideals of manhood and notions about achieving these ideals are inextricably linked to specific organizational frameworks set up and maintained by the Israeli state.

This rite—as we saw, actually a series of tests—is carried out within the context of a highly rational, goal directed bureaucratic organization. The significance of this situation may be clarified through the body practices to which men/soldiers are subjected. From an organizational point of view, a prime problem for military frameworks is to assure the control and predictability of the body (Frank 1991: 51). Put somewhat differently, because of the kinds of environments the armed forces are deemed to perform in, soldiers' actions must be liable to calculation, planning, and prediction (Vidich and Stein 1960: 496). Given the stressful conditions of combat the aim is mastery of their bodies (or more correctly bodies-in-use) and control of their emotions so that the military—a large bureaucratic organization—can operate on its environment. To make the reasoning here explicit, in order to achieve military goals—such as destroying enemy personnel and equipment, or taking and holding territory—men must be trained and prepared to be able to manage their emotions and actions.

The overall imagery here is of the military as an organization characterized by "machine-like" characteristics. Within the IDF (as in all modern armies) the dominant metaphors used by officers and soldiers in describing military "work" are related to machinery and to industrial production. Indeed, analyses of the military have long underscored its claim to professional competence: the management of violence (Lang 1972: 29; Shay 1995: 17). This claim alerts us to the fact that underlying much of modern military structures are certain notions of organization. The Israeli military, like all modern armies is characterized by strong mechanistic assumptions and images: units of the armed forces are thought to operate and have the qualities of machines.[3]

The point here is that the creation, management, and understanding of men's bodies within service as a rite of passage is part of, and subordinated to, the overall logic of the military. In other words, the IDF uses ideals of manhood to motivate many of its recruits to undergo the harsh regimen by which they learn to control their bodily actions and feelings so that they will be able to carry out different types of military missions.[4] For example, unlike the quests undertaken by initiates in Indian-American societies, the night-time navigations carried out in the Israeli army are explicitly linked to the bureaucratic rationality of a modern military establishment and are harnessed to its goals. Similarly, harassment and punishment are meted out not only as part of the rite whose aim is to create men, but also as part of making disciplined, obedient soldiers whose actions can be predicted.

But notions of manhood—through their inextrable link to military organizations—are also related to national, state mandated goals. David Gilmore (1990) proposes that at its core, warrior-hood involves accepting the notion of men's expendability: the idea is that to be men, individuals must accept the fact that they are dispensable. In fact, in many cultures the acknowledgment of one's expendability constitutes the basis of the "manly pose." Recognizing this social truth often "constitutes the measure of manhood, a circumstance that may explain the constant emphasis on risk taking as evidence of manliness" (Gilmore 1990: 121). From a social point of view, such ideals of "real manhood" should be seen as inducements to carry out collective goals because of the universal urge to flee from danger. In this sense, manhood ideals make an indispensable contribution to the continuity of social systems by encouraging men to act according to a code of conduct that advances collective interests by overcoming inhibitions" (Gilmore 1990: 121).

In terms of the Israeli case, the point is that manhood ideals prepare young men to sacrifice themselves for the nation-state. While I did not find any explicit examples of this theme in Lieblich's book, let me provide two illustrations from reports in contemporary Israeli newspapers. Such portrait reviews are quite common in the

Israeli media and often give public voice to prevailing notions and assumptions. The first example is an article about an officer—Uri Azulai—who was killed in Lebanon during a skirmish with the Hizbullah. In the account, one of Azulai's commanders relates a conversation he had with him about death. The commander says:

> All during my military service I was in dangerous activities and I scraped with death. The role necessitates this. You have to take this into account. Uri looked at me and answered unhesitantly: "I tell my family and my girlfriend that I will never be hurt, but I know that I may die. I am ready for this, but if it happens, I want to be killed in battle." (Abramowitz 1996)

Another article is a sketch of officers's training course in Israel's southern desert (Bernheimer 1996). In it the battalion commander says very bluntly (and according to the journalist with no pathos): "our role [tafkid] is to die." One of the company commanders explains:

> When I was in Lebanon we had a number of engagements. It takes 15 seconds. Once I stood up facing a terrorist. It was him or me. Then we returned to Israel and I went into a settlement and you see the local kindergarten, and everything is quiet... And you think "This is great" they didn't even feel that anything happened. It brings a smile to your face. You are here for that... I am still ready to die. I told my soldiers, we came here to die instead of the civilians. Everyone here is willing to die so that everyone can live.

Such ideas clearly resonates with notions of sacrifice found in the central national narratives of Zionism. We are led back to our starting point of how, in contemporary Israel, notions of warriorhood are *still* explicated and legitimated through a profusion of stories, sites, and rituals devoted to heroism and sacrifice. In this sense, such passages are still related to wider foundation myths of the Zionist state as a solution to the Jewish problem of the diaspora (Aronoff 1993); to the intentional cultivation of a "new" active and combative Jew differing radically from the passive and nonresistant Jew of the diaspora (Lomsky-Feder 1992; Katriel 1986: chap 2); and, to the birth and preservation of the nation-state in war and on the basis of sacrifice (Sivan 1991).

Consequently, my argument is for understanding the situation in terms of mutuality, of recursive feedback between the military and civilian parts of people's experiences (see Lomsky-Feder and Ben-Ari,1999). It is not just a matter of the military preparing men, or disposing them to think and act in a certain manner once they are civilians. To be successful, ideologies must appeal to and activate preexisting cultural understandings which are themselves compelling. In the IDF, Israeli myths of heroism and notions of mastery and rationality all reverberate against each other and are actualized in the specific contexts of military units. Thus the army invokes and actuates wider cultural understandings about manhood and nationhood, and in this manner it feeds back into and energizes these very ideals. It is in this light that the accounts in Lieblich's study demonstrate how ideals of masculinity influence soldiers' attitudes towards military service and combat even before entering the army.

Conclusion: Body-Building, Character-Building and Nation-Building

In this paper, I have explored the relationship between ideals of manhood and military service in Israel. I have carried out this exploration through a re-reading of the accounts found in Amia Lieblich's seminal book about transition to adulthood during military service. The "logic" of military service as a rite-of-passage is predicated on a series of tests—involving physical and psychic tribulations—that boys must undergo in order to become men. The power of these tests lies in the combination of public scrutiny and private self-discovery that they entail. While the prime trial is that of combat, military service provides a host of other situations for testing individuals in terms of characteristics that are gendered male: composure under pressure, endurance, perseverance and adherence to goals. In all of these situations, the body and the emotions are crucial: becoming a man is thus both a project of "body-building" and "character-building." Alongside the paradigmatic image of a man there are other kinds of manhood. Thus another version of manhood—man-as-organizer—exists in the IDF but is subordinated to the ideal of man-as-warrior. The manly ideals associated with this organizational version are based on civilian notions of responsibility, power and rationality. The warrior ideal, in other words, is not only the hegemonic version of masculinity (Morgan 1994), but is crucial for organization building within the military (through motivating men to serve) and for nation-building (by readying them for sacrifice).

By way of conclusion, let me make a number of points about the wider implications of my analysis: three about the exclusion of women, and one about contemporary Israel. The exclusion of women is firstly, and perhaps most apparently, effected through the "simple" fact that by centering attention on men and their interconnection, military service as a rite-of-passage places women outside a central arena in Israeli society (Izraeli 1997). But the exclusion of women is carried out in much more subtle ways. My second point is that gendered social practices change the meaning and character of people's bodies by actually altering them physically. The

> physical sense of maleness experienced by many men comes not just from the symbolic significance of the phallus, or even simply from the images of power frequently attached to the male body by popular culture. It also derives from the transformation of the body through social practices. (Shilling 1993: 110)

As I showed earlier, the social definition of men as holders of power is translated not only into mental body images and fantasies, but into muscle tensions, postures, the feel and texture of the body. This important point here is that this "translation" is one of the main ways in which power becomes "naturalized," that is, seen as part of the order of nature (Connel 1987: 85, 1993, 1995). Women then, according to this reasoning, are assumed to be inferior (and therfore to merit exclusion) because they do not have the kinds of bodies which have been fashioned through military service and have not actively participated in this "bodily" transformation.

On a third level, the control demanded of men within the army reverberates with—that is, it tends to be reinforced by, and to reinforce—wider notions of self control as a central element of manhood in western industrialized societies (Brandes 1980: 210). This reverberation is not surprising, since in such societies—and Israelis belong to such a society—there is a strong link between rationality and masculinity: rationalization is gendered. As Haste (1993: 90) observes, "a central metaphor of rationality is mastery—over the environment, over others, and over ideas." This link is, moreover, intensified by the social contexts of the Israeli army: men must display control and mastery within public arenas of small, relatively cohesive, groups and (often) close relations with commanders. Thus by performing as soldiers, individuals are implicitly actualizing the gendered division between rational and irrational, between men and women.

My final point is related to contemporary Israel. One may well ask whether the IDF and Israeli society are not changing in ways that have made the model I have explicated—based as it is on Lieblich's material from the 1980s—rather dated? By way of an answer, I would suggest a complex model. I propose that Israel is the site of emerging developments toward peace *and* the continued importance of the military in Israeli society. My point is hence not a simplistic argument about the "inevitable" transformation of Israeli society from one based on a "war footing" to one characterized by a "peaceful" existence. Rather, we should recognize an emerging complexity in which existing considerations related to security will continue to hold sway alongside newer factors. I assume that at the same time as Israel is moving toward long-term accords with some of its neighbors, that its armed forces and military service will *continue* (in the foreseeable future) to be socially and culturally significant (Cohen 1995; Schiff 1995). Thus, the links between manhood and military service will continue to hold sway for at least part of Israel's Jewish population in the coming few years.

Notes

*   Thanks are due to Efrat Ben-Ze'ev, Danny Kaplan, Amia Lieblich, Pamela Lubell, Edna Lomsky-Feder and Vered Vinitzky-Seroussi for comments on earlier drafts.

1.   A word about the "methodology" of this study. In basing my analysis on the testimonies found in Amia Lieblich's book, I follow what is now an established practice in the social sciences: the utilization of existing works in order to further explore or refine theoretical issues. Out of a range of such studies that have been carried out in the past one may cite Strauss's (1978) reexamination of monographs written about negotiations, Lyman and Scott's (1976: chap 7) reappraisal of the dramaturgical significance of resistances, or Handelman's (1976) reconsideration of a study of symbolic interaction in a work setting. Such attempts may admittedly lead to a dilution of what is often a "thick description" (Geertz 1973) in the original work. Yet the contribution of such studies may lie precisely in their theoretical potential: in their potential to extend the hints and insights that are often found in sensitive accounts and to relate them to broader analytical issues.

2.   To revert back to a point made earlier, as Edna Lomsky-Feder shows (1992, 1998), it is easy for young people to identify with stories of heroism propagated on the national cultural level, since many of them incorporate attributes of masculinity such as grit, unyielding adherence to goals, ability to withstand physical and emotional pressure, and loyalty to friends.

3.   At the individual level, the image is similar to one found in sports or parts of medicine, the body-as-machine: "the body is seen as a complex machine whose performance can be enhanced, and which can break down and be repaired, just like any other machine" (Shilling 1993: 37).

4.   This type of stress is found both in tribal (Roscoe 1996: 662) and complex societies: "That the military organization seeks to capitalize on these tenden-

cies in American life is seen in such perennial recruiting slogans as "The Marine Corps Builds Men" and "Join the Army and Feel Like a Man" (Moskos 1970: 154).

References

Abramowitz, Shlomo 1996. "A Legend in His Time." *Yediot Aharonot,* November 1. (Hebrew).

Abu-Lughod, Lila and Catherine A. Lutz 1990. "Introduction: Emotion, Discourse and the Politics of Everyday Life." In *Language and the Politics of Emotion,* edited by Catherine A. Lutz and Lila Abu-Lughod, 1-23. Cambridge: Cambridge University Press.

Adams, Abigail E. 1993. "Dyke to Dyke: Ritual Reproduction at a U.S. Men's Military Academy." *Anthropology Today* 9, 5: 3-6.

Aronoff, Myron J. 1993. "The Origins of Israeli Political Culture." In *Israeli Democracy Under Stress,* edited by Ehud Sprinzak and Larry Diamond, 47-63. Boulder, CO: Lynne Reinner.

Aronoff, Myron J. 1989. *Israeli Visions and Divisions.* New Brunswick, NJ: Transaction Publishers.

Arkin, William and Lynne R. Dobrofsky 1978. "Military Socialization and Masculinity." *Journal of Social Issues* 34, 1: 151-68.

Badinter, Elisabeth 1992. *XY: On Masculine Identity.* New York: Columbia University Press.

Barrett, Frank J. 1996. "The Organizational Construction of Masculinity: The Case of the U.S. Navy." To appear in *Gender, Work and Organization.*

Ben-Ari, Eyal 1998. *Mastering Soldiers: Conflict, Emotions and the Enemy in an Israeli Military Unit.* Oxford: Berghahn Books.

Ben-Ari, Eyal 1997. *Body Projects in Japanese Childcare: Culture, Organization and Emotions in a Preschool.* London: Curzon.

Ben-Ze'ev, Efrat and Eyal Ben-Ari 1996. "Imposing Politics: Attempts at Creating a Museum of 'Co-Existence' in Jerusalem." *Anthropology Today* 12, 6: 7-13.

Bernheimer, Avner 1996. "A Bomb Full of Motivation." *Yediot Aharonot,* 27 September.

Boene, Bernard 1990. "How Unique Should the Military Be? A Review of Representative Literature and Outline of Synthetic Formulation." *European Journal of Sociology* 31, 1: 3-59.

Bourdieu, Pierre 1977. *Outline of a Theory of Practice.* Cambridge: Cambridge University Press.

Bourne, Peter G. 1967. "Some Observations on the Psychological Phenomena Seen in Basic Training." *Psychiatry* 30: 187-96.

Brandes, Stanley 1980. *Metaphors of Masculinity: Sex and Status in Andalusian Folklore.* Philadelphia, PA: University of Pennsylvania Press.

Bruner, Edward M. and Phyllis Gorfain 1983. "Dialogic Narration and the Paradoxes of Masada." In *Text, Play and Story,* edited by Edward M. Bruner, 56-79. Washington DC: Proceedings of the American Ethnological Society.

Cameron, Craig M. 1994. *American Samurai: Myth, Imagination and the Conduct of Battle in the First Marine Division, 1941-1951.* Cambridge: Cambridge University Press.

Cockerham, William 1973. "Selective Socialization: Airborne Training as Status Passage." *Journal of Political and Military Sociology* 1: 215-29.

Cohen, Stuart 1995. "The Israel Defense Forces (IDF): From a 'People's Army' to a 'Professional Military'—Causes and Implications." *Armed Forces and Society* 21, 2: 237-54.

Connel, R.W. 1995. *Masculinities*. London: Polity Press.

Connel, R.W. 1993. "The Big Picture: Masculinities in Recent World History." *Theory and Society* 22, 5: 597-624.

Connel, R.W. 1987. *Gender and Power*. Cambridge: Polity Press.

Dornbusch, Sanford M. 1955. "The Military Academy as an Assimilating Institution." *Social Forces* 33, 4: 316-21.

Frank, Arthur W. 1991. "For a Sociology of the Body." In *The Body: Social Process and Cultural Theory*, edited by Mike Featherstone, Mike Hepworth, and Bryan Turner, 36-102. London: Sage.

Gal, Reuven 1986. *A Portrait of the Israeli Soldier*. New York: Greenwood Press.

Geertz, Clifford 1973. *The Interpretation of Cultures*. New York: Basic Books.

Gilmore, David D. 1990. *Manhood in the Making: Cultural Concepts of Masculinity*. New Haven, CT: Yale University Press.

Handelman, Don 1976. "Rethinking 'Banana Time': Symbolic Integration in a Work Setting." *Urban Life* 4, 4: 433-48.

Handelman, Don and Elihu Katz 1995. "State Ceremonies of Israel: Remembrance Day and Independence Day." In *Israeli Judaism: The Sociology of Religion in Israel, edited by* Shlomo Deshen, Charles Liebman and Moshe Shokeid, 75-85. New Brunswick, NJ: Transaction Publishers.

Handelman, Don and Lea Shamgar-Handelman 1996. "The Presence of Absence: The Memorialism of National Death in Israel." In *Grasping Land: Space and Place in Contemporary Israeli Discourse and Experience*, edited by Yoram Bilu and Eyal Ben-Ari. Albany: State University of New York Press.

Haste, Helen 1993. *The Sexual Metaphor*. New York: Harvester/Wheatsheaf.

Herdt, Gilbert 1981. *The Sambia: Ritual and Gender in New Guinea*. Fort Worth, TX: Holt, Rinehart and Winston.

Horowitz, Dan and Moshe Lissak 1989. *Trouble in Utopia: The Overburdened Polity of Israel*. Albany: State University of New York Press.

Izraeli, Dafna 1997. "Gendering Military Service in the Israel Defense Forces." *Israeli Social Science Research* 12,1: 129-66.

Kalderon, Nissim 1988. *The Feeling of Place*. Tel-Aviv: Hakibbutz Hameuchad. (Hebrew)

Katriel, Tamar 1986. *Talking Straight: Dugri Speech in Israeli Sabra Culture*. Berkeley: University of California Press.

Kimmerling, Baruch 1993. "Patterns of Militarism in Israel." *European Journal of Sociology* 34: 196-223.

Kratz, Corinne A. 1990. "Sexual Solidarity and the Secrets of Sight and Sound: Shifting Gender Relations and their Ceremonial Constitution." *American Ethnologist* 17, 3: 449-69.

Lang, Kurt 1972. *Military Institutions and the Sociology of Law*. Beverly Hills, CA: Sage.

Levy-Schrieber, Edna and Eyal Ben-Ari forthcoming. "Body-Building, Character-Building, and Nation-Building: Gender and Military Service in Israel." *Studies in Contemporary Judaism*.

Lieblich, Amia 1989. *Transition to Adulthood During Military Service: The Israeli Case*. Albany: State University of New York Press.

Lieblich, Amia and Meir Perlow 1988. "Transition to Adulthood During Military Service." *The Jerusalem Quarterly* 47: 40-76.

Liebman, Charles S. and Eliezer Don-Yehiya 1983. *Civil Religion in Israel: Tra-*

ditional Judaism and Political Culture in the Jewish State. Berkeley: University of California Press.

Lomsky-Feder, Edna 1998. *As If There Was No War: The Perception of War in the Life Stories of Israeli Men.* Jerusalem: Magnes. (Hebrew).

Lomsky-Feder, Edna 1994. *Patterns of Participation in War and the Construction of War in the Life Course: Life Stories of Israeli Veterans from the Yom Kippur War.* Ph.D dissertation. Department of Sociology and Anthropology, Hebrew University of Jerusalem.

Lomsky-Feder, Edna 1992. "Youth in the Shadow of War—War in the Light of Youth: Life Stories of Israeli Veterans." In *Adolescence, Careers and Culture,* edited by Wim Meeus et al., 393-408. The Hague: De Gruyter.

Lomsky-Feder, Edna and Eyal Ben-Ari 1999. "Introductory Essay: Cultural Constructions of War and the Military in Israel." In *The Military and Militarism in Israeli Society,* edited by Edna Lomsky-Feder and Eyal Ben-Ari. Albany: State University of New York Press.

Lyman, Stanford and Marvin B. Scott 1976. *The Drama of Social Reality.* New York: Oxford University Press.

Meiron, Dan 1992. *If There Is No Jerusalem... Essays on Hebrew Writing in Cultural-Political Context.* Tel Aviv: Hakibbutz Hameuchad. (Hebrew)

Meisels, Ofra 1995. "An Army in the Process of Liberation." Paper presented at the conference: "An Army in Light of History: The IDF and Israeli Society." Hebrew University of Jerusalem. June. (Hebrew).

Morgan, David H.J. 1994. "Theater of War: Combat, the Military and Masculinities." In *Theorizing Masculinities,* edited by Harry Brod and Michael Kaufman, 165-82. Thousand Oaks, CA: Sage.

Moskos, Charles C. Jr. 1970. *The American Enlisted Man: The Rank and File in Today's Military.* New York: Russel Sage Foundation.

Mosse, George L. 1990. *Fallen Soldiers: Reshaping the Memory of the World Wars.* Oxford: Oxford University Press.

Schiff, Rebecca L. 1995. "Civil-Military Relations Reconsidered: A Theory of Concordance." *Armed Forces and Society* 22, 1: 7-24.

Schwarz, Barry, Yael Zerubavel and Bernice M. Barnett 1986. "The Recovery of Masada: A Study in Collective Memory." *The Sociological Quarterly* 27, 2: 147-64.

Shay, Jonathan 1995. *Achilles in Vietnam: Combat Trauma and the Undoing of Character.* New York: Touchstone.

Shilling, Chris 1993. *The Body and Social Theory.* London: Sage.

Sion, Liora 1997. *Images of Manhood Among Combat Soldiers: Military Service in Israel's Infantry Brigades as a Rite of Passage.* MA Thesis. Department of Sociology and Anthropology, Hebrew University of Jerusalem.

Sivan, Emmanuel 1991. *The 1948 Generation: Myth, Profile and Memory.* Tel Aviv: Ministry of Defense. (Hebrew).

Strauss, Anselm 1978. *Negotiations: Varieties, Contexts, Processes and Social Order.* San Francisco, CA: Jossey-Bass.

Turner, Victor 1982. *From Ritual to Theatre: The Human Seriousness of Play.* New York: Performing Arts Journal Publications.

Turner, Victor 1974. *The Ritual Process.* Harmondsworth: Penguin.

Vidich, Arthur J. and Maurice R. Stein 1960. "The Dissolved Identity in Military Life." In *Identity and Anxiety,* edited by Maurice Stein, Arthur Vidich and David White, 493-506. New York: The Free Press.

8

The Meaning of War through Veterans' Eyes: A Phenomenological Analysis of Life Stories*

Edna Lomsky-Feder

> *"When I think about war, half of my life has passed since then. I was 19 then and now I'm 34. It is almost insignificant. It seems to me truly anachronistic. Almost. (...) The war didn't change me at all. I always wanted to travel and I always loved to travel. I never took things too seriously. That's all."*
> —Haggai, tanks brigade

> *"It is so much more dramatic than life, so crazy, dramatic and intense. I have no doubt that . . . that my life is divided into before the war and after the war."*
> — Zohar, tanks brigade

These two quotes illustrate two different outlooks on the meaning of war as a factor in the personal biography of the speaker. They were culled from life stories of two Israeli men who participated in the 1973 Yom Kippur War as part of their regular military service. Both were combat soldiers, from a similar sociocultural background. The two men look retrospectively on their war experience of fifteen years ago, trying to evaluate its place in and importance to their life stories.

Which of these quite different perceptions characterizes the personal meaning the Israeli man lends the war experience in his life

269

story? What is the nature of the difference, and how can it be explained?

The research literature that focuses on the influence of war on veterans states that the view of war as a crisis or transition (as expressed by Zohar, the second subject) typifies the place of war in the veteran's course of life (see, for example, Card 1983; Figley and Leventman 1980; Laufer 1988a, 1988b; Lifton 1973; Milgram 1986; Modell and Haggerty 1991). This approach uses psychological models to explain the difference between the two speakers. The first veteran, Haggai, who asserted that his life was not affected by the war, would be said to lack self-consciousness, or to be compelled to repress or deny his awareness of the influence of war on his life.

Is the underlying assumption that war is a traumatic experience a necessary one? Is the psychological explanation of Haggai's statement the only way to explain his remarks? And is there a sociocultural context in which the non-crisis outlook can characterize the meaning the individual gives to war in his life story? Can the differences between the speakers be explained through cultural rather than psychological models?

The decisive answers found in the existing literature express assumptions about the predominantly traumatic effect of war on the individual. This approach views war as foreign to the course of normal life. War is perceived as a difficult and stressful experience with far-ranging and traumatic repercussions (Danish et al. 1980; Reese and McCluskey 1984).

Because of these assumptions, the prevailing research does not ask whether and how far war experience is integrated into the world of the individual, or whether the experience necessarily distorts and disturbs his life. The researchers' certainty about the severe impact of war and their neglect of the social context have largely overshadowed veterans' responses to the question of whether or not they interpret the war in traumatic terms. The phenomenology of war is perceived as unimportant, and the authors are hardly, if at all, attentive to the meaning the war receives in the eyes of those who fought it.[1]

This is especially problematic when we wish to examine the meaning of war among veterans in societies such as the Israeli society, where the wars are frequent and are given social and cultural centrality.

Given this background, the aim of this study is twofold. First, to deal with the phenomenology of the war experience, i.e., to discover the personal interpretation of war, as revealed in the Israeli veterans' life stories, and to understand the interpretive mechanisms through which the war experience is integrated into the course of their lives. Second, to explain the war's personal meaning within the cultural context in which the veterans live and construct the reality of their lives.

The central thesis here is that the phenomenology of war is to be understood in context. More specifically, my claim is that because war is not a shocking and transformative factor in Israeli society at the macrosocial level, it is not so at the individual level. It seems as if war is institutionalized and normalized into the Israeli social order, and so the individual also integrates and co-opts it into his personal biography.

Life Stories and the Phenomenology of War

The meaning of war was examined through life stories of war veterans. A "life story" is the individual's description of the course of his life. This story is a collection of events and experiences that the narrator chooses (consciously or unconsciously) to present as his or her personal story along a time axis. According to this perception, a life story does not represent a random collection of events, but is "narrative rather than a chronicle; in it resides the evaluation of one's own existence" (Frank and Vanderburgh 1986: 186).

In researching the life story, it is necessary to examine not only the contents of the narrator's biography, but also the formal or "literary" aspects through which he or she shapes the story. The researcher must therefore also relate to the rhetoric of the biography and the manner in which life is articulated as a narrative (Bertaux and Kohli 1984; Bruner 1987; Cohler 1982; Corradi 1991; Crapanzano 1984; Denzin 1989; Sarbin 1986). The goal of the researcher studying the life story is not to reconstruct actual reality as such, but rather to extract the *story* of a particular narrator concerning this reality, and to expose his or her perceptions of it.

In general, the phenomenological approach to life stories focuses on the individual's interpretation of his or her world, rather than on his or her behavior or personality (Berger and Luckmann 1975;

Schutz 1970; Schutz and Luckmann 1974). The life story as interpretation takes a personal story and embeds it in a larger sociohistorical context. The sociohistorical context provides narrative content for the life story; at the same time, the life story is a significant sociocultural document, as the individual constructs a personal version of the common social-historical context. Life stories are a mediating mechanism between society and the individual; between the cultural meaning attached to events and experiences and their personal meaning; between collective memory and personal memory. For this reason our interest in the life story stems from the phenomenological constructions of the narrator (Ryff 1984).

The life story, or, according to Schutz and Luckmann (1974) the "biographical articulation" of the individual, is one of the means used by the individual to construct his or her world. In particular, it is a tool by means of which the individual *cruises* through different levels of time and links them to his or her world. According to Schutz and Luckmann, biographical articulation is conditional upon society: "The categories of biographical articulation are not really categories of inner duration as such, but rather are categories which are formed intersubjectively and established within relative-natural world view. They are basically imposed upon the individual and become interiorized by him" (1974: 57). The social categories of biographical articulation are part of a world view assumed to be self-evident. Moreover, the individual experiences these categories as part of his or her world, as part of the reality of his life, and they define the operational limits for him (Schutz and Luckmann 1974: 94). Schutz and Luckmann are aware of the unique nature of the individual, and maintain that it is important to examine how personal experience is reconstructed in the life story. They posit that the unique nature of the individual is expressed in the way he organizes and arranges his life experiences on the continuum of time, and in the relative weight he lends these experiences (p. 58).

Although Schutz and Luckmann (1974) place emphasis on the way the subject perceives his world, their analysis focuses on the inter-subjective construction. They therefore argue that the individual's perception of his life is largely determined by the sociocultural context. The individual can do no more than choose one from among a number of available "typical biographies" (pp. 94-5). The degree of openness and variety with regard to the selection of typical biographies depended on sociohistorical and class contexts; it also varies according to gender.

Watson (1976) attempts to forge a systematic link between the phenomenological approach and the "life story" approach. He claims that the situation in which a subject recounts his life story to the researcher is a special situation of biographical articulation. According to Myerhoff (1980), by complying with the request to tell about himself, the subject is compelled to contemplate his life, thereby formulating conscious into a more coherent articulation. In this situation, the narrator departs from the spontaneity of his existence in the world, becomes more aware of his "natural points of view" concerning his life which he has taken for granted, and adopts a more phenomenological position regarding the world (Natanson 1962). In other words, he examines and investigates his natural assumptions about the world and makes his own experiences the object of examination. This situation emphasizes the significance and meaning of events and experiences in the narrator's life—such as the war experience.

The experience of war is of special significance in the life-story. In his seminal article "The Homecomer," Schutz (1945) argues that war is a unique experience in the context of the phenomenology of man. This is because war has the potential to shatter "taken for granted social knowledge." The emphasis in this article is on the detachment from home and the return home with a different perspective. The subjects are soldiers who spent long periods at war, i.e. a situation in which the dimension of time is of central significance in the construction of the soldier's memory of home and family. The act of coming back home confronts these constructs of memory with the reality of the encounter. To a large extent, the return is also a transition from one reality of life—war—to another reality—home. This transition locates the soldier in a position of "phenomenological strangeness"—a position that is largely unexpected, since this strangeness occurs within what used to be familiar. This position inevitably shatters what used to be taken for granted. Returning home disrupts the natural and spontaneous continuum according to which the soldier used to perceive his home reality. Largely against his will, he is pushed from natural observation to phenomenological observation of the reality of home life. The emphasis in Schutz's analysis is therefore not on the essence of the war experience as one which undermines that which has been taken for granted, but on the quality of the return from this experience. In this respect, each return home has the potential to disturb that which has been taken for granted. Even returning home from a vacation, in Schutz's opinion, unsettles to a certain extent what used to be seen as something that did not even require consideration. From this perspective, the longer the absence from home, the more significant the experience of return; the more different the "other reality" is from the "home reality" to which one returns, the more dramatic the new encounter. It thus follows that return from war carries the potential for crisis, since it cancels the rules of the game to which the soldier has become accustomed in the war reality and forces him to confront the radically different rules of the home reality.

I would argue that this mechanism of challenging that which has been taken for granted is potentially present in the preceding biographic chapter—the moment the soldier is "thrown" into the war reality. Thus, not only the return from war is significant in terms of the perception of everyday life, but also the actual encounter with war as a unique phenomenon. This particular point is not emphasized by Schutz, who deals with the nature of the experience in the context of the process of homecoming. Given its powerful emotional and moral impact, the experience of war carries the potential to challenge the social presumptions that structure the home reality. The war experience forces the individual to confront extreme and abnormal situations in which the fighter is pushed toward a phenomenological position regarding his life-world. War is so different from what was previously familiar, known and expected in the home reality; experiences in war also usually differ from the soldier's expectations of war itself. These gaps—between the soldier's social knowledge of the reality of his life and of the reality of war on the one hand and his actual experiences in war on the other—may potentially lead to the breakdown of the individual's natural assumptions about his life.

War, then, is an event that carries the potential to break down the models according to which one interprets the everyday world. Is this potential always realized? Does the social context in which this experience takes shape influence the realization of this potential?

The goal of this study is to examine the meaning of the war experience in the life story within its cultural context and by means of a phenomenological framework. More specifically, through the life story I seek to clarify whether war is presented as part of the "typical biography." I examine the connection between war and other events perceived as part of the typical biography; the "natural" assumptions concerning war and the "phenomenological" points of view; whether war is taken for granted in social terms, or is it seen as an experience that challenges natural suppositions concerning the world, leading to a phenomenological perception of life.

My analysis of the stories stemmed from the term "interpretive relevance" as developed by Schutz and Luckmann in their book *The Structure of the Life-World* (1974). According to them, when a subject or event receives conscious attention and is the object of focused discussion (i.e., when it is of thematic relevance), the narrator does not usually confine himself to presenting the matter, but also interprets it and determines its significance and meaning on the basis of the interpretive schemes available to him. "Interpretive relevance," then, is the encounter between the individual's experience and his social knowledge, by means of an appropriate interpretive scheme (Schutz and Luckmann 1974: 207). As long as events, incidents and experiences are familiar and do not create "noise," they are interpreted in a routine and almost automatic manner. The interpretation will be in accordance with familiar and approved categories of meaning, which can be seen as the norms in the construct of reality. When an experience shatters and violates the ongoing and routine continuum of how the individual "reads" the environment, to the extent of undermining norms, it receives more attention.

I use the conceptual framework set by Schutz and Luckmann to examine how the speaker conceives and interprets war in the context of his pre-war knowledge which was taken for granted. Is this an experience that has led to a "disturbance" in routine interpretation, to the point of destabilizing knowledge that has been taken for granted, requiring a purposeful, conscious and reflexive grappling with its meaning? Against this background, what meaning does the experience receive in the context of the veteran's perception of the continuum of his life?

The objective is to understand the way in which war relates to the world of "the taken for granted," and to unravel the interpretive solutions, proposed by the story-tellers, to confront the experience. These outlooks expose an important aspect of the repertoire of cognitive strategies, which the subjects use to incorporate the experience into their lives. The assumption is that the more the war experience is presented as an event that did not break down "taken for granted" social knowledge, the more normalized it has become in the course of life.

Methodology

My research is based on sixty-three life stories, gleaned from in-depth, open-ended interviews. Israeli society is engaged in a constant struggle and offers an especially interesting case for studying the meaning of the war experience among war veterans. There are several reasons for this: (1) war occupies a special place in Israeli culture and ideology, since the idea of being a "besieged" nation is a central component in the country's self-image; (2) war is one of the primary factors in building social institutions, developing the economy, and maintaining social solidarity; (3) within a relatively short period of time, Israeli society has been involved in many wars, each of which had its own character and set of social implications; and (4) military service in Israel is compulsory, universal, and long (both regular service and reserve duty). As an experience shared by almost all Jewish men and by many women, it transcends differences of social class and ethnicity. Because Israel has been involved in a number of wars, most Israeli men fought in at least one war and have been forced to deal with this experience (Horowitz and Lissak 1989; Kimmerling 1985, 1993; Mintz 1984).

Given these factors, it is clear that military service and the war experience are deeply rooted in the reality of life in Israel and are central to defining the identity of the Jewish Israeli and the constructs of his or her day-to-day existence. The interviewees are men born between 1952 and 1954, who fought in the 1973 Yom Kippur War as part of their regular military service. Half held combat positions and half were support staff.

For our purposes, the 1973 Yom Kippur War is especially pertinent to the phenomenology of war in Israel. This war is universally regarded as the most traumatic of Israel's wars. Absolute faith in the power of the Israel Defense Forces (IDF) was shattered and feelings of impermanence and insecurity that were put to rest in the 1967 (Six Day) War, and supposedly banished from individual and collective consciousness, were reawakened. This crisis of faith in the power of the IDF forced Israelis to re-examine the meaning of war on the individual and the societal level.

The interviews took place between 1987 and 1989, about fifteen years after the Yom Kippur War. The men told their life stories and talked about this from a retrospective viewpoint. This cohort is interesting because those involved entered adolescence when Israel was basking in the euphoria of the 1967 war; they were brought up with the values, images and symbols that had become entrenched as a result of the 1967 war. There is a special significance to the fact that this generation came fully of age around the time it had to fight in the Yom Kippur War—a war that, from a military and morale point of view, was the antithesis of the 1967 war. Later, aged around thirty years, the members of the cohort served as reservists in the Lebanon War (1982-4) and have since taken part in the IDF response to the Palestinian *Intifada* (uprising) which broke out in 1987. Both the Lebanon War and the Palestinian *Intifada* have further divided national opinion and shattered collective faith in the role of the IDF. These additional experiences with war doubtless influence the way in which war is perceived—specifically the Yom Kippur War—and lend an additional dimension to the veterans' retrospective interpretation of the meaning of this experience.

The stories were chosen from a clearly defined social group: middle-class, educated, secular men—the group from which most of the Israeli "elite" comes. This point is extremely important because, to a great extent, the interpretations presented here express a hegemonic ideology, i.e., the cultural models of the dominant groups in Israeli society. Even the "other voice," critical of the IDF, is phrased in conceptions that are meaningful for dominant Israeli groups and are not the interpretations typical of a person from the social periphery of the society, such as criticisms developed among marginal groups like women, lower-status second-generation immigrant groups, and Israeli Palestinians.

Findings

Before presenting the different meanings of the war experience in the life stories, it is important to stress that this study is concerned with the interpretive level of the individual, not with behavior or personality. In other words, the focus is on the way in which the subjects interpreted and explained the effect war had on them. It does not claim to answer questions such as "does war have an impact on life?" or "what sort of an impact does war have?"

An analysis of the life stories reveals a number of interpretations of the war experience. In the following section, the different interpretations and the relationships between them are discussed. These various interpretations are not necessarily mutually exclusive; generally, the interviewee presents a number of interpretations in the course of a single story. The relationships between the different interpretations are an additional factor in the analysis of the material and in understanding the interpretive mechanisms through which the individual integrates the war into his life story.

The "Dominant Voice"—Interpretation of Continuity

The interpretation of continuity integrates war into the course of life by means of previous social knowledge. This interpretation is the dominant voice, expressed in some 75 percent of the life stories. Whether they served in combat or non-combat positions, most of the veterans perceive the war as a significant experience, but one that did not undermine their taken-for-granted knowledge of the world. Life is perceived as a linear continuum; war has a localized place on this line but did not alter it. Even if the interpreter wonders about social or personal truths, the experience of war did not substantively change his understanding and feeling of control in the external and internal environment. This interpretation of continuity receives many meanings, and several variations can be identified within the dominant voice of continuity. As mentioned earlier, these meanings do not necessarily preclude one another, and some tellers simultaneously present a number of interpretations to establish the interpretation of continuity.

The most routine interpretation of war is heard in the stories in which war is perceived as an event integrated into a period of conscious lethargy, which the tellers claim is characteristic of youth, and specifically of regular military service. An example of this is the following comment by Amnon, a paratrooper:[2]

> We didn't have the faintest idea what was happening to us, we were jerks. Regular soldiers aren't... their lives don't belong to them. I think it's partly a

matter of age, and partly the context of being a regular soldier, that makes you not think in terms of profits and losses.

These stories portray the army as a totalitarian institution, which fosters mechanical, automatic, and herd-like behavior. According to these veterans, military service silenced their emotions and vulnerability, and dulled their ability to think independently and critically. They claim that "being in the army" took control of how they dealt with war experience. Expressions such as "brain washing" and "deadening" are used to describe this outlook.

Another expression of the interpretation of continuity originates in acceptance of and fundamental identification with the hegemonic Israeli ideology of war. For example, Koby, a pilot:

> I was into the whole business of Zionism and heroism—the same things that now seem a bit stupid to me. But I remember that because I was too young, they didn't let me fly for the first two days. I went on hunger strike. I refused to shave and I went around barefoot, until they let me fly.[3] I was injured slightly once, and some of my friends died and some were taken prisoner, and it was... but it wasn't even traumatic, because it all happened completely under the banner of Israeli heroism... that's it.

While Koby's comments have a dimension of self-irony, this is combined with his nostalgia for the enthusiastic, believing and committed young man he used to be. The war experience was shaped by the cultural models that led him to enlist and according to which he acted during his service. The war did not challenge his interpretative patterns; indeed, he claims that it even reinforced them. A further example is Chaim, a personnel officer in the armored corps:

> All those people who say that it destroyed things for them... that it destroyed myths for them... what did it destroy? (...) On the contrary, this was *the* just war... we didn't start it, we didn't want it, we didn't have any goal. These evil guys came and surprised us on Yom Kippur. At that time, the mere fact that they came on Yom Kippur was considered a serious sin. So I'm not convinced that myths were really destroyed, I think that in some ways it strengthened the myth (...) Certainly the war strengthened the connection with all these things. Look, this is our war. Each generation has its own war, and this was our war.

These subjects speak about identification, rather than a sense of "deadening" or the absence of criticism. Most proponents of this interpretation view the war as justified and just. The tellers accept the cultural assumption that as long as a war is justified, it should be seen as a political means based on rational and utilitarian considerations. The Yom Kippur War did not destroy this basic conception and these individuals justify the war in defining it as an experience of continuity. Avi, a flight supervisor, was asked how the war influenced his life:

> I'm not sure how to isolate the war. Because there is no doubt that we were attacked in the Yom Kippur War, and we had to defend ourselves, which is what we did (...) We were educated to fight in an unavoidable war, an ethical war, a war that couldn't be prevented (...) But the war took place and had to take place. What can we do? This is part of life. Death is part of life.

According to Avi, a justifiable war should be seen as part of life. The Yom Kippur War fulfilled this criteria, but the war in Lebanon—which undermined Avi's conception of war in Israeli society—did not. Many of the stories that present the Yom Kippur War as an experience of continuity through identification compare this war to the war in Lebanon. In this they differ from the subjects who viewed the war as an experience of discontinuity and who often speak about the 1967 war as a heroic war in collective memory.

Stories in which the identity-based interpretation is particularly strong are ones where the narrator presents himself as a young man who was, before military service, rather alienated from the ethos of the army. In these stories, the war is portrayed as an event that reinforced his links and identification with such an ethos. This type of interpretation is particularly evident in the life stories of interviewees who spent long periods of their youth outside Israel and felt somewhat marginal in Israeli culture, especially given its militaristic element. Army service was a particularly significant experience for these young men in the process of their induction and integration in Israeli culture.

Another voice in the interpretation of continuity through identification is expressed in the stories presenting war as a rite of passage to the fighter role. Yuval, an officer in the infantry, expresses this view:

> This was the first time I tested myself in truly difficult situations. All the time you ask yourself how you will react under fire, if you will be able to assist others in difficult circumstances. On the one hand, this is a difficult situation, and all that, but on the other hand, I ... it's not nice to say, but I felt a certain personal satisfaction because I was pleased with myself.

This outlook sees war as an experience that binds the individual with the cultural models addressing the nature of the good soldier.

According to this interpretation, war accentuates the connection between "common knowledge" and action on the battlefield. This interpretation of continuity is latent in most of the stories but is especially prominent when the teller feels a certain dissatisfaction with his military service or the war. The emphasis on functioning successfully on a personal level compensates for feelings like "my role wasn't sufficiently combat oriented" or "I wasn't able to integrate into the army" or "I didn't fight in the war with my buddies."

Until this point, the tellers have presented themselves as passive interpreters who adopt cultural models without criticizing them. Some interviewees present themselves as more active interpreters who take part in more conscious negotiations with their interpretive schemes. In this case they are critical – but the criticism existed before the war and was reinforced by it. War is an experience that confirms and promotes the process of undermining the social norms which previously existed.

An example of this is given by Rafi, a soldier from the intelligence corps. Later in his life, Rafi became one of the leaders of the Peace Now movement.[4] In his story, the Yom Kippur War is perceived as an experience that strengthened the way he challenged things that were taken for granted, an experience that reinforced a separation and detachment from hegemonic cultural models:

> At a very young age, my opinions were very unmilitaristic (...) In these matters, I was very different, even before we went to the army. I mean, I always expressed very clear views on this matter. My big brother, let's say, was in a combat unit, but I wasn't. I already had objections to this even at this stage. (...) I never identified , I mean when I was very young I didn't identify really with what they were doing. Even when the Labor Party was in power, I was very much in opposition. I mean, that they weren't doing enough, and that they were putting us in danger, etc., etc. Yom Kippur was the proof of this.

Rafi's comments convey his basic interpretation of the war as an event that confirmed his opposing stance. The war was additional evidence for an already consolidated viewpoint.

While in some cases the war brings confirmation of a critical standpoint that had taken shape before military service, in other stories it is the nature of the encounter with the IDF during military service that leads to a significant challenging of social norms. The war experience becomes integrated in a process that had begun earlier.

The last variation of the interpretation of continuity appears in stories which define war as an experience that raised awareness and put an end to youthful naivete. Here, the war is integrated into the natural process of maturation. The subjects do not perceive war as an event that caused a dramatic change, a shock, reform; war is presented as a catalyst—a meaningful experience that propelled a process that would have taken place in any case. War did not cause a conscious revolution but raised questions and thoughts that characterize the maturation process.

In stories characterized by highly routine interpretations, the sobering up that comes with maturity is presented as occurring after the period of service. Here, however, the war serves as a catalyst, accelerating processes that are characteristic of the natural process of maturation.

To sum up, for the time being, the stories illustrate that interpretation of continuity is dominant in the life stories. According to these veterans, the war did not undermine the models that interpret the world and did not interrupt the flow of life.

The "Other Voice"—Interpretation of Discontinuity

Despite the dominance of the continuous voice, we must also consider the less common voice, which presents the war as an experience that disturbed the ongoing interpretation, destroyed understood assumptions about the world and interrupted the continuum of life. This is termed the "interpretation of discontinuity."

This interpretation was heard in about 25 percent of the life stories, is far less common and can be seen as the "other voice" in the stories. At the same time, it is presented with emotional and dramatic power. It is voiced more often but not exclusively by combat soldiers.

Haggai, an operations sergeant in the air force, relates:

Suddenly I couldn't identify with anything happening here. Everything looked awful. Awful, shocking. In one fell swoop, the Garden of Eden became Hell.

And Ze'ev, a pilot:

-What did you do in the war?
-I was a pilot in the Air Force. The Yom Kippur War was a trauma. It was a turning point in my life, in all respects, in all respects. And within a few hours of the war beginning, I really felt that I was a totally different person.

In response to an innocent and conventional question in the Israeli context, Ze'ev presents the war as an immediate, direct and spontaneous turning point in his life. His first association with his military service is the war and its being a transforming event. The revolution was immediate and absolute—he was "a totally different person."

Stories which present war as an experience that interrupted the course of life have a common narrative core: a youth who is obligated and believes in hegemonic cultural models goes to war. During the war he undergoes difficult experiences but meets the demands of the Israeli fighter ethos and functions as dictated on the personal level. His encounter with war is a source of great disappointment to him and he acknowledges the gap between reality and image. Because of this insight, he seriously questions these images and takes a critical stance, sometimes even adopting alternative models. The shock lies in the interpretive schemes that construct the relationship of the teller with his society. The disturbance in the flow of life is translated into sociopolitical terms more often than into psychological-existential ones. This approach *undermines* the interpretive schemes dealing with the relationship with the society, and *purifies* the schemes dealing with the "self."

In the interpretation of purification, war is defined as an event that shocked the taken-for-granted accepted dogma; later, one tries to recover what is truly important and significant. For example, Ofir, an infantry soldier:

> At that point, the whole world of values crumbled. All that was left were the simple questions that I don't see now I'm quoting to you from my life then. At that point, there was no Schopenhauer or Hegel or Nietzsche, no "relative good" or Socrates. There was just one desire, and in the end one feeling: fear. Then you ask yourself "why am I afraid" and the answer is real simple and not at all sophisticated: "Because I want to live." So then you ask the question that led to all the philosophies in history: "Why do you want to live." And the answer is real simple: You want to live because, and I remember the answer, because you want to see Tali [his future wife] and the family, and I want a red-roofed house in a green field in Australia. Now, after you return, part of the answer becomes your Ten Commandments, because these are the things that were the very kernel of your life then, in these sacred and honest moments. Later, you can argue with this, and ask whether these minutes really represent something in life, or whether they are actually removed from life. But when I came back from the war, my feeling was that these feelings bind me.

This interpretation neither undermines nor combats accepted norms but consciously deals with the same knowledge as a way of separating the wheat from the chaff. According to this interpretation, war enables the veteran to construct a "cleaner," "truer," and "purer" interpretive scheme for the meaning of existence.

The undermining of the interpretive scheme focuses on the model that deals with the social meaning of war. The interviewees assert that the process begins with a difficult encounter with the IDF during the war, engendering feelings of helplessness and despondency. The gap between expectations and reality was so enormous that some refused to acknowledge the reality and interpreted the situation as a tactic, an exercise in camouflage.

Their stories express the shock, despair and pain in the meeting with war. The war was so different from what they had been promised, so different from what they believed it would be, so out of line from what they were trained to do.

They were deeply disappointed by the IDF's performance and arrogance. According to these veterans, the IDF was neither "strong" nor "good." The stories present a different picture of the IDF, one far removed from the traditional image and the Israeli fighter ethos. Yigal, in the tanks brigade, expresses this view:

> Myths like, you don't leave the wounded, you don't leave them behind on the battlefield. The commanders are like that . . . all that disappeared . . . when you see how they leave the wounded, how they leave bodies, how . . . hysterical commanders do not command, don't have control, when you see how the soldiers don't believe in their commanders, how the soldiers avoid their responsibilities, so as not to be wounded. So . . . maybe myths aren't broken, they simply fade out.

Belief in the IDF, and the idea that it is a strong, advanced, and ethical army were shattered. The interviewees present the IDF as a physically and morally battered army.

It is no accident that the 1967 (Six Day) War is always present. The comparison with this war is interwoven (at both explicit and implicit levels) in all the stories that present this interpretation. The interviewees undergo profound soul searching when confronted with militaristic world-views and with the feeling of collective arrogance that became deeply rooted after 1967.

The undermining interpretation also incorporates the alternative interpretation that was more strongly voiced in the Lebanon War (Helman 1993; Lieblich 1989) and which continues to be heard during the *Intifada*. Although the undermining interpretation is not the dominant voice in the life stories of the Yom Kippur fighters, the manifestations of this interpretation are powerful in the overall context of the "war voices" heard in the social group that we investigated. The obvious question is: who is doing the undermining? And can these subjects be characterized in terms of their patterns of involvement in the war?

Three characteristics are typical of this group which, as stated, comprises about one-fourth of all the interviewees: First, most of them are combat soldiers who do not express any doubts about the extent to which they were combative participants in the war. Second, most of these soldiers were involved (through combat experience or otherwise) in the battles and events that have come to symbolize the Yom Kippur War and constitute its collective memory. Third, most of them finished their army service shortly after the end of the war and did not spend a prolonged period serving in the military.[5]

The "Accompanying Voice"—Interpretation of Isolation

Although the interpretations of continuity and discontinuity lend a different meaning to war, they both acknowledge the direct tie between war and life and both view war as part of the course of life. In addition, there is a third interpretation as well: the "interpretation of isolation." Here, war is seen as a separate, unrelated experience, not part of the social models that construct the reality of daily life. War is presented outside the continuum of life.

If the interpretation of continuity is the "dominant voice" and the interpretation of discontinuity is the "other voice," the interpretation of isolation is the "accompanying voice." To differing degrees, this voice appears in most of the stories. This interpretation appears beside and always accompanies the interpretation of continuity or discontinuity. The interpretation of isolation does not stand alone.

This interpretation has two central meanings. In the first, war is defined as a "different world" and therefore irrelevant to daily reality; the models that interpret the reality of war are neither valid nor relevant to constructing the course of life. Arnon, a soldier in the infantry, describes this well: "Life and war are parallel, and do not intersect."

Ilan, an infantryman, concurs: "A different planet (...) You can't apply it at all to any other area of life."

The second interpretation of isolation defines war in psychological terms as a repressed experience, hence one that is not present in daily life. The veteran does not consciously deal with the experience and therefore it is not part of the daily interpretive "game." This psychological meaning is clearly expressed by Moshe, a member of the armored corps:

> It passed for a simple reason—my repression mechanisms are very strong, very sophisticated. I didn't even have one nightmare at the time I saw all the horrors of war, all the horrors!

Moshe uses psychological jargon of "repression mechanisms," as do many other interviewees.

The two meanings of the isolation of the experience from life are fundamentally different. There are two types of narrators present here: the "phenomenological" narrator who defines the war as "other reality" and the "psychological" narrator who defines the war as "suppressed reality." The "phenomenologist" focuses on the war, emphasizing the nature of the experience as a social entity; the "psychologist"concentrates on the fighter who has been through the war and is busy coping with memories of the experience. The phenomenologist examines the relevance of this experience to the everyday social environment of the actor; the psychologist relates to the presence of war in the inner environment of the actor. The phenomenologist defines the war as a totally different experience from normal life; the psychologist defines it as a particularly difficult experience. According to the phenomenologist's interpretation, the experience was outside life, since these are two social entities which do not meet; the psychologist assumes that there is an encounter between war and life, but that this is on the unconscious level, since war is so difficult to cope with. The isolation of war from the life course is mainly evident when veterans seek traces of the experience in their private lives. In this regard, the notion of isolation—especially in the psychological definition —offers an easily accessible cultural model for veterans, who explain why it is difficult for them to place, identify, and articulate the influence of war on their lives.

The "Voices of War"—Dual Interpretation

On the face of it, the interpretation that isolates war from life runs counter to the other interpretations—of continuity and discontinuity—which tie it to life. How can war simultaneously be defined as part of life and outside it? It appears that lack of consistency is one of the means of dealing phenomenologically with complex situations, including the war experience. I would argue that the internal contradiction in presenting the war in the life stories is what enables the veteran to live with and "normalize" war in his daily life.

The different interpretations should be seen as a complex of "voices of war" which construct a dual perception of experience—simultaneously part of life and outside life. The dual perception is an additional interpretive mechanism used to integrate war into the course of life.

The dual interpretation coalesces war into the course of normal life, but at the same time remains a "lacuna." War is a meaningful event in an autobiography but one which generally leaves no obvious traces in the life story. This experience constructs the veterans' reality of life but is at the same time defined by them as a different reality, neither relevant nor present. The dual interpretation of war is a paradoxical concept—the normal in the abnormal. War is integrated in life by removing it from life; war is normalizes by delimiting and isolating it.

I claim that the tension and the lack of consistency between the interpretations allow the individual interpretive flexibility and degrees of freedom in coping with the experience. This dual interpretation does not present a continuum from isolation to connection. These are two important facets of the same outlook—sometimes one is referenced and sometimes the other is. The question is: in what context does the teller employ the interpretation which connects war to life and in what context does he use the interpretation which isolates war from life?

Analysis of the life stories demonstrates that interpretation is dependent upon context: the connection of war to life (in a continuous or a discontinuous way) comes to the fore when the teller is concerned with the public sphere of the experience and his role as a citizen. The isolating aspect is more common when the veteran deals with the experience as a person and speculates about the meaning of war in his private life—career, family, friends, and so on. The interpretation of continuity appears when the veteran faces the fact that every few years he is exposed to a war situation and sometimes is sent to the front. In this context, he cannot ignore the presence of war and its relevance to his public life. War must be seen as part of life, whether by accepting or rejecting it. In contrast, removing war from private life is an extremely effective interpretive model for explaining why war did not affect the personal biography and why it is difficult to distinguish its marks on the course of daily life.

The more difficulty the veteran had in identifying the vestiges of war in his private life, the greater his need for the interpretation of isolation. What is striking and interesting is that in the stories which presented the interpretation of continuity, the teller had greater difficulty in pinpointing the impact of war on the private life story and he expressed the interpretation of isolation in a more powerful fashion. And so, paradoxically, the more a person integrates the war into his life, the more he removes it from his life. When he strongly expresses the dominant voice—the interpretation of continuity—the dual interpretation plays a greater role in his life story.

Discussion: The Personal Meaning of War Within the Cultural Context

An analysis of the "interpretative relevance" of war in the life stories of Israeli soldiers shows that war is not perceived by the interviewees as a traumatic experience, but rather as one that is integrated in the flow of their lives. War is thus a "normalized experience" in their lives, contrary to the findings reported in the general literature relating to the impact of war on demobilized fighters (see also Syna-Desivilya 1991). The life stories reveal two interpretative mechanisms used by interviewees in integrating and normalizing the war experience in their lives. The first is the contiguous interpretation of war: i.e., integrating the war into the flow of life by means of interpretative schemes which were existent prior to the war. The second mechanism is the dual interpretation of war: i.e., representing it as an experience that was simultaneously part of life and removed from it.

It is precisely this perception of war, as a "normal" experience integrated in the flow of expected life without any "noise," that expresses the centrality of war in Israeli life. This perception reflects the definition of war as an integral part of life in Israeli society, as well as an additional 'voice' in the construction of this meaning. Subjects' interpretation of war is an expression of what Kimmerling

(1993) calls civil militarism or cognitive militarism, which characterizes Israeli society—i.e., a state in which the members of a society take the centrality of war in life for granted. The expectation that war will break out sooner or later is part of the collective, as well as personal identity and conscience.

The interpretive mechanisms that were seen in the life stories—the interpretation of continuity and the dual interpretation—are symbolic expressions on the individual level for the routinization of war into the life-world. These cognitive "strategies" can be seen as mediation mechanisms that translate the collective meaning of the war experience as a "normal experience" into personal meaning. These interpretations enable the individual to satisfy the social demand that war be smoothly integrated into his life. Put differently, when war does not cause social change, it will probably not affect the personal biography. In the Israeli context, coping with war means living as if the war had not taken place—on both the societal as well as the individual levels.

I would contend that the personal meaning of war, as voiced by these interpretations, allows veterans to cope with the frequent demand to participate in war as well as to live their lives as if the war had not taken place. This outlook on the experience does not only serve the society; it also allows the individual to live with war, without shocks or extreme changes, in the demanding context of Israeli society. It therefore recruits the commitment and conscription of members of society and controls the disruptive effects of crucial changes for the individual that arise in a war situation.

It is hard not to be impressed by the way subjects integrate the war into their life and by their commitment to the social meaning attached to war. It appears that this is linked to a large extent or even dependent upon the institutionalization of relations between society and the army (Horowitz and Lissak 1989); or in the characteristic nature of Israel's wars (in close proximity and short term) which enable a rapid return to routine on the general societal level as well as the personal level (Kimmerling 1985, 1993). However, there are also macrosocial arrangements that construct the normalization of this experience on the individual level and serve as mechanisms for social control. Several such arrangements can be suggested.

The first arrangement is the entire range of socialization processes in Israeli society during childhood and adulthood, which prepare the youngster for the role of soldier and for a state of war. The recognition of war and of the army as part of reality is consolidated in light of the repeated picture of a father going off to the army and accompanies the life cycle until the stage when the youngster himself becomes a father accompanying his own sons as they go off to army service (Katriel 1991; Lieblich 1989). The war and all that goes with it are among the most central themes reflected in formal and informal curricula, overtly and latently (Bar-Tal and Zoltak 1990; Bet-El and Ben-Amos 1995); and the presence of the army and of security are daily and central features in the media (Herzog and Shamir 1994; Katz et al. 1992). The Jewish Israeli prepares for war and faces war throughout his life: the internalization of the war as part of life is the result of socialization processes.

An additional social arrangement structuring the normalization of war on the individual level is military service. The army is apparently a central social institution which functions as an agent for socialization for fighting situations through regular service and reserve duty. However, in terms of its significance in the construction and preservation of the war experience, as part of life and beyond life, military service is much more than just another socialization agent. Regular service is a central element in the social and cultural construction of the stage of youth in Israeli society (Azarya 1989; Lieblich 1989; Rapoport and Lomsky-Feder 1994). The frequency of wars and the need to recruit youngsters for the sake of the collective have obliged the Israeli state since its establishment to shape what Mosse (1990) terms the "myth of participation in war." In other words, society has systematically and deliberately reinforced the noble facet of military service and war. The heroism of war is culturally structured through the myth of the warrior hero, which is a key element in the ethos of the *sabra* (native-born Israeli) and has served as a model of identification for generations of young Jewish Israelis (Gal 1986; Horowitz 1988). Israeli society has succeeded in channeling the aspirations of its youngsters into military service and in structuring the characteristics of youth through the role of the soldier.

The institution of reserve duty is also central in terms of the social structuralization of the significance and normalization of war within life. This is one of the key mechanisms focusing male discourse around the army and war (Ben-Ari 1992; Helman 1993). Reserve duty is an "island" of the army within everyday life, exposing participants to fighting situations in an interrupted continuum over decades of their life.

A third social arrangement that fosters the normalization of war on the individual level is the continual and intensive preoccupation with the war experience in the different "cultural fields." Israeli society is preoccupied with war—particularly with its painful and critical aspects, in the various channels of its culture. Social time is dotted with wars—from the Memorial Day ceremonies immediately before Independence Day, through the anniversaries of the various wars (Handelman and Katz 1991). Relative to other societies, the culture of commemoration in Israeli society is exceptional with regard to the number of memorial sites and the extent of memorial literature (Levinger 1993; Sivan 1991). War is a central theme in Israeli literature, drama and poetry (Katz et al. 1992; Meron 1992; Shaked 1988). It occupies artists (Ofrat 1991; Tzalmona 1993) and appears as a central theme in popular culture, on the radio, in the movies, and in rock music (Katz et al. 1992).

This intensive and public preoccupation removes the subjects from the private domain and implies an acceptance and institutionalization of the "naturalness" of war in the Jewish Israeli's life. Addressing the cost of war, from a position of empathy for pain or from a critical standpoint, through so many channels and on an almost daily basis, leads to the expropriation in the collective arena of the personal experience of coping with these subjects. This may also allow the pain, tension, anger, and criticism to be channeled into the public arena enabling the individual to continue life as usual. These "cultural fields" can be seen as arenas for a form of catharsis that is at once collective and individual.

Notes

*   This article was first published in *International Sociology*, 10, 4: 463-482.
1.   Exception to this is, for example, the studie by Schutze (1992), that discusses the way in which the collective guilt of the German people affects the individual memory of German soldiers and civilians who lived through the Second World War.
2.   The following analysis of the material includes excerpts from life stories. The excerpts are transcribed with minimal editing, to ensure optimal proximity of the reader to the manner in which the narrator expressed himself. Where the narrator's comments included breaks in speech, the sign ... is used. Although the excerpts were not edited, the very choice of excerpts includes a significant dimension of editing. The sign (...) marks an omission.

3. The narrator refers to the Jewish mourning customs which are also part of rituals for Yom Kippur (Day of Atonement), a highly significant day for observant Jews. The war began on this religious holiday. This description is interesting and unique, because the narrator is not religious; this is an instance of the absorption of religious symbols into a secular context.
4. Peace Now is a social protest movement established in the 1970s to push forward the peace process between Israel and the Palestinians.
5. A detailed discussion of the connection between the personal, crisis meaning carried by war among this group, and patterns of participation in war appears in Lomsky-Feder (1994).

References

Azarya, Victor 1989. "Civil Education in the Israeli Armed Forces." In *Education in a Comparative Context,* edited by E. Krauz, 119-47. New Brunswick, NJ: Transaction Publishers.

Bar-On, Dan 1993. "Peace pre-Traumatic Disorder (PPTSD): the Israeli Experience." Paper presented during a symposium of the Israeli Society for Family Therapy, May 19, Tel-Aviv University.

Bar-Tal, Daniel and Shmuel Zoltak. 1989. "The Reflection of the Character of the Arab and Jewish-Arab Relations in Readers." *Megamot*, 32, 3: 301-17.

Ben-Ari, Eyal 1992. *Conflict in the Military World View: An Ethnography of An Israeli Infantry Battalion.* Jerusalem: H. S. Truman Research Institute and Shain Center.

Berger, Peter and Thomas Luckmann 1975. *The Social Construction of Reality.* Harmondsworth: Penguin.

Bertaux, Daniel and Martin Kohli 1984. "The Life History Approach: A Continental View." *Annual Review of Sociology*, 10: 215-37.

Bet-El, Ilana and Avner Ben-Amos 1995. "Rituals of Democracy: Ceremonies of Commemoration in Israeli Schools." Paper presented in the Annual Meeting of the Israeli Anthropological Association, January 4.

Bruner, Jerome 1995. "Life as Narrative." *Social Research*, 54, 1: 12-32.

Card, Josefina 1983. *Life After Vietnam.* Toronto: Lexington Books.

Cohler, Bertram 1982. "Personal Narrative and Life Course." In *Life-Span Development and Behavior*, edited by Paul Baltes and Orville Brim, 206-41. New York: Academic Press.

Corradi, Consuelo 1991. "Text, Context and Individual Meaning: Rethinking Life Stories in Hermeneutic Framework." *Discourse and Society*, 2, 1: 105-18.

Crapanzano, Vincent 1984. "Life History." *American Anthropologist*, 86: 953-9.

Danish, Steven, Michael Smyer and Carol Nowak 1980. "Developmental Intervention: Enhancing Life-Event Processes." In *Life-Span Development*, edited by Paul Baltes and Orville Brim, 339-66. New York: Academic Press.

Denzin, Norman 1989. *Interpretive Biography.* London: Sage.

Figley, Charles 1978. "The Psychology Adjustment Among Vietnam Veterans: An Overview of Research." In *Stress Disorders Among Vietnam Veterans*, edited by Charles Figley, 57-70. New York: Brun Mazzel.

Figley, Charles and Seymour Levantman (eds.) 1980. *Strangers at Home.* New York: Praeger Publishers.

Frank, Gelga and Rosamaund Vanderburgh 1986. "Cross Cultural Use of Life History Methods in Gerontology." In *New Methods For Old Age Research*, edited by Christine Fry and Jennie Keith, 185-212. South Hadley, MA: Bergegnan Garey.

Gal, Reuven 1986. *The Portrait of the Israeli Soldier.* New York: Greenwood Press.

Helman, Sara 1993. *Conscientious Objection to Military Service as an Attempt to Redefine the Contents of Citizenship.* Ph.D. Dissertation, Hebrew University of Jerusalem. (Hebrew).

Handelman, Don and Elihu Katz 1991. "State Ceremonies In Israel: Remembrance Day and Independence Day." In *Models and Mirrors: Towards an Anthropology of Public Events,* edited by Don Handelman, 191-233. Cambridge: Cambridge University Press.

Herzog, Hanna and Ronen Shamir 1994. "Negotiated Society? Media Discourse on Israeli Jewish/Arab Relations." *Israel Social Science Research,* 9, 1-2: 55-88.

Horowitz, Dan 1993. *The Heavens and the Earth: A Self-Portrait of the 1948 Generation.* Jerusalem: Keter. (Hebrew).

Horowitz, Dan and Moshe Lissak 1989. *Trouble in Utopia.* Albany: State University of New York Press.

Katriel, Tamar 1991. "Picnics in a Military Zone: Rituals of Parenting and the Politics of Consensus." In *Communal Webs,* edited by Tamar Katriel, 71-91. New York: State University of New York Press.

Katz, Elihu et al. 1992. *Leisure Culture in Israel: Changes in Patterns of Cultural Activity, 1970-90.* Jerusalem: The Louis Guttman Israel Institute of Applied Social Research.

Kimmerling, Baruch 1993. "Patterns of Militarism in Israel." *European Journal of Sociology,* 2: 1-28.

Kimmerling, Baruch 1985. *The Interrupted System.* New Brunswick, NJ and Oxford: Transaction Publishers.

Laufer, Robert 1988a. "The Aftermath of War: Adult Socialization and Political Development." In *Political Learning in Adulthood,* edited by Roberta Sigel, 415-57. Chicago: University of Chicago Press.

Laufer, Robert 1988b. "The Serial Self." In *Human Adaptation to Extreme Stress,* edited by John Wilson, 33-53. New York: Plenum Press.

Levinger, Ester 1993. *War Memorials in Israel.* Tel-Aviv: Hakibbutz Hameuchad. (Hebrew).

Lieblich, Amia 1989. *Transition to Adulthood During Military Service: The Israeli Case.* Albany: State University of New York Press.

Lifton, Robert 1973. *Home From the War.* New York: Simon and Schuster.

Lissak, Moshe 1971. "The Israel Defence Forces as an Agent of Socialization: A Research in Expansion in Democratic Society." In *The Perceived Role of the Military,* edited by M.R. van Glis, 327-339. Rotterdam: Rotterdam University Press.

Lomsky-Feder, Edna 1994. *Patterns of Participation in War and the Constructions of War in the Life-Course: Life Stories of Israeli Veterans From the Yom Kippur War.* Ph.D. Dissertation, Hebrew University of Jerusalem. (Hebrew).

Meron, Dan 1992. *Facing the Silent Brother.* Jerusalem: Keter. (Hebrew).

Milgram, Norman 1986. "An Attribution Analysis of War-Related Stress: Modes of Coping and Helping." In *Stress and Coping in Time,* edited by Norman Milgram, 9-25. New York: Brunner/Mazel.

Mintz, Alex 1984. "The Military Industrial Complex: The Israeli Case." In *Israeli Society and its Defence Establishment,* edited by Moshe Lissak, 103-27. London: Frank Cass.

Modell, John and Timothy Haggerty 1991. "The Social Impact of War." *Annual Review of Sociology*, 17: 205-24.

Mosse, George 1990. *Fallen Soldiers: Reshaping the Memory of World Wars*. Oxford: Oxford University Press.

Myerhoff, Barbara 1980. "Life History Among the Elderly: Performance, Visibility and Remembering." In *Life Course: Integrative Theories and Exemplary Population*, edited by Kurt Back, 133-53. Boulder, CO: Westview Press.

Natanson, Maurice 1962. *Literature, Philosophy and the Social Sciences*. The Hague: Martinus Nijhoff.

Ofrat, Gideon 1991. "The Fading of the Khaki." *Studio*, 27: 6-11. (Hebrew).

Rapoport Tamar and Edna Lomsky-Feder 1994. "Israel." In *International Handbook of Adolescence*, edited by Klaus Hurrelman, 207-33. Westport, CT: Greenwood Press.

Reese, Hayne and Kathleen McCluske 1984. "Dimensions of Historical Constancy and Change." In *Life-Span Developmental Psychology*, edited by Kathleen McCluske and Hayne Reese, 17-45. New York: Academic Press.

Ryff, Carol 1984. "Personality Development From the Inside: The Subjective Experience of Change in Adulthood and Aging." In *Life-Span Development and Behavior*, edited by Paul Baltes and Orville Brim, 243-79. New York: Academic Press.

Sarbin, Theodore 1986. "The Narrative as A Root Metaphor for Psychology." *Narrative Psychology*, edited by Theodore Sarbin, 3-21. New York: Praeger.

Schutz, Alfred 1970. *On Phenomenology and Social Relations*. Chicago: University of Chicago Press.

Schutz, Alfred 1945. "The Homecomer." *American Journal of Sociology*, 50, 5: 369-75.

Schutz, Alfred and Thomas Luckmann 1974. *The Structure of the Life-World*. London: Heinemann.

Schutze, Fritz 1992. "Pressure and Guilt: War Experiences of a Young German Soldier and Their Biographical Implications." *International Sociology*, 7, 2: 187-207 and 7, 3: 347-67.

Segev, Tom 1991. *The Seventh Million*. Jerusalem: Keter. (Hebrew).

Shaked, Gershon 1988. "Shamir, Shaham, Mosenson—Three Authors on The War." Paper presented at a seminar on "Reading War in Literature, Film and the Media." Everyman's University and Tel-Aviv University. (Hebrew).

Sivan, Emmanuel 1991. *The 1948 Generation: Myth, Profile And Memory*. Tel-Aviv: Marachot. (Hebrew).

Syna-Desivilya, Helena 1991. *War Experiences In the Israeli Veterans' Lives: A Life Course Perspective*. Zikhron Ya'akov: Israeli Institute For Military Studies.

Tzalmona, Ygal 1993. "Hummus with Whipped Cream?" *Mishkafayim*, 19: 9-14. (Hebrew).

Watson, Lawrence 1976. "Understanding a Life History as a Subjective Document." *Ethos*, 4, 1: 95-131.

9

Citizenship Regime, Identity and Peace Protest in Israel*

Sara Helman

On March 1997, two helicopters carrying Israeli troops to routine military tasks in South Lebanon collided, killing 73 soldiers. The heavy toll exacted by the accident threw the country into deep grief and mourning; expressions of national solidarity as well as of a momentary unity could be felt all over. The event, dubbed in public discourse as the "helicopters' disaster," evoked two seemingly contradictory responses.

By the end of the week, expressions of national solidarity gave way to what is usually termed in Israel as the "blood account" (*chesbon ha'dam*). The printed media was swamped by a continuous flow of letters to the editor, editorials and articles referring to the disaster. Talk shows turned into an arena in which the meaning of the disaster was heatedly debated. A common theme could be discerned from the heated and charged debate conducted in and through the media: who were the fallen soldiers and to what groups in Israel they belonged. By explicitly stating to what groups the fallen soldiers belonged (kibbutz and moshav members, city dwellers, veteran Israelis, new immigrants, Bedouins and Druzes), media discourse established by default what groups were missing: the ultra-orthodox Jews, the Palestinian citizens of Israel and some spoiled middle-class youngsters that refused to be drafted to the Israeli army.

The colliding helicopters became a metonymy of the nation. The bodies of the fallen soldiers signaled its boundaries as well as the criteria of membership in it.[1] The intensive public debate—artfully orchestrated by the media—capitalized on one of the symbols of Israeli-Jewish national identity, the Israeli army. Participation in military service was mobilized as the pristine expression of membership in the nation and the Israeli army represented as the main mechanism forging national solidarity. The "helicopters' disaster" turned into a macabre celebration of a national identity that seemed to be fading away after the 1996 elections.[2]

The second response to the "helicopters' disaster" highlighted the heavy toll exacted by the Israeli occupation of Southern Lebanon. Two months after the collision, a group of women, claiming to represent a wide public of soldiers' mothers, most of them living in the Northern part of Israel, established a movement called "Four Mothers."[3] The movement launched a campaign for the withdrawal of the Israeli forces from Southern Lebanon. The nurturance and care associated with republican motherhood was diverted and imbued with a new content: the well-being of the soldiers and of the nation was to be protected by bringing the sons back home. Republican motherhood was transformed into an oppositional symbol.

While both responses to the helicopters' disaster capitalized on consensual symbols and mobilized them to challenge the legitimacy of the ruling coalition and its policies, the following event displayed the opposite dynamics. The Prime Minister of Israel mobilized the same symbolic repertoire to delegitimize challengers of both governmental policies and the ruling coalition. In October 1997, during Succoth (the Tabernacles Feast), Netanyahu visited Kabalist Itzchak Kaduri. While sitting in the Tabernacles Feast booth, Netanyahu whispered to Kaduri: "the left has forgotten what to be a Jew means. They want the Arabs to defend us." The whisper was recorded by a journalist of the Israeli Broadcasting Authority, and became the hottest news in the country. Netanyahu's statement reasserted the intimate link between military might and Jewishness. The ongoing readiness to defend Jewish sovereignty over the Land of Israel and an uncompromising attitude towards the Palestinians were considered as the main criteria of membership in the nation. The willingness to compromise with the Palestinians—so Netanyahu's statement implied—casts doubts on the loyalty, masculinity and military prowess of the "left."

This statement aroused an immediate and angry reaction of rank and file citizens, most of them belonging to the "left." Letters were sent to newspapers' editors, members of Netanyahu's military unit published a petition in a daily newspaper, and by the end of the week a group of reserve soldiers marched from Tel-Aviv to Jerusalem and joined other demonstrators to protest against what they considered an insulting statement.

What is the common denominator of the former events? The main protagonists of the different events—be those staged by the media and those in the demonstrations and petitions—are individuals and groups belonging to the Israeli peace movement. The "left" alluded to by Netanyahu comprises an array of organizations that since the 1970s has conducted a sustained challenge to the military policies of the state, and has striven for the de-escalation of the Arab-Israeli conflict. However, the intriguing aspect of the phenomena is that the main organizations of the Israeli peace movement (Peace Now, Mothers against Silence, There is a Limit [*Yesh Gvul*], Four Mothers, etc.) display militaristic symbols and skillfully appeal to the "security culture" (Kimmerling 1998) in order to frame their actions and challenges against this very culture. Another outstanding feature of the Israeli peace movement is the social composition of its leadership and constituency: it appeals mainly to a particular ethno-class (middle class Israeli-Jews of Western origin or Ashkenazis). Despite the mobilization of consensual codes—such as the loyalty to the IDF, the military record of its leadership and its commitment to the Zionist character of the Israeli state—the Israeli peace movement has not succeeded in broadening its constituency beyond the limits of the Ashkenazi middle classes. Its attempts to mobilize wider constituencies (such as the Israeli-Jews of Oriental origin or Mizrahim and the Palestinian citizens of Israel) had not only been limited but also unsuccessful, generating animosity and alienation (Feige 1995; Reshef 1996).

Moreover, despite the fact women have been prominent amongst the movements' constituency, they have been marginal in leadership positions, performing mainly administrative and technical tasks (Sasson-Levi 1995). The close association between the fighter's role and peace activity has pushed women to the periphery of its organizations and networks. Therefore, women's peace activity mobilizes republican motherhood and womanhood (Helman and Rapoport 1997; Hermann and Kurtz 1995; Zuckerman-Bareli and Benski 1989), within the framework of single gender organizations. However, republican motherhood—and even less womanhood—have not been attractive frames to women other than those belonging to the Ashkenazi middle classes.

Why is it that a peace movement celebrates a national identity marked by militaristic undertones while peace and peace protest had turned into the exclusive asset of an ethno-class? Moreover, why does the Israeli peace movement still adhere to soldiering and motherhood as mobilizing frames while the conditions that gave raise to these frames are precisely those that the Peace movement's leadership and constituency are protesting against?

My contention is that in order to understand this seeming paradox it is not enough to examine the class bases of the Israeli peace movement (see Levy 1997) nor the pattern of national identity it has promoted in its collective action frameworks (Kimmerling 1985, 1998). Each of these perspectives is valuable, yet limited. The class-based analysis grants changing material interests (however unconscious they maybe) a weight that ignores the ways in which institutions and mainly political institutions (such as the state and its agents) shape not only material interests but also social identities. The political culture analysis, while successfully mapping the oppositional symbols and versions of national identity elaborated by social movements, tends to ignore the ways in which class interests converge in different versions of national identity. While both frameworks of analysis prioritize state-making and war-making as explanatory factors, they disregard the ways in which the state through war shapes the citizenship regime. The citizenship regime is more than the mode of incorporation of social categories into citizenship. For as Tilly (1995: 9) has observed: "states often use previously existing ties ... for forming ties of citizenship or as grounds for the exclusion from citizenship."

I thus propose that the citizenship regime may constitute a linking mechanism between class structure and national identity and may explain how social classes develop different versions of national identity. Modes of inclusion (full, partial, differential, or exclusion) are constitutive of socio-political identities. In turn the identities constituted by state's policies in the area of citizenship may be embraced to challenge state's policies and institutional arrangements (Helman 1999a; Marx 1995). Moreover, the mode of inclusion into citizenship not only shapes socio-political identities but also the coalitions formed in the process of mobilization for collective action.

I sustain that peace politics in Israel should be understood with reference to the ways in which the state through war has shaped citizenship regime. My claim is that war-making and conflict management have ethnicized and genderized citizenship in Israel through the formation of a hierarchical structure of citizenship. This hierarchy was constructed through the differential incorporation of social categories in missions through which the Zionist national project was composed. This differential incorporation has shaped the class position of different groups as well as the versions of national identity represented in their collective action frameworks. In turn, this differential incorporation may explain the riddle I began with, namely, why peace protest is framed in militarist symbolism and is confined to the Ashkenazi middle class.

This article is composed of four sections. In the first I analyze the citizenship regime in Israel and how war-making and routine conflict management have been constitutive of its different positions. In the second section, a short outline of the challenges to the citizenship regime is presented. This section outlines the opportunity structure that enabled the emergence of the Israeli Peace movement. The third and four sections discuss four organizations: Peace Now, There is a Limit, Mothers against Silence, and Women in Black. The final part of the article presents a discussion of the limited mobilizatory potential of the Israeli peace movement.

Citizenship in Israel: Zionism, the State, and War Management

National projects strive for a separate political representation for a collective and establish through narratives, practices and institutions, the boundaries of the community of citizens as well as the place of different categories within that community. However, national projects take place and are shaped by historical forces and structural constraints. These forces and constraints explain the type of ties between citizenship and national identity that national projects promote. The Zionist national project, "imagined" the boundaries of the citizens' community and its links to the nation in ethnic-cultural terms, rather than in relation to the administrative and territorial boundaries of the state. The Zionist character of the state of

Israel can explain how citizenship rights and obligations have been nationally framed, but it cannot explain the different types or forms of citizenship that comprise the community of citizens, its institutions, practices, standards of civic value and, consequently, the flow of resources and rewards. Moreover, Zionism can explain who may be singled out as a potential candidate to be granted effective membership, but not how effective membership is instituted and reproduced.

Zionism as a national project that sought a separate political representation of Jews took its shape within the framework of a colonising enterprise. The Zionist colonisation and settlement of Palestine was characterised by a protracted state of conflict between Jews and Palestinians, a conflict that Kimmerling (1983), Peled and Shafir (1996), Shafir (1989) and others have identified as featuring settler or frontier societies.[4] The management of conflict shaped the institutional makeup, infrastructural capacity and legitimacy of the Jewish community in Palestine.

War and routine conflict management, and its construction in terms of a struggle for survival, has been central to the consolidation and enhancement of the power and autonomy of the Israeli state since its establishment in 1948. As such, it structured the relationship between citizenship, ethnic origin, gender and nationality which gave rise to a stratified and fragmented structure of citizenship in which different forms or types of citizenship coexist, and are legitimated by different discourses (Peled and Shafir 1996).

War and routine conflict management played a central role in the shaping of the community of citizens, as well as in the organisation of membership and participation in the political community. Moreover, war and the army (based on universal conscription) have been the main mechanisms for the construction of what Ben-Eliezer (1995: 280-308) has conceptualised as the "ethnic nation," i.e., in imagining and constructing the bonds and solidarity between settler-immigrants in terms of a common project, the defence of Jewish sovereignty. Within the boundaries of the ethnic nation, full and effective citizenship has been constructed in republican terms, i.e., with an emphasis on the individual's contribution to the fulfilment of collective goals (see Oldfield 1990; Peled 1993; Van Gunsteren 1994). However, the republican principle was differentially implemented, by assigning social categories to the different missions the nation and state building project generated. This differential implementation gave rise to different types or forms of membership and participation, ordered in a hierarchical structure. Moreover, this hierarchical ordering of national missions shaped the social identity of the different categories, and created links and overlaps between ethnic origin, class position, and interpretations of national identity.

Participation in war and military service was identified as the ultimate token of political obligation as well as the highest contribution to the enhancement of collective goals. The institutions of war making, and especially the military, were instituted as the prime arenas of political integration and as the signifiers of full and effective membership in the political community (see e.g., Horowitz 1982; Horowitz and Kimmerling 1974; Kimmerling 1979). Thus, civic virtue was constructed in terms of, and identified with military virtue. However, military virtue was differentially distributed, thereby reflecting the ways in which the state in Israel reshaped the links between gender, class, ethnic origins, nationality, and citizenship. The fighter's role—the ultimate expression of civic virtue—came to be "naturally" associated with Ashkenazi-Jewish males, whereas Mizrahi Jews (males and female) were relegated to the periphery of civic virtue, thereby strengthening through the idiom of a mass and egalitarian army their already peripheral position in society (see e.g., Levy 1996).[5]

The formation of the citizenship regime went hand in hand with the formation of the class structure of Israel. The state's immigration absorption and settlement policies (Seguev 1984), the preferred pattern of economic growth and development (Carmi and Rosenfeld 1979; Swirski and Bernstein 1980) shaped an ethnicized class structure. In this structure, Ashkenazi Jews were highly represented in the state formed middle classes, whereas the state shaped Oriental-Jews as the working class.

This class structure and the positions of the different groups within it were legitimized by the republican principle, the pillar of the statist ideology (*Mamlachtyut*). Thus, the unequal allocation of material resources and the differences of prestige accruing from the different positions were legitimized both in terms of the relative contribution of the different groups to the enhancement of the collective good as well as in terms of the cultural capital of each of them (Carmi and Rosenfeld 1979; Levy 1997; Swirski and Bernstein

1980). Cultural capital and contribution to the enhancement of the common good were closely related. For in order to effectively participate in the political community and capitalise on their contribution, Oriental-Jews were required to transform themselves. Jewishness was a necessary but not a sufficient attribute to progress through the nation (Cohen 1989). However, class marginality reinforced those very attributes that were considered as hindrances to effective participation in the political community.

Whereas military service was mandatory for both men and women, the equation between civic and military virtue was gendered, and applied exclusively to men (Izraeli 1997). Women were accorded a different mission within the national project, despite their being drafted. The defence of Jewish sovereignty was also conceived in terms of the continual growth of the Jewish population, to counteract the presence of a different national-ethnic community (the Palestinian citizens of Israel) within the framework of the settler state (Berkovitch 1997). Whereas the national mission of Jewish men was that of "military defenders of the nation," the national mission of Jewish women was to be "physical reproducers of the nation" (Berkovitch 1997; Walby 1992; Yuval-Davis 1997). Thus citizenship for Jewish women meant their inclusion as mothers within the boundaries of the citizens' community (Berkovitch 1997).[6]

The ethnic-cultural understanding of nationhood, as well as the constraints and imperatives emerging from the settler colonial character of the Israeli state, brought about the inclusion within the republican project of categories that did not comply with the equation civic virtue = military virtue. These categories ranged from opposition to ambivalence in their stance towards the state of Israel: i.e., the ultra-orthodox and religious groups. In order to incorporate them within the nation building project, the state in Israel created special legal arrangements that postponed the military service of ultra-orthodox males and exempted religious Jewish women from it (see Hofnung 1991: 238-47; Kimmerling 1979, 1994). The pattern of incorporation of religious men and women into citizenship was constructed in terms of their role in preserving the "cultural survival of the collectivity." The religious sectors of society provided the legitimation needed by a settler state and society in a state of protracted conflict with its geopolitical environment (Kimmerling 1994; Peled and Shafir 1996).

Whereas the religious and ultra-orthodox groups were incorporated into the community of citizens by exempting them from military service, the same principle turned into an exclusionary practice towards the Palestinian citizens of Israel. Their exclusion from military service, as well as from the access to the institutions that were signalled as furthering the collective good, shaped their membership and participation in the community of citizens. As the Palestinian citizens of Israel are debarred from the furthering of collective goals, this has granted them a differential access to the rights of citizenship (see e.g. Haidar 1987; Kretzmer 1987). Practices such as differential treatment by welfare agencies (Rosenhek 1995), differential access to the labor market (Lewin-Epstein and Semyonov 1993), and military and political surveillance (Lustick 1980) were legitimised by means of their exclusion from military service. In contrast to the republican type of citizenship characteristic of Jews, the Palestinian citizens of Israel are accorded what Peled (1993) has conceptualised as liberal citizenship, political rights on an individual basis.

The inclusionary drive of nation building should not conceal that the republican principle gave rise to a stratified community of citizens. The republican principle and the relative importance assigned to its different missions, differentiated between men and women, and between Ashkenazi and Mizrahi Jews. The ethno-cultural framing of the "enhancement of the common good," in turn, differentiated between Israeli-Jews and the Palestinian citizens of Israel (Peled 1993, Peled and Shafir 1996). Concomitantly, the citizenship regime shaped social identities of the different social categories. These social identities were mobilized to alter institutional arrangements, as the political opportunity structure changed for each of them (Tarrow 1994: 77-8).

Identities and Social Mobilization:
Challenging the Citizenship Regime

The legitimacy of the different positions within the community of citizens, as well its dominant discourses began to be challenged at the beginnings of the 1970s. In 1971, a social movement composed mainly of second generation lower class Mizrahi Jews– The Black Panthers—questioned their peripheral role in Israel. [7] The movement—important as it was—was short-lived, due to co-optation moves by the political establishment, a weak organizational structure and the 1973 war (see e.g., Bernstein 1984). According to Cohen (1989), the Black Panthers rejected the republican principle (the granting of rewards according to the contribution to the collectivity) and demanded access to equal citizenship in terms of the ethnonational principle, i.e., solely by their Jewishness. However, Oriental Jews also claimed access to the rewards of citizenship by relying on their participation in military service. Thus they mobilized the warfare-welfare nexus in order to challenge their position in the community of citizens as well as the unequal social structure (Levy 1997: 149-54).[8]

Another precedent of the challenge of the position within the community of citizens came from the side of the Palestinian citizens of Israel in 1976. The "Land Day"—a widespread wave of protest against land expropriations in the Galilee—turned into a symbolic media to protest against the deprived position of Palestinians in Israel (see Yiftachel 1999).

It was only by the end of the 1970s that an open questioning of the war policies of the state emerged. The occupation of the West Bank and the Gaza Strip and the 1973 war (Yom Kippur War) engendered divisions in Israel around issues of national security.[9] The change in government in 1977 (the decline of the Labour Party hegemony and the rise to power the right wing Likud Party), signalled the rising dominance of new socio-political coalitions (mobile Oriental Jews, religious groups and parties) (Kimmerling 1993; Shapiro 1989). Concomitantly, due to the changed political hegemony, the Ashkenazi middle classes, traditionally identified with the Labour party, began to develop more autonomous stances on issues of national security. The peace treaty with Egypt (1978) furthered the disagreements over issues of national security, and promoted the crystallisation of protest groups (left and right, doves and hawks).[10]

The emergence of organizations (such as Peace Now and the Block of the Faithful) that openly questioned the state's policies in the area of national security represented a shift in state-society relations. These protest groups organized around the issue of how to enhance or further the defense of Jewish sovereignty and disclosed the different discourses—the republican and the ethno-national (mixed with a strong religious and even messianic component)— legitimating the state of Israel (see e.g., Kimmerling 1993). These different interpretations of the legitimating principles of the state in Israel unveiled controversies over the role of the state, its boundaries (social and physical), and the criteria for the distribution of rights and obligations (see e.g., Kimmerling 1985). However, this challenge to state-society relations was not matched by a significant change in patterns of political obligation as expressed in military service. Moreover, the republican principle—as it applied to the intimate relationship between military service and citizenship—preserved its hegemonic position. Nevertheless, the cycles of protest that slowly developed from the end of the 1970s, mainly with the emergence and growth of Peace Now, bear witness to the changing meaning of the fighter's role. The sectors constituted as prime carriers of civic virtue increasingly mobilized the fighter's role to challenge the war policies of the state, and as a lever to claim participation and voice in security policies (Feige 1995; Reshef 1996). However, the mobilization of the fighter's role as a collective action framework could generate consensus against the system that created it only amongst those groups and individuals constituted as carriers of civic virtue. It marginalized women, alienated the Palestinian citizens of Israel and antagonized significant sections amongst the Israeli Oriental Jews.

From Fighter to Citizen: "Peace Now" and "There is a Limit"

"Peace Now": The Fighter's Role as an Oppositional Symbol

Groups and organizations that conducted a continuous challenge to the state's war policies existed in Israel prior to the end of the 1970s. However, they did not succeed in mobilizing broad constituencies. The symbols and ideological messages they displayed were an anathema to the majority of Israeli-Jews. It was only by the end of the 1970s when the political opportunity structure changed that a massive peace movement began to develop in Israel.

On March 7, 1978, a petition signed by 348 reserve soldiers and officers was published in daily newspapers. The petition—known as the "Officers' Letter" was issued by a group of university students, most of them veterans of the 1973 war. The petition was rich in military symbolism. Its opening paragraph stated the signers' status as a group of citizens in their capacity as reserve commanders

and soldiers. The next sentence referred to the new hopes and opportunities opened by the peace initiative with Egypt and ended with
the request to avoid any step that could endanger the path to peace.
The petition closed with the following sentence: "true security will
be achieved with the coming of peace. The I.D.F strength dwells in
its soldiers' identification with the policy of the state of Israel. We
urge you to choose the path of peace and in so doing to strengthen
our faith in the rightness of our mission."[11]

The signers mobilized their identities as citizen-soldiers to legitimate their claims. However, the emphasis put on citizens in their
capacity as commanders and soldiers (some of them holders of the
highest military decorations), was more than a tactical move. The
social identity mobilized to support the Peace initiative and warn
against its stalemate, was expressive of the way the signers were
incorporated into citizenship and interpellated as ultimate carriers
of civic virtue.

The petition mobilized a semiotic code (Swidler 1995), the
fighter's role, but altered its meaning. It stated the signers' loyalty
to the IDF, but conditioned it to the implementation of peace policies. Moreover, the petition adopted a consensual and sacred symbol in Israeli-Jewish political culture—security—but offered alternative policies and practices to achieve it. It promoted the idea
that peace—and not war-preparations and escalation—is the way
to true security.

The petition succeeded in the formation of consensus across networks and groups all over the country. It was also successful in
mobilizing consensus across networks and groups; the emerging
movement—known as Peace Now—organized massive demonstrations and public events in which the symbols and slogans inscribed
in the "Officers' Letter" were displayed in banners, buttons and elaborated in speeches. The collective action framework capitalized on
the war experience of the 1973 cohort (Feige 1995; Lomski-Feder
1994; Reshef 1996) while simultaneously creating and articulating
cultural meanings and narratives amongst members of a specific
social group.

The successful mobilization of a wide constituency as well as
of allies amongst the political, intellectual and artistic elites, should
not obscure the fact that the collective action framework was the
expression of the social identity and of the culture of a very spe-

cific strata in Israel. One of the central leaders of Peace Now consciously stated this when featuring what he termed "the sociological profile" of the founding group: "Our parents were university professors and professionals. We trained ourselves to follow the same occupational path. Most of us graduated from prestigious high schools, we were members of youth movements, and we served in combat units. We grew up in veteran families and had a complacent and confident sense of belonging. We belonged to the heart of Israeli society, and we were confident that all the paths were open to us. We had the political know-how..." (Reshef 1996: 69).

The potential constraints emerging from the class and ethnic composition of the movement as well as from its symbolic package were not felt during the protest cycle initiated to enhance the peace treaty with Egypt (1978-1979). Although an open struggle over the meanings of security—and through it over national identity—were already under crystallization (see e.g., Cohen 1989; Kimmerling 1985, 1993), they consolidated into a clear antagonism during the 1982 war and its aftermath.

Israel's war in Lebanon was a highly contested war, with differing interpretations coming into contact and conflict. It was the first time a war was conducted under a Likkud government and was perceived by the center and left wings of Israel's politics as a "political war," waged as a political instrument, rather than a defensive war aimed at countering a threat to the nation's existence. The state's elites were unable to construct the friction with the PLO[12] (after a year of cease-fire in the northern area of Israel) as an existential danger, and to universalize it. This war turned into a landmark in the relationship between the home-front and the battlefront; from its first week, small demonstrations were staged, and as the war progressed, massive waves of protest swept the country.

The 1982 demonstrations' motto, was that the war in Lebanon was a "war of choice" (*milchemet breirah* or the Israeli equivalent to the "unjust war"). Such a motto contained an open questioning of the legitimacy of the war and the authority of the state's elites to declare and wage war. Moreover, the protest against the war also challenged the right of the state to command its male citizens to kill and be killed under any circumstances. Thus, in a newspaper add published by Peace Now the following slogan could be found: "What are we killing and being killed for." This slogan was unprecedented in Israel until 1982.

The protest cycle against the war begun to gain momentum only by the end of its second week. While small scale protests were mounted in the central cities (by small networks and groups, such as feminist organisations, non-Zionist groups, etc.), Peace Now was caught in a conflict of interests between activists and constituencies in the home front and some of its leading members taking an active role as commanders and soldiers in the war. The tension between the fighter's role as an oppositional symbol and war participation— a tension that was inherent in the collective action framework of Peace Now—reached its peak during the war in Lebanon (Helman 1994, 1999b; Reshef 1996). Peace Now members acted both as mobilized and de-mobilized soldiers. The movement considered military service as a necessary condition for political participation and opposition. The attempt to remain within the limits of legitimate political discourse, by participating in war and simultaneously protesting against it, brought about the estrangement of the militant sectors of the constituency. The protest cycle created opportunities for new groups and movements of war challengers. The protest against the war turned into an arena in which social identities were mobilized, negotiated, and renewed.

"There Is a Limit": The Split Between Soldiering and Citizenship

The tension between war participation and war protest, brought about the emergence of a movement of selective conscientious objectors (Helman 1999a, 1999b; Linn 1986, 1994): "There is a Limit" (*Yesh Gvul*). During the war's three years (and mainly between 1982- 1984), 130 reserve soldiers were court-martialed for refusing to participate in the war. The sociological profile of the conscientious objectors was not different from that of Peace Now's constituency. The movement drew mainly from highly educated, middle class Ashkenazi males kibbutz members and city dwellers.[13]

Although Yesh Gvul did not reject military service altogether, the practice it promoted—refusal to participate in the war—demystified the unconditional obedience to and compliance with military service. It prioritized civilian values to military commands in cases in which they came into open conflict. Moreover, Yesh Gvul signalled the possibility that citizenship and soldiering can no longer be considered as complimentary and harmonizing roles. Yesh Gvul disclosed the political character of the sphere of national security and legitimized opposition to the state's policies in terms of the political ideology soldiers as citizens may hold. In this sense, the movement of conscientious objectors demystified the "a-political" veil of the nation in arms. Furthermore, it demanded the extension of democratic practices—such as the active and even autonomous participation of individuals and groups in the shaping of political society—to the sphere of national security. As the sphere of national security is closely connected to the fashioning of collective and individual fates, Yesh Gvul claimed that it must be opened up to participation according to criteria that are autonomous and even different from the interests represented in the practices of the state and its elites (Helman 1999a, 1999b). The movement imbued political cultural codes with new meanings. It split the taken-for-granted association between soldiering and citizenship, and promoted the idea that active and effective political participation may be pursued in arenas other than military service.

The fighter's role was skillfully monopolized by the above organizations to launch opposition against war policies, but also to signal their movement away from the state. However, it was this monopolization and the attempt to change the meanings of the fighter's role that turned the protest for and against the war into an arena in which a struggle over national identity was conducted. The struggle took place between those constituted as carriers of civic virtue (mainly Ashkenazi middle class males) and between the eternal candidates to civic virtue (mainly Oriental Jews). The opposition to the war was interpreted by the latter as an attempt to devaluate their war participation (Levy 1997: 179) precisely in a period in which they could capitalize on it. This struggle over national identity reached its zenith in February 1983, when one Peace Now demonstrator was killed and four wounded in the course of a demonstration.

The *Intifada* (the Palestinian popular uprising) ignited again the struggle over issues of national security and through it the debate over national identity. Peace Now renewed its activities, through mass demonstrations at first in Jerusalem and Tel-Aviv and then in the occupied territories. The demonstrations in the occupied territories set a precedent, since they were the joint effort of Israelis and Palestinians (Reshef 1996: 148-69). Internal changes at the leadership level (Feige 1995: 288), such as the access of women to leadership positions, the partnership with Palestinians and occasionally with Palestinian citizens of Israel (Reshef 1996), brought about changes in the cognitive framing of protest. Human rights discourse coupled with a discourse emphasizing the damage the caused by the oppression of another people to the moral fiber of Israeli society, substituted for the fighter's role and the symbolic package of citizen militarism (Feige 1995: 391). Nevertheless, Peace Now remained loyal to the formula establishing that military service was a precondition for political participation and opposition. As during the protest cycle of 1982-1985, this adherence to the limits of legitimate discourse, albeit reformulated, estranged the radical elements in the constituency. "There is a Limit" renewed its activities, and as much as 190 soldiers were court martialed for refusing to serve in the Occupied territories. Moreover, considerable numbers of soldiers found alternative arrangements and refusal (declared and undeclared) turned into a viable option for opponents of the military policies of the Israeli state at that period.

From Motherhood to Citizenship: "Mothers against Silence" and "Women in Black"

"Mothers against Silence": The Politicization of Republican Motherhood

Although women were prominently represented in Peace Now and supported the conscientious objectors movement, the framing of protest against the war in terms of the fighter's role, marginalized them. It was only in May 1983, almost a year after the war begun that a new movement, Mothers Against Silence, joined the protest cycle (Barzilai 1992; Zuckerman-Bareli and Benski 1989). At the end of April 1983, a soldier's mother published a letter in three daily newspapers. The letter clearly stated the contradictions republican motherhood was beset with. It emphasized the mounting tension built in Zionist-Left wing education between the pursuit of coexistence and peace with the Palestinians on the one hand and on other hand the loyalty to the IDF and to the role of the fighter. The mother called upon parents concerned for their sons' lives to interrupt their silence, to take responsibility, and to protest against "this damned war." The letter succeeded not only in the formation of consensus across networks of parents across the country but also in mobilizing them for action. The movement was called "Parents against Silence" but mainly mobilized women to action, most of them Ashkenazi and highly educated.[14] By mobilizing motherhood, the movement made use of the social identities constituted by the incorporation of women into citizenship. Republican motherhood was turned into an opposi-tional symbol and mobilized to challenge the war.

"Mothers against Silence" created a discursive space in which motherhood was politicized and turned into a legitimate resource to engage on issues of peace and security (Atzmon 1997). Coupling the narrative of republican motherhood to anti-war activity was met with expressions of hostility and did not succeed in overcoming the divisions within those interpellated as republican mothers. The move-ment did not succeed in broadening the constituency beyond the limits of the ethno-class. Coupling the narrative of republican moth-erhood to antiwar activity not only limited the appeal of the move-ment but also was met by hostility by the wider public. While the politicization of motherhood challenged the war politics of the state, it did not challenge the militaristic gender order.

"Women in Black": Challenging the Political and Gender Orders

The movement away from the militaristic collective action frame crystallized in Women in Black. The Women in Black movement was inaugurated by a small group of women in Jerusalem in Janu-ary 1988 (a month after the *Intifada* broke out) and continued uninterrupted for six years, disbanding officially in June 1994. The purpose of the movement was to protest the Israeli occupa-tion of Palestinian territories and the cycle of Israeli-Palestin-ian violence, and to prevent the issue from being swept under the national rug. The opposition to the occupation was itself two-fold: to decry the oppression of the Palestinian people and at the same time to warn against the concomitant moral corruption of Is-raeli society.

The central and virtually exclusive activity of the movement was the weekly demonstrations held every Friday between one and two p.m. in fixed locations throughout Israel. Throughout the six years of its existence, the movement maintained six minimal rules which defined the demonstration anew each week: the time, the site, the silent protest, the black attire, the all-woman makeup and the sign "Stop the Occupation."[15]

The weekly protest vigils of Women in Black were met by harshly negative reactions. Passersby would shower the women with sexist catcalls and curses, in which national conflict was wedded to gender and ethnic origin (Helman and Rapoport 1997). The act of protest of Women in Black represented both a challenge and an alternative to the identities constituted through the incorporation of Israeli-Jewish women into citizenship. The symbolic package displayed by Women in Black challenged Israeli-Jewish women's role as embodied symbols of the boundaries of the nation. The symmetry the movement posited between the suffering of both Jews and Palestinians was considered as a breach in national solidarity. Moreover, by claiming access to the public sphere and voice on issues of peace and security Women in Black challenged male monopoly on participation in the public realm and the gender division between the spheres. Their demand to voice their opinions on issues of peace and security as citizens of equal standing—not as the mother or wife of a soldier—undermined the male monopoly on this realm and the logic that governs gender relations in Israel.

Women in Black offered an alternative form of political participation to women, particularly in juxtaposition to the two accepted forms of non-establishment political participation in Israeli society. One entails a secondary behind-the-scenes role, subordinate to their male peers, who represent the group to the general public, appropriating for themselves the role of fashioners of the ideology and the mode of protest (Sasson-Levi 1995). The second involves the protest of the mothers and wives of male warriors, which does little more than extend traditional female roles beyond the domestic realm (see Rapoport et al. 1994).

Women in Black constituted a new symbolic type of political woman, whose struggle for her place in the political realm is couched in new terms: not in terms of her domestic identity or any other identity anchored outside the public sphere and the weekly event they performed. The symbolic type of political woman promoted by Women in Black challenged both the political and the gender orders.

Conclusions: Citizenship Regime and the Limits of Mobilization

This article began with a paradox: how to explain why a peace movement displays militaristic symbols while peace and peace protest turn into the assets of an ethno class. To disentangle this paradox I claimed that we should examine the ways in which the state through war has shaped the citizenship regime. Moreover, my claim was that the mode of incorporation into citizenship is constitutive of political identities and frames the kinds of contestations and claims of individuals and groups vis-a-vis the state. Concomitantly, the mode of inclusion into citizenship and the identities constituted may influence the coalitions formed in the process of mobilization for collective action.

The Israeli peace movement that began to crystallize slowly from the end of the 1970s mobilized the fighter's role as the main asset to claim peace. Women on the other hand mobilized republican motherhood. Soldiering and republican motherhood, as the main identities constituted by the citizenship regime, became resources that could be mobilized to protest against the excesses in national missions and in the process of rising alternative policies vis-a-vis the state. However, the subversion of motherhood and soldiering were almost monopolized by the Ashkenazi middle and upper middle classes. This is the common class and ethnic-national background of the members of the Israeli peace movement. It appeals mainly to members of this strata, and it reflects its changing interests vis-a-vis war and conflict management.

The 1982-1985 protest cycle against the war can be characterized as condensing the conflicts and contradictions emerging from the stratified and hierarchical structure of citizenship in Israel. The interests and social identities constituted by the differential inclusion of social categories into citizenship came into open conflict during the war. This conflict constrained the mobilizatory potential of the war challengers, despite the fact that their collective action frames lavishly drew on and displayed symbols of what Kimmerling (1998) termed the security culture.

The collective action framework could not, however, turn peace into the interest of other social categories in Israeli society. Whereas the fraction of the Ashkenazi elite belonging to the peace movement was signaling that war and escalation were becoming a hindrance in the pursuit of its interests, it was unable to universalize its claims to other social categories in Israel. Peace Now emerged when other social categories and bearers of social identities (such as the Mizrahim) began to capitalize on national missions. In other words, while the middle and upper middle classes in Israel were moving towards civil society, other groups were advancing towards the state and its missions. This movement—away from the state and towards the state—was the product of the political opportunity structure created by the new ruling coalition (the Likud-led ruling coalition). While the Likud ruling coalition did not create new patterns of national identity (Kimmerling 1993), it favored a national identity that de-emphasized the weight of the republican principle. By emphasizing ethno-national boundaries of the effective political community, the ruling coalition could mobilize the legitimation of social categories that were marginalized by the republican principle. This does not mean that the republican principle lost of its dominance, but it was tempered—thus allowing for a sense of advancement through national missions, formerly closed mainly to the Orientals (both mobile and non-mobile) and to religious groups.

It is against this background that the different patterns of national identity displayed by different ethno-classes should be understood. Social identities constituted by the differential incorporation of social categories in the framework of citizenship, turned into constraints to form a wider coalition for peace. Whereas during the 1982-1985 protest cycle the struggle was conducted by mobilizing identities constituted by citizenship, the Intifada protest cycle may be considered as an attempt to transcend these identities, mainly by challengers radicalizing the collective action frames of the Peace movement.

Women in Black and There is a Limit (*Yesh Gvul*) split the taken-for-granted association between citizenship and motherhood, and between citizenship and soldiering. They created alternative political cultural meanings and identities. However, these meanings and identities were minority options, even within the Peace movement itself. Furthermore, as the power of social movements rests on cultural innovation (Johnston and Klandermans 1995; Melucci 1989; Swidler 1995), the opposition met by the Israeli Peace movement disclosed how the cultural innovations it introduced reflected the emerging cultural, political and economic interest of an ethno-class.

The more collective action frames radicalized and moved away from the social identities and patterns of national identity constituted by citizenship regime, the more they emphasized the constraints this regime put on the formation of coalitions for peace.

Notes

*   I would like to thank Lev Grinberg and Uri Ram from Ben Gurion University and Zeev Rosenhek from the Hebrew University for their comments and remarks on a previous version of this article.

1.   One journalist (Meir Shalev, *Iedihot Achoronot*) dramatized the boundaries of the nation by skillfully reminding his readers in what areas of Israel the phone lines collapsed and lights never turned off. Bnei Brak (a town heavily populated by ultra-orthodox Jews) was set apart in his article as behaving as if nothing happened.

2.   The ruling coalition was widely termed as a coalition of minorities or of outcasts. This ruling coalition in which new immigrants from the former Soviet Union sit side-by-side with ultra-orthodox Jews (belonging and representing the oriental ethno-class) and nationalistic religious Jews was considered as an anathema by the dominant classes of Israeli society.

3.   The movement's name alludes to the biblical mothers of the Israeli-Jewish nation: Sarah, Rivka, Rachel and Leah.

4.   Following Lamar and Thompson (1981), Peled and Shafir (1996: 395) characterize conflicts typical of colonial frontier societies as conflicts over land resources and the control of the land.

5.   For a detailed analysis of the processes that led to the close association between the fighters' role and the Ashkenazi middle class males, see Levy 1997.

6.   Berkovitch (1997) observes that Palestinian women were silenced in discourse, and therefore no reference to their role as mothers can be found in the different laws establishing special rights to women as mothers, nor in the different debates at the Israeli Parliament.

7.   The 1967 war brought about an extension of economic activity in Israel. Whereas most groups "gained" from the war, these gains did not alter the main contours of the ethnicized class structure.

8.   Levy (1997) indicates that changes in the organizational structure of the IDF as well as in the civilian sector of Israeli society constrained the opportunities of Oriental Jews to capitalize on their military service. Changes such as the higher value granted to managerial and technological skills were taking place when mobile Orientals entered fighting units.

9.   For an extensive analysis of the socio-political and economic changes brought about the 1973 war, see Kimmerling 1993; Levy 1997: 143-65.

10. For a detailed analysis of the political opportunity structure that facilitated the emergence of protest movements, see Kimmerling 1993; Levy 1997; Peled and Shafir 1996.

11. For a full version of the petition, see Reshef 1996: 13-4.

12. The Lebanon War was defined by the then Chief of Staff of the Israeli army, Rafael Eitan, as a war waged on behalf of the Land of Israel or the territories occupied during the 1967 war. As controversy over the occupation began to gain salience in Israeli politics, to wage a war over the Land of Israel could hardly contribute to the construction of the war in existential terms.

13. The entire population of conscientious objectors included 130 individuals who were court-martialed and imprisoned for periods ranging from 21 to 35 days. Conscientious objectors to the war in Lebanon can be characterized as belonging to the dominant or elite group in Israel. Most of them were of Ashkenazi origin (81.8 percent); residents of major cities (Tel Aviv, Haifa, and Jerusalem) or kibbutz members (25 percent); and former members of youth

movements. They comprised a highly educated group: 36 percent held a bachelor's degree; 23 percent have obtained a master's degree; and 7.5 percent held Ph.D.s (mainly in the natural and exact sciences). The rest were undergraduate students or high school graduates. Their military service was in combat, sometimes elite units. After their refusal, most of them (86 percent) continued their yearly tour of duty in the reserves. Most of the conscientious objectors belonged to the center-left continuum of Israeli politics (Labour Party and the Citizens Rights Movement); a minority belonged to the radical left (the Communist Party and Trotskyst organizations).

14. According to Zuckerman-Bareli and Benski (1989), 77 percent of the movement's constituency were women with a post-secondary education and from Ashkenazi background.

15. For an extensive analysis of Women in Black, see Helman and Rapoport 1997.

References

Atzmon, Yael 1997. "War, Mothers and Girl with Braids: Involvement of Mother's Peace Movements in the National Discourse in Israel." *Israel Social Science Research* 12, 1: 109-28.

Barzilai, Gad 1992. *A Democracy in Wartime: Conflict and Consensus in Israel.* Tel Aviv: Sifriat Hapoalim Publishing House. (Hebrew).

Ben-Eliezer, Uri 1995. *The Emergence of Israeli Militarism 1939-1956.* Tel Aviv: Dvir Publishing House (Hebrew).

Berkovitch, Nitza 1997. "Motherhood as a National Mission: The Construction of Womanhood in the Legal Discourse of Israel." *Women Studies International Forum* 20: 605-619.

Bernstein, Deborah 1984. "Conflict and Protest in Israeli Society: The Case of the Black Panthers." *Youth and Society* 16, 2: 129-52.

Carmi, Shulamit and Henri Rosenfeld 1979. "Appropriation of Public Resources and the State's Middle Class in Israel." *Makhbarot Le'Mechkar U'Bikoret* 2: 43-84. (Hebrew).

Cohen, Erik. 1989. "The Changing Legitimations of the State of Israel." *Studies in Contemporary Jewry* 5: 148-65.

Feige, Michael 1995. *Social Movements, Hegemony and Political Myth: A Comparative Study of Gush Emunim and Peace Now Ideologies.* Ph.D. dissertation, Department of Sociology and Anthropology, Hebrew University, Jerusalem.

Haidar, Azziz. 1987. *Social Welfare Services for Israel's Arab Population.* Tel Aviv: International Center for Peace in the Middle East.

Helman, Sara 1999a. "Redefining Obligations, Creating Rights: Conscientious Objection and the Redefinition of Citizenship in Israel." *Citizenship Studies* 3, 1: 45-70.

Helman, Sara.1999b. "Yesh Gvul." *Theory and Criticism,* Special Issue: 50 to 48: Critical Moments in the History of the State of Israel: 313-19. (Hebrew).

Helman, Sara 1994. *Conscientious Objection to Military Service as an Attempt to Redefine the Contents of Citizenship.* Ph.D. dissertation, Department of Sociology and Social Anthropology, Hebrew University, Jerusalem.

Helman, Sara and Tamar Rapoport 1997. "Women in Black: Challenging Israel's Gender and Socio-Political Order." *British Journal of Sociology* 48, 4: 681-700.

Hermann, Tamar and Gila Kurtz 1995. "Prospects for Democratizing Policymaking: the Gradual Empowerment of Israeli Women." *Middle East Journal* 49, 3: 447-66.

Hofnung, Menachem 1991. *Israel - Security Needs vs. The Rule of Law*. Jerusalem: Nevo Publishers. (Hebrew).

Horowitz, Dan 1982. "The Israeli Defense Forces; a Civilizianized Army in a Partial Militarized Society." In *Soldiers, Peasants and Bureaucrats*, edited by Roman Kolkowitz and Andrzej Korbonski, 77-105. London: George, Allen and Unwin.

Horowitz, Dan and Baruch Kimmerling 1974. "Some Social Implications of Military Service and the Reserves System in Israel." *European Journal of Sociology* 15, 2: 262-76.

Izraeli, Dafna N. 1997. "Gendering Military Service in the Israeli Defense Forces." *Israel Social Science Research* 12, 1: 129-66.

Johnston, Hank and Bert Klandermans 1995. "The Cultural Analysis of Social Movements." In *Social Movements and Culture*, edited by Hank Johnson and Bert Klandermans, 3-24. Minneapolis: University of Minnesota Press.

Kimmerling, Baruch 1998. "Political Subcultures and Civilian Militarism in a Settler-Immigrant Society." In *Security Concerns: Insights from the Israeli Experience*, edited by Daniel Bar-Tal, David Jacobson and Aharon Kllieman, 395-416. Stamford, CT: JAI Press.

Kimmerling, Baruch 1994. "Religion, Nationalism and Democracy in Israel." *Zmanim* 13: 116-31. (Hebrew).

Kimmerling, Baruch 1993. "State Building, State Autonomy and the Identity of Society: The Case of the Israeli State." *Journal of Historical Sociology* 6, 4:396-428.

Kimmerling, Baruch 1985. "Between Primordial and Civil Definitions of the Collective Identity: *Eretz Israel* or the State of Israel." In *Comparative Social Dynamics*, edited by Erik Cohen, Moshe Lissak and Uri Almagor, 262-83. Boulder, CO: Westview.

Kimmerling, Baruch 1983. *Zionism and Territory: The Socio-territorial Dimensions of Zionist Politics*. Berkeley: Institute of International Studies, University of California.

Kimmerling, Baruch 1979. "Determination of Boundaries and Frameworks of Conscription: Two Dimensions of Civil-Military Relations." *Studies in Comparative International Development* 14: 22-41.

Kretzmer, David 1987. *The Legal Status of the Arabs in Israel*. Tel Aviv: International Center for Peace in the Middle East.

Lamar, Howard and Leonard Thompson 1981. *The Frontier in History: North America and Southern Africa Compared*. New Haven, CT: Yale University Press.

Levy, Yagil. 1997. *Trial and Error: Israel's Route from War to De-Escalation*. Albany: State University of New York Press.

Levy, Yagil. 1996. "War Politics, Interethnic Relations and the Internal Expansion of the State: Israel 1948-1956." *Theory and Criticism* 8: 203-23. (Hebrew).

Lewin-Epstein, Noah and Moshe Semyonov 1993. *The Arab Minority in Israel's Economy: Patterns of Ethnic Inequality*. Boulder, CO: Westview Press.

Linn, Ruth. 1994 *Conscience at War: The Israeli Soldier as a Moral Critic*. Albany: State University of New York Press.

Linn, Ruth 1986. "Conscientious Objection in Israel During the War in Lebanon." *Armed Forces and Society* 12, 4: 489-511.

Lomski-Feder, Edna 1994. *Patterns of Participation in War and the Construction of War in the Life Course: Life Stories of Israeli Veterans of the Yom Kippur War*. Ph.D. dissertation, Department of Sociology and Anthropology, Hebrew University, Jerusalem.

Lustick, Ian 1980. *Arabs in the Jewish State*. Austin: University of Texas Press.

Marx, Anthony 1995. "Contested Citizenship: The Dynamics of Racial Identity and Social Movements." *International Review of Social History* 40, 3: 159-83.

Melucci, Alberto 1989. *Nomads of the Present: Social Movements and Individual Need in Contemporary Society*. London: Hutchinson Radius.

Oldfield, Adrian 1990. *Citizenship and Community: Civic Republicanims and the Modern World*. London: Routledge.

Peled, Yoav 1993. "Strangers in the Utopia: The Civic Status of Israel's Palestinian Citizens." *Theory and Criticism* 3: 21-38. (Hebrew).

Peled, Yoav and Gershon Shafir 1996. "The Roots of Peacemaking: The Dynamics of Citizenship in Israel, 1948-93." *International Journal of Middle East Studies* 28: 391-413.

Reshef, Tzali 1996. *Peace Now: From the Officer's Letter to the Peace Now*. Jerusalem: Keter Publishing House. (Hebrew).

Rosenhek, Zeev 1995. *The Origins and Development of a Dualistic Welfare State: The Arab Population in the Israeli Welfare State*. Ph.D. Dissertation, Department of Sociology and Anthropology, Hebrew University, Jerusalem.

Sasson-Levi, Orna 1995. *Radical Rhetoric, Conformist Practices: Theory and Praxis in an Israeli Movement*. Shaine Working Papers No. 1. Department of Sociology and Anthropology. Hebrew University of Jerusalem.

Seguev, Tom 1984. *1949 - The First Israelis*. Tel-Aviv: Domino. (Hebrew).

Shafir, Gershon 1989. *Land, Labor and the Origins of the Israeli Palestinian Conflict*. Cambridge: Cambridge University Press.

Shapiro, Yonathan 1989. *The Road to Power: Herut Party in Israel*. Albany: State University of New York Press.

Swidler, Ann 1995. "Cultural Power and Social Movements." In *Social Movements and Culture*, edited by Hank Johnston and Bert Klandermans, 25-40. Minneapolis: University of Minnesota Press.

Swirski, Shlomo and Deborah Bernstein 1980. "Who Worked in What, For Whom and For What?" *Machbarot LeMechkar U'Bikoret* 4: 5-66. (Hebrew).

Tarrow, Sydney 1994. *Power in Movement*. Cambridge: Cambridge University Press.

Tilly, Charles 1995. "Citizenship, Identity and Social History." *International Review of Social History* 40, 3: 1-17.

Van-Gunsteren, Herman 1994. "Four Conceptions of Citizenship." In *The Condition of Citizenship*, edited by Bart van Steenbergen, 36-48. London: Sage Publications.

Walby, Sylvia 1992. "Woman and Nation." *International Journal of Comparative Sociology* 33, 1-2: 81-100.

Yiftachel, Oren 1999. "The Land Day: Palestinian Protest in the Israeli Ethnocracy." *Theory and Criticism*, Special Issue: 50 to 48: Critical Moments in the History of the State of Israel: 279-89. (Hebrew).

Yuval-Davis, Nira 1997. *Gender and Nation*. London: Sage Publications.

Zuckerman-Bareli, Chaia and Tova Benski 1989. "Parents Against Silence." *Megamot* 32, 1: 27-42. (Hebrew).

Part 4

The Notion of "National Security"—
Institutions and Concepts

10

The Link Between the Government and the IDF During Israel's First 50 Years: The Shifting Role of the Defense Minister

Amir Bar-Or

Introduction

The characteristics of civil-military relations in any given state are influenced by the components of its national security policy. Professor Moshe Lissak (1991) published an article in 1991 on this subject showing that every security doctrine, has an effect on civil-military relations in general and on the military elite and political elite in particular. In the case of Israel, where there is no formal national security doctrine, and the conventional security concept is "a rule partially written and partially intuited" (Tal 1996: 12), the nature of the civil-military relations might lead one to draw the conclusion that Israel is a *sui generis* among modern states. Many factors are involved and explanations for Israel's uniqueness can be found in numerous studies dealing with the development of Israeli civil-military relations since 1948.

Perlmutter (1977) referred to the State of Israel as a society in military uniform, suggesting that the country's innovative battlefield tactics evolved from the need to deal with perpetual security challenges. This in turn produced a special type of modern revolutionary soldier trained to protect the pioneering efforts of the Jewish population prior to the establishment of the state. Finer (1977) also claims that Israel's uniqueness dates from the early years of the state-in-the-making under the British Mandate. He distinguishes Israel from most other new states because Israel's political system was conceived before its political framework was created, and its democratic political culture was influenced by the organizational structure of Jewish communities in the Diaspora. Furthermore, the Israeli military was established by a political leadership that insisted on maintaining the principle of civilian hegemony and political supremacy. This also contributed to the singular style of political administration over the military branches already in the period preceding independence. That is to say, in the formative years of the state, the manner in which the political leadership's directives were channeled to increase its hold on the military apparatus was intended to prevent any obfuscation in the para-military organization's relations with the highest level of political leadership. This approach, according to Ben-Meir (1995: 100, 170), was designed to limit misunderstandings arising over the primacy of the political leadership which, he observed, benefited from the informal nature of relationships prevalent in Israeli society and the close-knit character of Israel's political system.

Peri (1983) attempted to answer the question whether Israel should be considered a Third World nation or a developed one, by referring to Israel's uniqueness from another perspective, basing his observation on standard scientific criteria in the field of civil-military relations. Thus, the Israeli example can be investigated in similar fashion to laboratory research by discerning the evolution of military force in its formative stages, observing its objectives, and pinpointing its early conflicts. Ben-Meir (1995), too, views Israel as a laboratory case, although for different reasons. He sees Israel as exceptional because it has "... one foot simultaneously in the Third World and the other in the Western World." According to Ben-Meir, Israel "...shares with the western nations a deep-rooted tradition of solid commitment to democratic government where the constitutional principle of civilian supremacy over the armed forces is firmly grounded both in law and custom" (1995: xi).

Ben-Meir admits that these observations offer only a partial explanation. "Although there is no question that, from a formal point of view, the Israeli government exercises civilian control over the IDF [Israel Defense Forces], new evidence... appears to indicate that behind the scenes the IDF calls the shots" (1995: xi). Tal, also, points out that while " . . . no one disputes the military command's subordination to political leadership, the IDF has become over the years a self-ruling, autonomous, dominating institution whose authority and responsibilities extend far beyond the limits defined in the early days of the state" (Tal 1996: 108).

This situation completely contradicts the principle canonized by Ben-Gurion in a document he dispatched in October 1949 to Lt. Gen. Yadin upon his appointment as the IDF's second chief-of-the-general-staff (CGS). Ben-Gurion stipulated, *inter alia*, that "the army determines neither the policies, administration, nor the laws in the state. The army does not even decide its own structure, missions, or methods of operation, and it most certainly does not wage war or sign peace treaties. The army is nothing more than the executive arm for the defense and security of the State of Israel. Organizing the army...[is] the sole responsibility of the civilian authorities who form the government, the Knesset and the voters...The army is unconditionally subordinate to the government" (1971: 82). These principles were not only formulated by Ben-Gurion, but they became his guiding-light for as long as he shouldered the double role of Prime Minister and Defense Minister during the state's first fifteen years.

The aim of this article is to re-examine the reasons which caused the situation, described by Ben-Meir and Tal, which placed the army in such a distinctive position vis-a-vis the political echelon despite Ben-Gurion's stated principle regarding civilian authority over the military. This study, therefore, focuses on the manner in which the Defense Minister has acted as the link between the Israeli government and the army's high command. The Defense Minister in Israel deals with security issues in the broadest definition of the term. The holder of office is not in charge of the army as minister of war or secretary of defense, but he wields tremendous power extending beyond the formal parameters enshrined in parliamentary legislation. For this reason, nominees for the job have traditionally been loyal supporters of those Israeli Prime Ministers who have chosen not to carry the dual burden. Of the eleven Prime Ministers, (including the current chief executive, Ehud Barak), six have assumed the role of Defense Minister while in office. Their range of perspectives and methods of handling this dual capacity will, in effect, be our prism for examining the uniqueness of civil-military relations in Israel.

To fully grasp the complexity of the issue, it is not enough to study the Israeli government's legal, formal *modus operandi* as commander-in-chief of the armed forces. The informal characteristics inherent in the Israeli political system will be taken into consideration throughout this work. Special attention will be placed, therefore, on recalling and analyzing the informal nature of relationships in the power triad of Prime Minister, Defense Minister, and CGS. This informal network has been the result, on the one hand, of the absence of formal legislation regulating their relationships and, on the other hand, the need to maintain working ties during the first three decades of the state's existence. Only in 1976 did *The Basic Law: The Army* officially incorporate the recommendations of the Agranat Inquiry Commission which had investigated the early stages of the Yom Kippur War (1973). Nevertheless, the complexities embedded in the power triumvirate continued to maintain, as before, informal working arrangements for guaranteeing three-way co-operation.

The Defense Minister's Role as a Prism for Examining Civil-Military Relations in Israel: Major Trends

1947-1967: Combined Roles

In 1946, in the course of discussions at the 22nd Zionist Congress, David Ben-Gurion, the head of the Jewish Agency, the highest political institution in the Jewish state-in-the-making, first conceived of the idea of a defense portfolio. This initiative resulted from Ben-Gurion's assessment that immediately after British evacuation full-scale war was likely to erupt with an invasion of Arab regular armies.

Security became the foremost priority of the Jewish community, and the need to prepare for a life-and-death struggle led to the creation of a defense department separated from the political department within the Jewish Agency. All military and civilian considerations were to be organized by the new department. Sensing the urgent need to gain an integrated feel of events, Ben-Gurion agreed personally to handle military contingencies and simultaneously deal with the vast array of political matters so that the flood of decisions would be directed in as thorough and comprehensive a manner as possible.

Upon establishment of the state, Ben-Gurion was appointed Prime Minister of the provisional government, and retained the defense portfolio. He held onto many of the authoritative powers, in order to secure his control of both the new governmental apparatus and the army (whose structure was still based primarily on the underground, para-military organization: the "Haganah"). In this way, as head of the new political entity, he neutralized rival political figures and other institutions that threatened to interfere with his efforts to attain full command of mobilizing the country for war. His "almost dictatorial" style in handling defense matters was not lightly tolerated by his political colleagues, yet Ben-Gurion insisted on keeping both roles despite the harsh criticism. This enabled him to concentrate on establishing an army at the end of the War of Independence to serve the young state's national and social objectives.

According to Ben-Gurion, the army had a vital role to play in strengthening the fragile social fabric of Israel's burgeoning community of immigrants, and the army's image as a "melting pot" would set an example for all sectors of society to emulate. Thus, the security threats at the end of the war notwithstanding, Ben-Gurion did not hesitate to markedly reduce the size of the fighting forces so that most of the country's meager resources could be channeled towards the absorption of masses of immigrants. At the same time, he invested much effort in constructing the defense ministry, and tried to make a clear distinction between the civilian and military branches by defining their respective roles. The concept of the defense ministry and Ben-Gurion's ability to exert supremacy over it, combined with his personal stature and authority, enabled him to formulate the relationship between Israel's political and military echelons in a way appropriate to his singular character and work style.

The efficacy of the joint civil-military relationship set up by Ben-Gurion was put to the test when he resigned from government between 1953 and 1955. Two ministers took over the roles he had previously fulfilled, and a number of clashes occurred, both on professional and personal levels, between civilian and military officials. These flare-ups, caused by the lack of clear boundaries of authority, had been kept in check as long as the dominant figure of Ben-Gurion was riding in both saddles. Upon returning to the prime-minister office and as a result of an intelligence fiasco in Egypt, while he was out of office, for the first time public criticism was voiced alluding to the surfeit of power enjoyed by Ben-Gurion, and the presence of undemocratic procedures overtaking the security sector as well as in social and political institutions. These developments exploded into a full-blown political crisis during the Lavon Affair in the beginning of 1960 when personal confrontations, partisan ramifications, and international implications, in addition to other factors, led Ben-Gurion to retire for good from his dual role as Prime Minister-Defense Minister in 1963.

His replacement, Prime Minister Levi Eshkol, also chose to serve both offices, but his limited military experience forced him to design a different set of relationships with the army and its high command. The appointment of a dominant personality like Itzhak Rabin to CGS in 1964 and the creation of the best general staff the country had known, served to reduce Eshkol's day-to-day involvement with the IDF, and thus strengthen the CGS's position. Eshkol, unlike Ben-Gurion, was dependent solely upon the CGS.

This was in marked contrast to Ben-Gurion's dual performance, when the Defense Minister was practically the IDF commander-in-chief, and the general staff operated on behalf of the Defense Minister. The role of chief military commander could be carried out by Ben-Gurion because during the War of Independence the heads of the various branches in the general staff, including the CGS, served the Defense Minister as if they were his staff officers. "Based on his political savvy and formidable experience, Ben-Gurion could weigh matters and make decisions while working directly with the general staff's branches. He had the required experience for military administration whereas the general staff lacked the same outstanding confidence and ability. This resulted in their need to win the Defense Minister's assurance of its decisions on matters of organization, weapons acquisition, and combat strategy, as well as the appointment of senior officers" (Tal 1996: 110).

Eshkol's credibility as Defense Minister was impaired on the eve of the 1967 Six-Day War. Pressured by public opinion and toiling under an ominous threat to the existence of the state, Eshkol was compelled to relinquish the portfolio of Defense Minister and, against his will, appoint Moshe Dayan. This separation of roles substantially changed the way the Minister of Defense functioned, and left its mark on civil-military relations in Israel for the following twenty-five years, until 1992.

1967-1992: Separation of Roles

The dazzling IDF victory in the Six-Day War illuminated a halo of fame over the CGS, Lt. Gen. Itzhak Rabin and the Defense Minister, Moshe Dayan who, it will be recalled, was deputized only a few days before the outbreak of war. Dayan's appointment placed a top-rate security personality, the IDF's popular fourth CGS, right in the middle of the now weakened Prime Minister and the reigning CGS, Rabin. In the absence of a cut-and-dry definition or a allocation of powers and jurisdictions between the Prime Minister and the Defense Minister, Eshkol and Dayan decided upon an informal appropriation of responsibilities in what came to be known as "the constitution."

Another factor that bolstered Dayan's status as Defense Minister after the war was his commitment to institute a formal military government, under the auspices the defense ministry, over the occupied territories. Shortly afterwards, Lt. Gen. Bar-Lev replaced Lt. Gen. Rabin, and Golda Meir succeeded Eshkol who died in 1968. These events completed the process of Dayan's elevation to "defense czar" in the years between the Six-Day War and the Yom Kippur War (1973). Ironically, Dayan's heightened stature germinated the seeds of his own decline and set in motion the elevation of the CGS's position. The implications of this process would also increase the CGS's independence in later years (Tal 1996: 112).

The convoluted dynamics of the three-way interaction between Prime Minister, Defense Minister, and CGS became manifest during the Yom Kippur War in view of the manner in which operations were conducted by the political and military echelons. *The Basic Law: the Army*, whose purpose was to correct the blatant shortcomings revealed during the war, stated that the CGS is subject to governmental authority and is subordinate to the Defense Minister. The government, as a collective body, is the undisputed commander-in-chief and may rescind any orders issued by the Defense Minister, whose job is to execute government policy and who, according to the law, has been granted control of the army on behalf of the government. Vague legal phraseology concerning relationships within the chain of command and an unprecedented allocation of authority between the government and the Defense Minister left a wide breach open for interpretation that did little to alter the rules of the three-way game.

Golda Meir's resignation following the Agranat Inquiry Commission's recommendations, which had refrained from naming either her or the Defense Minister as key figures responsible for the short-comings of the war, led to the inevitable dismissal of Dayan. Itzhak Rabin was appointed Prime Minister and Shimon Peres became Defense Minister. According to the Commission's suggestions, Lt. Gen. David Elazar was replaced by Lt. Gen. Mota Gur. But political rivalry between Rabin and Peres exacerbated interpersonal relationships within the unholy trinity. Peres had previously held several key positions in the defense ministry and was skilled in wielding power and influence within the military establishment, especially with regard to advancing his own position in the race for Prime Minister. The exhaustive rivalry between Peres and Rabin hastened the process that led to the CGS's increased independence.

In 1977 Ezer Weizman became Defense Minister after the Likud party won the elections (following twenty-nine years of Labor Party rule), and found himself facing a very sovereign CGS. Weizman wanted to dismiss Lt. Gen. Gur, but the Prime Minister, Menachem Begin, vetoed the idea. During Weizman's tenure the first attempt was made to establish a National Security Advisory Staff within the defense ministry. This body's two-fold task was to assist the Defense Minister by acting as a counterbalance in his breach with the general staff, and tempering his over-dependence on the CGS (Tal 1996: 111). Weizman's dispute with Begin over the scope of his authority as Defense Minister led to his resignation from the government in 1980. Despite Begin's meager experience in national security matters, he became acting Defense Minister in addition to his role as Prime Minister until the 1981 elections. This move provided Lt. Gen. Rafael Eitan, Gur's replacement, with a golden opportunity to assume additional power as CGS.

Sharon's appointment as Minister of Defense, after the 1981 elections, brought a significant change to the role. He increased the use of the National Security Advisory Staff, now operating out of his office, and aimed at severing the ties between the army and the government, thus strengthening his own power in the triangle of forces. Sharon was also the first Defense Minister who attempted to change Israel's traditional security concept by employing military force to achieve political goals beyond Israel's borders even though there was no threat to the country's existence.

This novel concept was milked to the bone as a justification for the Lebanon War in 1982. Sharon's involvement in strategic agreements with the United States, as well as his support of Prime Minister Begin at the first stage of the Camp David Agreements, catapulted him to a key position in the triad of forces and empowered him to dominate all military operations as a "super CGS." His brief, yet intense, term of office ended abruptly in 1983, following a public cry for his ouster based on evidence provided by the Kahn Inquiry Commission. The Kahn Commission differed from its predecessor, the Agranat Commission, by not distinguishing between the military's responsibility and the government's. This resulted, for the first time, in forcing both the military high command (the CGS and the chief of intelligence) and the political leadership to pay the price for involvement in the Lebanon War.

Moshe Arens replaced Sharon as Defense Minister for a short period until 1984. One of his first acts was the dissolution of the National Security Advisory Staff which, it will be recalled, had swollen out of proportion during Sharon's tenure and whose dismantling now meant the return of some of its members to the planning branch of the general staff. The chief of the planning branch was thus obliged to "serve two masters" concurrently—the Defense Minister and the CGS. Despite Arens' deft promotion of Lt. Gen. Moshe Levi to CGS by stipulating his appointment on the creation of a Ground Forces Command and the "Lavi" jet-fighter project, he soon came to reveal himself as a Defense Minister who was unable to project his authority over the army.

The National Unity Government, subject to a bi-partisan rotation of prime ministers, brought Itzhak Rabin to the office of Defense Minister for a six-year period (1984-1990). He immediately sized up the power now concentrated in his hands after having served as both CGS and Prime Minister. In spite of rigid decision-making mechanisms within the National Unity Government, Rabin's personal authority and professional credit enabled him (with the Prime Minister's backing) to pass the resolution for pulling IDF troops out of Lebanon. At a later stage Rabin terminated the "Lavi" project, siphoning its financial outlay towards the army's other needs. Half of his term in office was spent dealing with the *Intifada* that had erupted in the occupied territories in late 1987. Rabin partially neutralized the CGS, Lt. Gen. Dan Shomron, because he believed that the conflict in the occupied territories had major political ramifications. But the army's image, as well as that of the CGS, was substantially blemished by the military's lack of readiness for dealing with this worsening situation. From his vantage point, Rabin strove to alter long-held, narrowly-defined views of future relations with the Palestinians. In 1992, he succeeded in bringing the Labor Party back into power, thus receiving a mandate to realize his vision of advancing the peace process.

1992-1996: Return to the Dual Role

Rabin's objective might have been clear, but attaining it was an overwhelming task. The importance he attributed to the peace process as a key to achieving security, necessitated Rabin at the outset of his term in office to adopt Ben-Gurion's approach and assume the dual roles of Prime Minister and Defense Minister. The skill he brought to the job, combining political know-how with an exemplary security background, enabled him to steer the negotiations towards a breakthrough with the Palestinians. Rabin's decision to sign an agreement with the Palestinians was no less momentous than the decision to establish the State of Israel made by Ben-Gurion in 1948, on the eve of the British evacuation.

Rabin withheld all information from the army and its high command until the signing of the Oslo Accords in September 1993. Thereafter, according to Rabin, the army would have a major role to play in implementing the agreement, and its involvement in the post-Oslo negotiations with the Palestinians would be essential. Contrary to the CGS's advice, Rabin appointed the deputy CGS, Maj. Gen. Amnon Shachak, as head of the negotiating team. Thus, Rabin was able to maintain direct communications with his chief negotiator, while applying his problem-solving skill whenever necessary. Upholding the agreement at its early stages was made possible because the army had been harnessed to the political process, a move that offset military opposition to certain ground deployments that might otherwise have been seen as conflicting with military needs and interests.

Peres' short term in office after Rabin's assassination, and especially his decision-making during "Operation Grapes of Wrath" (April 1996), reflected Peres' political dependency on the General Staff and the CGS. Although Peres also held the double roles in the short period of time between Rabin's assassination in November 1995 and the May 1996 elections, he hesitated in making the right choices in Lebanon. This miscalculation nearly escalated to a full-blown war, and can been attributed to Peres' over-reliance on the IDF, as well as the absence of a National Security Advisory Staff which might have prevented the campaign in the first place (Tamir 1996).

1996—1999: The Netanyahu Government

The new political leadership that emerged as a result of the 1996 elections presented an opportunity for reshaping Israel's civil-military relations. Prime Minister Benjamin Netanyahu's intention to establish a National Security Council in his office as stated in *The Basic Law: The Government* could have created a new set of ground rules in civil-military relations. The formation of the council to be headed by Maj. Gen. (Res.) David Ivri, former commander of the air force and one of the most distinguished figures within the national security establishment, who had held the high-profile position of director general of the defense ministry for over ten years, was supposed to usher in a major overhaul in the Prime Minister's capacity to oversee military.

In addition, the National Security Council was designated to coordinate various inter-ministerial efforts such as the Anti-Terror Unit, and to deal with issues that came under General Ivri's direct responsibility in the defense ministry, e.g. strategic co-operation with the United States and heading the Arms Control Steering Committee.

Amassing all the above-mentioned roles in the Prime Minister's office could have enabled the Prime Minister, on behalf of the government, to exercise unprecedented control over national security policies by reducing the army's de facto influence on defense issues. Furthermore, the direct election of the Prime Minister, for the first time in Israel's history, was an additional unformal factor strengthening the Prime Minister's position. Direct elections provided him as assumed by the Prime Minister with greater authority than before when he was only "first among equals." The transfer of security responsibilities, in the past held by the government as a collective body, directly to the Prime Minister, as a result of direct elections, granted him only according to his understanding additional leverage that could undoubtedly erode the status of both the

Defense Minister and the CGS even further. This initiative met op-position from the army and the Defense Minister, who was concerned over his position and who had only recently joined the government's inner cabinet. Stiff pressure by the Defense Minister altered the de-cision and determined that the shape of the National Security Coun-cil remain within a limited forum with Maj. Gen. (Res.) Ivri at its head, and under auspices of the defense ministry and not in the Prime Minister's office.

The Prime Minister's volte-face a few weeks later drew public attention to the on-going power struggle within the triad of forces and was a clear victory for the Defense Minister, Itzhak Mordechai, still testing the depth of political waters after his thirty-five years in uniform. But there was nothing in this decision to change the basic views of the Prime Minister concerning the army's role in carrying out political directives. According to Netanyahu, the army should be separated from everything connected to political decision-mak-ing, and in its place Netanyahu would station his own close advisors in positions of authority.

This was, above all, a transparent attempt to remove high-rank-ing officers who had supported the previous government's policies from proximity to decision-makers politicians. Maybe this was the actual reason why Netanyahu adopted an entirely new channel: dis-missing the army from the political arena. This step also reflected the widening gap between the new political leadership and military echelons, as the top-brass were no longer called on to present their professional opinion in political colloquiums. Two examples illus-trate this point: (1) When the Hebron withdrawal plan was discussed by the cabinet, the military, and especially the chief of Central Com-mand, were kept in the dark about it. (2) Before the disastrous West-ern Wall Tunnel opening in the Old City of Jerusalem (September 1996), the military was neither consulted about the timing nor was it issued any direct orders once the decision was made. All it received was painfully short notice to get the troops ready. The fact that mili-tary personalities were no longer "in the game" put them in the awk-ward position of having to provide military solutions for political con-tingencies that they were not consulted on in the first place.

These two examples bore witness to Netanyahu's trend of side-lining the army and obstructing it from fulfilling its missions due to the reckoning of political accounts. This drift had the potential for obfuscating the orderly functioning of political institutions in Israel. Although certain developments continued to reinforce this orientation, a reaction of outspoken criticism set in, emanating from anonymous voices within the military establishment that found their way into the media. In vitriolic language they castigated the Prime Minister's relations with the IDF high-command and impugned his criticism of high-ranking brass during military events such as the graduation ceremony at the National Security College or during heated discussions on the IDF's 1998 budget where Netanyahu used blunt language bordering on political demagoguery to override professional estimates of the general staff (Gevirtz 1997).

This criticism was all the more valid as it came from retired officers protesting the insult and inattention towards the IDF's most competent professionals. The cumulative effect of constant downsizing led to a weakening and further polarizing of communication between the Prime Minister and the CGS (Oren 1997). The brunt of the Prime Minister's criticism was being directed, as mentioned, towards the CGS, Lt. Gen. Shachak, who was identified more than anyone in the military with the previous government's significant achievements at reaching the "Oslo Accords." Disregarding his military expertise from all aspects of the political process paralleled the relentless whittling down of his stature and authority.

Netanyahu's alienation was characteristic not only of his relations with the chief-of-staff, but also with the Defense Minister, Itzhak Mordechai, where a further dimension was added—the political one, stemming from the highly popular Defense Minister's potential threat to the Prime Minister's status. Moreover, Mordechai's political initiatives and his criticism of the cabinet's handling of major security issues led to his isolation within the government and his later removal from office (Caspit and Rahat 1997; Galili 1997). In contrast to his two predecessors, both of whom served the double roles of Prime Minister and Defense Minister simultaneously, Mordechai spent most of his time within the security establishment.

Taking all of the above-mentioned developments into account, it seems obvious why Mordechai made every effort to strengthen his control over the general staff. Many of the promotions in the IDF reflect this inclination as proven by the fact that top jobs have been handed to officers considered loyal to the Defense Minister. However, the majority of these appointments were not acceptable to the CGS (Rabin 1997) whose ability to oppose them was limited.

The Defense Minister's lack of political experience in a government where the Prime Minister was hell-bent on centralizing his own power and emulating the omnipotence of a feudal king, landed Mordechai in an extremely ticklish situation. He had to persuade the government to decide on complex and sensitive issues that once were solely within the jurisdiction of previous Defense Ministers. Although his political leverage was blocked by the Prime Minister, he was still able, because of the CGS's problematic relationship with the Prime Minister, to become a major figure representing the IDF's interests before the government and the public. This was one of the only cases where the CGS's status was truncated in a power struggle with the Defense Minister vying for influence and free passage-way to the Prime Minister's ear. In this scenario, the CGS found himself overshadowed by the two other players in the triad.

An Additional Paradox

A review of the major developments related to the Defense Minister's role under different governments and political arrangements offers a glimpse of how the IDF evolved into an autonomous, self-governing institution for so many years, as noted by Tal (1996: 108), despite formal governmental control over the military and parliamentary regulations. This review provides a background for better understanding of the complexity inherent in the informal processes and the role of political arrangement in the phenomenon under examination.

Lissak has examined the various aspects of civil-military relations in Israel in several publications (e.g. 1971, 1984). One study is dedicated to a discussion of the paradoxes typifying these relations. According to Lissak (1983), they originate in " . . . the adverse conditions under which the IDF emerged—first under autonomous Jewish national institutions during the [British] Mandate and later under the government of Israel—all of which left their mark on the IDF. These conditions created a number of paradoxes, or blatant incongruities, that have added a distinctive twist to civil-military relations in Israel."

The fifth and the last of the paradoxes introduced by Lissak (1983) is defined as follows: "Despite the fragmented and often permeable boundaries between the military and civilian sectors, their relations and divisions of labor are nonetheless based on the principle that the armed forces is subordinate to the political authority." This paradox stresses the army's implicit obedience to civilian authority, despite the existence of some "porous" zones that on occasion blur a formal demarcation of powers.

The definition of an additional paradox, dealing with the nature of informal, subordinate relationships versus formally defined ones, is offered now as an extenuation to Lissak's fifth paradox. Notwithstanding the specific turf outlined in *The Basic Law: The Army* which was designed to clarify chain-of-command relationships between the government, the Defense Minister, and the CGS, the wording is sufficiently ambiguous to permit the major protagonists *considerable maneuverability*. Their behavior is, in effect, a reflection of their own authoritativeness, their personal and political fortitude, and those precedents that have become canonized in relationships despite formal boundaries. These are the sources for the creation of innumerable informal arrangements that have been cemented over the decades.

This paradox focuses mainly on power-struggles between the three leading role-players entwined in a fluctuating relationship. It is necessary, therefore, to examine how the army, or its highest-ranking officer—the CGS—acquired a considerable degree of independence vis-à-vis the "army's chief civilian commander-in-chief," i.e., the government and especially the Defense Minister. The process of amassing power, that altered somewhat in the late 1980s, swerved onto a completely different track under Netanyahu's stewardship. On coming to power, the Netanyahu government enacted basic, destabilizing changes in the game-rules that had crystallized over the years.

It is an indisputable fact that national security policy is the sole responsibility of the government that holds the monopoly for mobilizing the military. The national leadership is responsible for formulating the security doctrine that is essentially the basic strategy for national preparedness. The political echelon is also supposed to determine the nature, scope and framework of the national defense system, as well as its political and military objectives (Tal 1996: 33, 51). It has been pointed out that planning a national security doctrine must be the task of the civilian leaders because the doctrine reflects their beliefs and perceptions of reality which in turn guide their political and strategic decisions. The government's success or failure is a yardstick for evaluating its particular contribution (Harkabi 1981). In the absence of a clearly defined, official policy in this complex area, the army has been forced to determine the parameters of its security plan, according to its understanding and basic needs, instead of following the government's national security policy.

The above observations rang true for the most part during the first years of the state when Ben-Gurion was both Prime Minister and Defense Minister. Later, during Israel's wars and in the intervals between them, the government had difficulty defining political objectives despite Rabin's claim that government defense policy is expressed primarily in the army's " . . . training, and its readiness to follow guidelines determined by the government." Rabin also stated that " . . . when the issue is the military's execution of the civilian authority's defense policy, then the army's main objective is to defend the state and its essential interests as well as to protect its territory and inhabitants." (1987). This definition indicates the ideal way in which the government directs the army, though the actual picture is somewhat different.

This is also true regarding the tension between political and military objectives. This "stems from the fact that political concepts during war-time take on non-military implications either because the political aspects undergo constant change (relative to the stable nature of military thinking), or because military rationale is basically different from political rationale" (Lanir 1985). The tangled process of defining objectives during the "Grapes of Wrath" Campaign (April 1996), as well as the incompatibility between the government's goals and the army's capabilities, hindered both the management and conclusion of that entire hapless operation.

The IDF's involvement in the campaign highlighted errors resulting from an omission of clearly defined political directives. The army was forced to bridge this gap by its own initiatives that were naturally influenced by the nature of the ground fighting. Directives, generally accepted by the political echelon, are essential for the smooth running of a sprawling military network.

The army and its commanding officer, assisted by the general staff and all its branches, usually perform their jobs in a systematic fashion that posits the CGS at an advantageous position vis-a-vis the government, Prime Minister and Defense Minister. The "Operation Grapes of Wrath" fiasco could have been prevented had the attempts to provide the Prime Minister and Defense Minister with professional advisory boards been successful. Even the appointment of senior military consultants never took place. The National Security Advisory Staff during Sharon's term in office was a serious effort to overcome the lack of an advisory board, but it too failed due to its pretentious bid to replace the general staff. Further attempts to establish a consultative agency were more limited in scope and treaded gingerly so as not infringe upon the army's turf.

Although the establishment of a "National Security Team" in the Prime Minister's office is a legal instruction within *The Basic Law: The Government*, it has never materialized in a way to guarantee success. Appointing "light-weight" personalities during Shamir's and Rabin's terms of office also came to nought because the work style of these Prime Ministers left little room for such advisory teams. Rabin preferred the military-like decision-making style of the army, and relied mainly on his own military assistants. Other decision-makers at the highest level are inclined to agree with ideas presented by the military, especially in war-time situations, due to the lack of other alternatives (Bar-Or 1989-90). This is most apparent when intelligence assessments, most of which are incomprehensible to civilian authorities, are presented by the IDF Intelligence Branch. This branch is thus responsible for national estimates that further enhance the army's stature.

Additional processes, as well as some sanctified behavioral precedents, have frozen into informal norms. Over the years, the top brass was awarded almost mythic admiration and unlimited professional respect from the political captains-of-state. The government and Defense Minister granted them almost boundless freedom of action, which gradually reduced the civilian authority's involvement in military affairs. Some Defense Ministers avoided direct control of the army, preferring to offer "ministerial advice," while others strove to become "super-CGSs," especially in wartime, although this did not necessarily win them mastery over the armed forces. One minister eschewed any intervention at all in military affairs (as a result of his own experience at having undesirable decisions forced upon him).

It is easier for a Defense Minister with a military background to challenge the army's hegemony if he so chooses. Defense Ministers did not always use their governmental backing to take full command of military matters. Exactly the same number of Defense Ministers came to the job from a military background as those who lacked military experience. While the complicated role of Defense Minister requires a deep understanding of military bureaucracy, it is not a prerequisite for guaranteeing success.

Since the separation of ministerial roles in 1967, power struggles between the Prime Minister and the Defense Minister have also contributed to the military's increased power. At times, government ministers intent on downsizing the Defense Minister's authority exerted all their influence in strengthening the CGS by boosting his leverage. Thus, it became possible for the CGS to appeal directly to the Prime Minister and government, outflanking the Defense Minister's office.

The National Unity Government's six years in power (1984-1990) was an additional stimulant to bolstering the CGS's position, despite the realization that the army's ability to handle the *Intifada* was limited. While the Defense Minister was trying to quell the uprising, it was the government, especially the cabinet, which actually hampered the army's efforts. The vacuum created by the National Unity Government's almost inveterate paralysis forced the army to take matters into its own hands and make critical decisions.

Notwithstanding Rabin's close supervision of the army while he was both Prime Minister and Defense Minister between 1992-1996, the CGS actually gained greater freedom of maneuverability. When Netanyahu ascended to office, the CGS's independent decision-making was abruptly curtailed and a reversal of game-rules characterized the Prime Minister's effort to concentrate power in his own hands while jettisoning the Defense Minister and the CGS to the sidelines. The result of these processes was a gradual shrinking of the government's ability, and the Defense Minister's authority, to supervise military performance. This fact does not alter the basic theme of this article that political supervision of and supremacy over the army rest on informal working-procedures wider in scope and greater in significance than formalized arrangements printed in government manuals.

A comprehensive summary of the ramifications of the Netanyahu Government is not attainable at this moment. Nevertheless, one can say almost with certainty that parliamentary supremacy over the army is an essential requirement for effective democratic government. Ben-Gurion and Rabin, both of whom held the double role of Prime Minister-Defense Minister, had such capabilities, albeit each came from a different background. Exerting supremacy, as defined by Dror, is the ability to handle security operations based upon an extended political overview, as well as the wisdom for creating original strategic options and imposing them on the army.[1] This definition enables one to differentiate between "the exertion of supremacy," and "control" (a term of restricted connotation). Assuming that the political leadership's activities are legitimate and represent the interests of society, its supremacy over the military is intended to guarantee that the army's behavior is kept within parameters that do not exceed the formal guidelines set down by law.

Conclusions

It seems that an analysis of the informal processes connected with the development of relations between the Prime Minister, Defense Minister, and CGS presented in this article strengthens the claim that this perspective is a valuable key to understanding civil-military relations in the State of Israel (Ben-Meir 1995: 106). While Ben-Meir has shown the importance of informal relationships among the three elite protagonists in Israel, this study has scrutinized the central, uppermost role played by informal processes and arrangements during the past fifty years in establishing the army and its highest-ranking officers as the dominant players in the menage-a-trois.

The clarification of informal processes at play in Israel's elite military power structure has introduced an additional paradox to Lissak's seminal list, and expands on Ben Meir's and Tal's observations in the beginning of this work. The manner in which the Defense Minister has functioned as the pivotal wedge between the civilian establishment and the military command structure throughout different governments can shed light on the paradoxical features of Israeli civil-military relations.

Due to a lack of clearly defined, formalized standards for the role of Defense Minister, informal arrangements evolved which lent ambiguity to his intermediary role as liaison between the government and one of its most important branches—the army. However, it should not be forgotten that the State of Israel, in its almost fifty years of existence, while bearing witness to the army's enhanced capability to influence the national agenda, has never faced the danger of a military takeover. While the army's aggrandized influence has often been used in power struggles with the civil authorities, it has never been manipulated against them. In spite of the army's dominance, Israel has remained a stable democratic society as shown by Lissak (1983).

The Israeli experience has proven that despite the obligatory nature of formal laws, the actual running of security matters derived from a vast complex of informal behavior patterns and arrangements which became canonized over the years. Attempts at changing these unwritten rules-of-the-game, whether by recommendations passed by committees investigating military disasters or by other means, have had extremely limited impact until now. However, there is a clear need to change the *The Basic Law: The Army* so that the ambiguity inherent in its imprecise sections will not continue to be a stumbling block at the highest level of the armed forces.

Note

1.   This definition was proposed by Professor Dror in his 1984 course "National Security Policy Issues," at the Political Science Department of the Hebrew University.

References

Bar-Or, Amir 1989-90. "Preemptive Counter-Attack and its Development in Alon's Security Thinking." *State, Government and International Relations*, 30: 61-79. (Hebrew).

Ben-Gurion, David 1971. *Yechud ve-Yeud* (Statements on the Security of Israel). Tel-Aviv: Ma'arachot. (Hebrew).

Ben-Meir, Yehuda 1995. *Civil-Military Relations in Israel*. New-York. Columbia University Press.

Caspit, Ben. and Menahem Rahat 1997. "Limor Livnat in a Tense Cabinet Meeting: 'Let's Get Things Straightened Out Around Here!'" *Yidiot Achronot*, June 22. (Hebrew).

Finer, Samuel 1977. *The Man on the Horseback*. Tel-Aviv: Ma'arachot. (Hebrew).

Galili, Orit 1997. "Locked and Loaded." *Ha'Aretz*, July 9. (Hebrew).

Gevirtz, Yael 1997. "Netanyahu—the Prime Minister of Virtual Reality—Interview with Maj. Gen. Res. Oren Shachor." *Yidiot Achronot*, August 15. (Hebrew).

Harkabi, Yehoshefat 1981. "Reflections on National Defense Policy." *Jerusalem Quarterly*, 19: 108-119.

Lanir, Zvi 1985. "Political Aims and Military Objectives in Israel's Wars." In *War by Choice*, edited by Aharon Yariv, 118-156. Tel-Aviv: Jaffe Center for Strategic Studies. (Hebrew).

Lissak, Moshe 1991. "The Civil Components of Israel's National Security Doctrine." *Iyyunim Bitkumat Israel*, 1: 191-210. (Hebrew).

Lissak, Moshe 1984. "Some Reflections on Convergence and Structural Linkages Between Armed Forces and Society." In *The Military, Militarism, and the Polity—Essays in Honor of Morris Janowitz*, edited by Michel Martin and Ellen McCrate. New York: Free Press.

Lissak, Moshe 1983. "Paradoxes of Israeli Civil-Military Relations: An Introduction." *Journal of Strategic Studies*, 6: 1-12.

Lissak, Moshe 1971. "The Israel Defense Forces as an Agent of Socialization and Education: A Research in a Democratic Society." In *The Perceived Role of the Military*, edited by M.R. van Gils, 327-339. Rotterdam: Rotterdam University Press.

Oren, Amir 1997. "On the Way to the Foreign Legion." *Ha'Aretz*, August 15. (Hebrew).

Peri, Yoram 1983. *Between Battles and Ballots: Israel Military in Politics*. Cambridge: Cambridge University Press.

Perlmutter, Amos 1977. *The Military and Politics in Modern Times*. New Haven, CT: Yale University Press.

Rabin, Eytan 1997. "Command Front General, and Friend." *Ha'Aretz*, August 28. (Hebrew).

Rabin, Itzhak 1987. "The Security Policy of Israel After the Six-Day War." *Skira Hodshit*, 34, 3-4: 4-11. (Hebrew).

Tal, Israel 1996. *National Security: The Few Against the Many*. Tel-Aviv. (Hebrew).

Tamir, Avraham 1996. "Interview." *Ha'Aretz*, June 28. (Hebrew).

11

A New Concept of National Security Applied on Israel

Henning Sørensen

Introduction

This article proposes a new understanding of the concept of national security after the end of the Cold War. The new concept consists of four ideal elements. The first is the political-military *position* of any country vis-a-vis other nations as either allies or enemies. In the latter case, the second element is the *perception of threat* of a country with enemies. The third element is the *purpose* of a country for its military involvement abroad, and the fourth element is the *point of time* for this action. In short, national security is defined by four Ps:Position, Perception of threat, Purpose, and Point of time for military intervention. This concept is then applied on the national security situation of Israel revealing the inconsistency between its incorrect and its proper (ought-to-be) foreign policy. Israel is today threatened by no nation from the outside and therefore Israel has more room for maneuvering in foreign policy than ever and than normally believed. But Israel acts as if it was not the case because Israel wrongly interprets the terror attacks of Hamas and Hizbollah as if they were actions of a hostile nation. It seems as if the present political leadership (December 1997) keeps Israel and the Israelis in a war-like situation instead of demonstrating political courage by redefining its security situation in a proper way: even though peace is neither close nor independent of terror-actions, peace in this region is today determined within Israel and by the Israelis, while twenty years ago it was decided by the Arabs outside Israel.

The New National Security Concept

The first element of the new security concept is the presence/absence of enemies or allies. It places any country in one of four security positions. A country with neither enemies nor allies has independent security. A nation with enemies, but no allies has isolated security. A country having enemies and, at the same time, being a member of a military alliance is in a collective security position. Finally, a country is in a selective security position having no enemies, but several formal allies, cfr. Figure 11.1.

In Figure 11.1 Denmark is used as an example for the illustration of the four security positions. In the Viking era 750 - 1050 a.c., Denmark had independent security with neither enemies nor allies. In the Middle Ages from 1050-1350 and from the Renaissance period and onwards, Denmark stood isolated with only enemies from either the south, i.e., German states, and or the east, Sweden. Only in two periods, during the Calmar Union 1397-1521 between Denmark, Sweden, and Norway, (with Norway the union lasted to 1814), and in the Cold War period with our membership of NATO, did Denmark rely on collective security. Today, Denmark experiences selective security with no enemy, several allies, and many security issues to select from.

Figure 11.1

The Presence/Absence of Enemies and Allies. The Case of Denmark

		Enemies	
		−	+
Allies	−	**Independent Security** 750 – 1050	**Isolated Security** 1050-1350, 1450-1949
	+	**Selective Security** 1990 -	**Collective Security** 1397-1521, 1949-1990

The shift from collective to a selective security position is new and common to all NATO countries after the end of the Cold War, while the former neutral nations such as Sweden, Finland, and Austria have reached the same selective security position from independent security. The new security position for all Western societies explains why most of them have demonstrated the will and capacity to participate in joint military operations to fight marginal security issues such as civil wars, etnic cleansing, catastrophes, starvation, riots, refugees, etc. in their regions or in the whole world. Here, even smaller nations can select among a number of foreign policy issues and thereby increase their role in world politics. For instance, Denmark was asked by the UN to deploy soldiers to Bosnia and did so, but declined the UN invitation to do the same in Somalia. Germany decided the opposite. But in both countries, Denmark and Germany, security is defined as an investment. They deploy soldiers "out of area" and may risk their lives for a good cause in the hope of avoiding higher casualties in the future.

In contrast, nations with enemies in either isolated or collective security positions perceive a *threat* from hostile neighbouring countries. The seriousness of this threat depends on the ability and willingness of the hostile adversary to conduct a military attack. This threat perception places both an isolated and a collective security country in one of four positions, cfr. Figure 11.2 below.

In Figure 11.2 Iraq is used as an example of an aggressor in the period up till and during the Gulf War of January-March 1991. At that time, Iraq had neither the will nor the ability to attack Denmark even if Denmark participated in the international maritime blocade

Figure 11.2

The Presence/Absence of Will and Ability of an Adversary to Attack. The Case of Iraq

		Willingness	
		−	+
	−	**Unrealistic** Denmark	**Potential** USA
Ability			
	+	**Unintentional** Iran	**Realistic** Kuwait

against Iraq during the Gulf War and therefore by the standards of International Law was at war with Iraq. So, an Iraqi attack on Denmark was unrealistic.

Neither had Iraq the ability to launch an attack on the U.S., but had most certainly the will to do so judged from the hostile statements of Saddam Hussein and the public demonstrations in Iraq against the U.S. So to the U.S, Iraq was/is a potential aggressor. In the war between Iraq and Iran (1980-1988), Iraq failed to win and both countries paid a high price of casualties for their warfare-experience; accordingly, Iraq at that time demonstrated its will and ability to attack Iran, but has today no intention of doing so. Thus, such an attack is considered unintentional. The Gulf War itself demonstrated the realistic ability and will of Iraq to attack Kuwait.

As stated above, this model is not applicable to Western countries having shifted from either isolated or collective security to selective security, because they at the same time changed their perceptions of threat from realistic to potential threats and from potential to unintentional threats. It means a reduced will among nations to use their military capacity. So, the low number of wars between nations seems better explained psychologically by the self-imposed control of the aggressor than militarily by the strength of the defender. Probably, this reluctance of an aggressive political elite to use violence against other countries or its own population is based on an increased world public opposition against violence/warfare now seen as uncivilized and inhuman behavior.

Countries in isolated or collective security positions can be placed in one of four threat positions. Nations with isolated security will normally not recognize a threat until the external aggression seems unavoidable or has taken place. In this case, the threat can best be met by a military build-up documented by high military expenditures, conscription, women in combat units, etc. In this case, the national security policy is to get sufficient military capacity to eliminate or reduce a (possible) violent attack from an aggressor. Security is here seen as a zero-sum game where increased security for one nation means decreased security for another.

The perception of a nation in a collective security position is not only based on the number and strength of hostile nations versus its own military capability, but includes as well the threat perception of the other members of the alliance. Therefore, a psychological dimension is here included as all the members of the alliance need not always share the same threat perception. So, the military capability of a specific country has to be combined with a mutual acceptance of using it. This psychological reluctance is explicitly formulated in the North Atlantic Treaty, Article 6, where the U.S. defines its right to decide whether or not to assist any NATO member having been unprovokedly attacked. For this and for economic reasons, the confronting blocs of NATO and the Warsaw Pact found common interest in reducing military budgets, hopefully in a balanced way. Here, security is like a prisoner's dilemma game with mutual and contrasting interests built on facts and bluff. Consequently, during the Cold War era the relations between the Warsaw Pact countries and NATO shifted between minor cooperation and major confrontation and the attempt of the opponent to exploit disagreement in the alliance.

The third element is the *purpose* of military intervention of any country under the presence/ absence of combat and/or of formal armed enemies. Thus nations can identify four ideal types of goals for their military intervention in conflicts around the world: alleviation, peace support operations (PSO), war prevention, and war, cfr. Figure 11.3 below.

In Figure 11.3, the U.S. illustrates the four purposes for military intervention and the category of people these official troops meet. Alleviation is deployment of military forces for humanitarian reasons in the event of natural catastrophes, disasters, rescue missions,

Figure 11.3

The Presence/Absence of Combat and Armed Enemy. The Case of the USA

| | | Formal Armed Enemies | |
		−	+
Combat Missions	−	**Alleviation** Humanitarian missions Original Somalia mission	**War Prevention** Deter opponent from violence Cold War
	+	**Peace Preservation** Neutral intervention Bosnia	**War** Gulf War

famine, etc. where no weapons actually are needed. Examples are
the Andrew Relief operation or UNOSOM, the original purpose of
the U.S. deployment of soldiers to Somalia to bring food from
Mogadishu to the starving rural population in 1992-1993.

Another purpose is PSO where combat takes place but not between two or more formally armed enemies. In this category we
find only one national or multinational army operating against terrorism, drugs, or in particular intervening in different forms of civil
wars from disarmament of informally armed gangs, deployment of
neutral soldiers between local fighting war lords—i.e., "the thin blue
line of the UN"—the establishing of cease fire, etc. By definition
peace still prevails as no external national enemy is identified. The
U.S. Army´s contribution to the IFOR/SFOR in Bosnia and the
KFOR in Kosova are examples of PSOs.

Yet another goal for deploying soldiers abroad is war prevention,
where two or more armed national armies stand hostile toward one
another. By definition, this situation is perceived as being more dangerous than that of PSO because no peace is here to be kept as the
national armies share a mutual intention of destroying the opponent. But war prevention will not always end up in a military confrontation, so often combat will not take place. The U.S. deployment of military forces in Europe as a reply to the USSR/WAPA
threat during the Cold War is one example of this type of military
intervention. Another example is the NATO air campaign of March-
June 1999 against Serbian troops terrorizing civilian Muslims in
Kosova.

The last purpose of military involvment is warfare itself. Here,
formally armed enemies meet in combat on the battlefield. Of
course, war is no goal in itself. It is not even the cause of a conflict
but only its outcome. The reluctance to wage war for its own sake
was shown by the U.S. in January 1991 in the Gulf War when the
U.S. withdrew its troops instead of taking advantage of the situation by destroying the political hegemony of Saddam Hussein in
Iraq.

So the basic change after the end of the Cold War is the shift from
international war/war prevention to intra-national PSOs and alleviation missions. This shift represents a more peaceful world among
nations probably due to the public demand for minimizing the use
of violence as mentioned above. Many indicators support the im-

pression of a more peaceful world. Today, we have almost 200 nations in the world, a generation ago around 160. In spite of this increase of nations by 20 percent, the number of wars between them has not risen accordingly. The India-Parkistan controversy over Kashmir is still the only registered international war right now. Moreover, many of the thirty internal wars in the world in the Cold War period were substituting conflicts for the ideological confrontation between the East and West. These civil wars have now come to an end and, consequently, most countries feel less threatened today than they once did.

However, the number of combat areas around the world has increased from around thirty to sixty (Seaquist 1996: 20). These "wars" are now waged more for ethnic and religious reasons than for political-ideological causes. They are fought by some four million men often neighbors, losely organized in armed gangs, shooting at one another occasionally. On top of that "there are, by one count, forty two private armies of the world" (Seaquist 1996: 25). Therefore, we are far from the large classic battles between well-defined enemies as WWI and WWII. The change in industrial production from mass production to flexible specialization is reflected in the change in warfare from mass production warfare to "demassified war-fare" (Toffler and Toffler 1993).

But the negative impression of more combat areas has to be seen in the light of growing ethnic and cultural self-consciousness among the more than 2,000 different cultural ethnicities organized in 200 nations, making only 5-10 percent of all countries nation-states where language/culture/religion and border coincide such as in Scandinavia.

The many more combat areas should moreover be compared with the increased democratization of around thirty former dictatorship countries since 1979, in particular in Latin America and Eastern Europe. These more than thirty nations have become democracies almost without the use of violence and by peaceful democratic means such as words, demonstrations, boycotts, strikes, etc. So, citizens inside a country have demonstrated a potential for peaceful political improvements.

The new combat situation characterized by local struggle between neighbors, occasionally fighting each other, sometimes using unorthodox weapons and causing outrageous violence against one another, and denying Western military "rules of the game" do, of course, influence the work of Western soldiers deployed into such combat areas.

So, since the end of the Cold War the armed forces of the U.S. and of most European countries have become increasingly involved in PSOs and alleviation missions. Over the last few years, U.S. forces have participated in more than twenty peace promotion/peace preservation operations (Volker 1997: 3) and the U.S. armed forces have moreover headed the war preservation missions during the Cold War period and warfare operations in "Desert Storm" in Iraq in January-March 1991.

Therefore, the most decisive change after the end of the Cold War is the increased involvement of Western countries in PSOs. The PSOs differ for several reasons from the other three missions. First, PSO is more difficult to conduct than the other operations because it often includes a twin purpose of military intervention in combat and humanitarian assistance to civil authorities and the local population. The latter includes such things as restoration of their public utilities like water stations, establishing fire brigades, organizing national elections, etc., as seen in Bosnia. Second, in PSOs a lot of civil actors other than governmental agencies are engaged such as Red Cross, Physicians for Peace, Christian Relief organizations, etc. In Somalia for instance, more than forty private NGOs served under a UN mandate along with the formal UN and governmental institutions. In Bosnia some 250 NGOs are accredited. Third, the opponents fighting are not common soldiers serving in formal national military units, but are often neighbors shooting at one another. Fourth, the rules and technology of violence used in local combat goes from household items (the Palestine Intifada 1987), to light weapons (the Brazilian guerilla terrorism 1970), to heavily motorized, land based weapons (the Afghan civil war), up to air- and seaborne weapon platforms (the Bosnian "inverse war 1994-1995") (Hjalte 1997). Fifth, public support at home for intervening abroad is rather labile and may end with a one-sided retreat. We saw that in 1982 when the U.S. withdrew its troops after the terror attack on the U.S. Army barracks in Beirut, Lebanon, where 182 U.S. soldiers got killed and again in 1993 in Somalia where seventy-six U.S. rangers were ambushed. So the casualty threshold at home for PSOs is rather low.

The last element of the new concept of national security is the *point of time* at which a country intervenes militarily. Based on the time when the challenge of an aggressor is met by the defender, three different responses can be identified. If the defender waits to stop the aggressor after his direct military attack, the defender only *reacts*. If the aggressor is stopped during his military build-up, the defender *acts*. If an aggressor is stopped even before that, cfr.the Clint Eastwood phrase as "Dirty Harry": "Don't even think about it," the defender *preacts,* cfr. Figure 11.4 below.

In Figure 11.4 NATO's strategy over the last forty years for when to meet a challenge is illustrated. Accordingly, NATO has pursued all three different types of action. In the early days of the Cold War, the Atlantic Alliance confined its security to be at stake only after a direct military attack by the USSR/WAPA, cfr. the Grand Strategy of nuclear retaliation in the 1950s even in case of a minor military attack from the USSR/WAPA.

From 1967 and onwards, NATO changed strategy to Flexible Response adapting its action to that of the attacker. Consequently, NATO told the USSR to have its middle range missiles, SS-20s, withdrawn or NATO would deploy the same type of missiles, Pershing, in Western Europe, cfr. the "Double Track Decision" of the 1980s. Today, the mass media and the public opinion of the NATO countries often demand anticipatory intervention, i.e., preaction, to avoid even minor security risks to expand/explode causing international crisis or humanitarian catastophes, as seen in 1997 in Albania when European NATO-countries decided to deploy soldiers down there to stop riots from escalating into a civil war.

So, NATO's military intervention policy has advanced from reaction via action to preaction while, at the same time, the national

Figure 11.4

Relation Between Aggressor's Initiative and Defender's Response. The Case of NATO

Aggressor´ initiative	Attack +	Risk +	Threat +
Defender's answer	React	Act	Preact (Anticipation)
NATO answer in	1950-60	1960s-1980s	1990s
Challenge	Massive retaliation	Double track	Albania intervention

security issue at stake for the intervening countries have diminished. So we now see faster military involvement for less important national security reasons. From the point of an aggressor, it means less time now for maneuvering than before. From the point of the defender/ mediator, he is caught between the public demand for anticipating military intervention to rescue human beings and the public expectation of this military intervention to cost almost no casualties.

Lessons Learned

Since the end of the Cold War, the concept of national security has changed from collective to selective security, from potential to unintentional threats, from international war prevention to intranational PSOs and alleviation missions, and from reactive to preactive military intervention. In short, the national security policy of Western industrialized countries have become more selective, courageous, humanitarian, and anticipatory. The old and the new elements of the concept of national security are illustrated in Table 11.1 below.

Table 11.1 sums up the changes of the elements of the national security concept: more nations have moved to a selective security position, are facing no enemy from the outside, are less threatened, have more foreign policy options to choose among, are more inclined to participate in military intervention in multinational missions for humanitarian reasons than before.

But still, there are some lessons to be learned from the implementation of such an active, anticipatory, and "all-around-the-compass" foreign policy. First, "the humanization of violence," i.e., the public demand for a minimum of violence more than for a maximum of victory, cfr. the Gulf War of 1991 where the world only

Table 11.1

Four Changes in National Security Concepts

	1977	1997
Position of Security	Collective Security	Selective Security
Perception of Threat	Potential	Unintentional
Purpose of Mil Intervention	War Prevention	Peace Preservation
Point of Time of Intervention	Reaction	Preaction

accepted military intervention to kick Saddam Hussein out of Kuwait not to destroy Iraq or its civil institutions. Moreover, the Western public support for military intervention depends on their low number of casualties. Otherwise they will be withdrawn, cfr. the retraction of American soldiers after the massacre in Lebanon in 1982 and in Somalia in 1993. The "humanization of violence" is furthermore supported by research in non-lethal weapons and the construction of target precision systems. A third indicator of this humanization of violence is the Nobel Peace price for 1997 awarded to the American Jody Williams working for an international ban of the use of landmines (ICBL). Finally, this support can be seen from the increased worldwide acceptance of the Geneva conventions excluding civilian populations from the struggle between organized military units.

Second, PSO is much more difficult for the intervening armed forces to conduct than war prevention/warfare due to the complexity of the former. This complexity is a result of the multiple tasks for the military organization to fulfill. In Bosnia, the OOTH (Operations Other Than War) included not only military intervention between fighting groups but also the repair and reopening of public facilities of electricity, water, and gas, assistance to a national election, policing, rebuilding of bridges, etc. The complexity is moreover a result of the diversity of weapons, of their prolieferation throughout the country, of the different types of soldiers from armed gangs, war lords, tribes, to rather well organized military forces, of the cause and rules for violence, and of the presence of NGOs and the medias in the area.

Third, a correct definition of the purpose of the military intervention is vital. An incorrect or changed identification in mid-stream may cost casualties. It was what happened to the U.S. in Somalia, expanding the purpose of their military presence from alleviation (bringing food to a starving population) to PSO (trying to capture the war lord Addeed in order to restore law and order in Somalia). So, it is only fair to state that the original U.S. alleviation operation was a success, even if it ended as a failure as a PSO.

The New Elements of the National Security Concept Applied to Israel

This analysis has concentrated on Western countries. But the new security concept is applicable on all nations; here, it shall be used on Israel to reveal the inconsistency of its national security policy. Over the last twenty years, Israel´s national security position has improved for several reasons. Twenty years ago Israel had potential enemies along all four borders. Today, Israel has relative peace on three of its four borders. Only at the border with Lebanon does Israel experience violent attacks from the Hamas/Hizbollah guerillas. Moreover, even though Israel was never a member of the Atlantic Alliance it was/is perceived as a member of the Western community by its Arab adversaries. So, by the end of the Cold War the automatic support of the USSR to the Arab countries ended and by the Oslo accord in 1993 and the peace negotiations with the PLO, the hostile rethoric from the neighboring Arab countries stopped as well.

This peaceful picture can be questioned. It can be argued that the terror of the Hamas guerillas initiated from Lebanon area is a threat toward Israels´s national security. It is not so. These attacks are no more a threat for the national security of Israel than those of the IRA are to Great Britain or of the ETA to Spain. So today, Israel has no hostile nation threatening its mere existence as was the case twenty years ago.

Second, the present Israeli-American alliance could be perceived as somewhat eroded. But no Arab country doubts the firmness of this alliance and both the U.S. and Israel have several times adapted their foreign policy to one another. For instance, during the Gulf War in 1991 Israel did not answer the Scud missile attacks from Iraq for the sake of continued Arab support for the U.S.-headed operation "Desert Storm." In the same way, the U.S. have critized the settlement policy of Israel for harming its foreign policy relations toward Saudi Arabia and Egypt without breaking close ties to Israel. So, tension between the U.S. and Israel has increased but not to a point of no return. Therefore, Israel has, as the other Western countries, moved from collective to *selective security* with no nation as enemy and with the U.S. as a firm ally.

But for many years Israel has denied itself the advantegous selective security position and thereby neglected its increased room for maneuvering in foreign policy. Of course, the Oslo Accord with the Palestinians can be seen as such a step. But since then, Israel has taken no further initiatives and has blamed the PLO for the same passivity. But here Israel makes a major mistake by comparing itself to the PLO. They are not actors at the same level. We should expect more bold foreign policy initiatives from Israel than from the PLO since Israel is a nation-state and the main power in this conflict and in this region. The PLO is neither. So Israel has the main responsibility for peace. At least, we should expect Israel not to abuse its strength and to allow settlements in occupied territories promised in the Olso Accord to be handed over to the PLO.

Instead, Israel exposes a threat perception of a nation in isolated or independent security, i.e., a nation surrounded by enemies and with no allies, cfr. the steady military build-up in the belief that sufficient military force can eliminate or reduce a presumed violent attack. Israel sees its security as a zero-sum game where increased security for herself means decreased security for another. But Israel is in a selective security position; therefore, we should expect quite another defence policy of downsizing its armed forces, of moving from conscription to an all-volunteer force, of reducing its military budget, etc. It would be a signal of good will, reason, and commitment to the peace process.

Nevertheless, the present political leadership of Israel sees its military intervention outside Israel as war or war prevention. Once, this was the case between Israel and its Arab neighbors. But Israel ought to see its occupation of Southern Lebanon and the West Bank as a PSO, the most complex of all four reasons for military intervention. As stated above, it is vital for any nation to define the purpose of its military intervention correctly. In the case of Israel, it is even more so because the official Israeli "Land for peace" policy will never succeeed. This is because the aggressor (Hamas) is no nation to whom territory means increased security by prolonged warning time and neutral battleground, as was the case for the USSR with its buffer zone of Eastern European countries. Actually, the Hamas attacks on Israel come from an area of Lebanon outside the control of this nation. So, Israel will never get peace with Hamas in exchange for land. The purpose of Israel´s military presence in these areas is to retaliate on the civil Arab population in the occupied territories whenever Hamas attacks in Israel. This hostage policy of Israel can only succeed if Hamas has more concern for the lives of innocent citizens than Israel, and probably it does not.

Instead, we should expect from Israel in a peace preservation position a "humanization of violence" policy, i.e., the minimum use of violence, and its military and civilian forces to participate in the repair and reopening of public facilities of gas, water, electricity, etc. So the present deployment of Israeli soldiers abroad is both too little, too much, and too bad.

Israel has for many years pursued an anticipatory military intervention policy (except in the Yom Kippur War of 1973) as is now the case for the Western countries. But for these countries engaged in mutual military UN-/OSCE-/NATO-missions, a humanitarian purpose has been included. But Israel´s fast military intervention has never had the purpose of alleviating human sufferings but was instead meant to retaliate. In fact, Israel has pursued a preactive military policy as found in the Western countries, but for quite another reason.

Israel´s present national security position is quite clear. Israel is in a selective security position with no national enemies and the U.S. as an ally. So, Israel should perceive no threat. But Israel reveals a threat perception of a nation in isolated or collective security as judged from its heavy military build-up. The official purpose of Israel´s military intervention is incorrectly defined as war/war prevention, but the proper definition is — as argued — PSO. This purpose is more complex to fulfill as it demands both military effort and civil assistance from Israel in the occupied areas.

So Israel´s foreign policy is inconsistent with its actual security position. The incorrect definition of Israel´s security position is used by the political elite in Israel to keep the armed forces and the military expenditures at a higher level than needed, to explain to the public why the peace talks do not succeed, to blame the PLO and the Arabs for that, and thereby to try to unite a divided Israeli population on this question. But this policy will neither succeed nor bring peace. A break-through in the Israeli-Palestine peace process starts with a correct definition of Israel´s security position.

Conclusions

After having identified the proper security position of Israel as one of selective security, of no threat, and of a humanitarian purpose and of fast military intervention, Israel should firstly recognize that there is neither a short cut to peace nor a military solution to the problem of terror. Second, Israel should realize that both military and civilain assistance is needed in the occupied territories. Third, Israel must revitalize the peace negotiations with the PLO instead of having them stopped. By doing so, Israel has indirectly supported the policy of Hamas of bringing the peace process to a standstill. This process should be reversed. A simple question may help the Israelis to reopen the peace process: "What will Israel look like in case of peace?"

The answers to that question are legion, but they will no doubt change the world's perception of Israel from the present passive, paralyzing and power-playing country to an active, assisting, and humanitarian nation in the Middle East region. Not only the Israelis but all the people in this region deserve this to happen.

References

Hjalte, Tin 1997. *A Typology of Civil Wars*. København: DUPI.

Seaquist, Larry (ed.) 1996. *The Venice Deliberations. Transformations in the Meaning of "Security"*. Paris: UNESCO.

Toffler, Alvin and Heidi Toffler 1993. *War and Anti-War — Survival at the Dawn of the 21ˢᵗ Century*. Boston: Little, Brown and Company.

Volker, Franke 1997. "Learning Peace:Attitudes of Future Officers Towards the Security Requirements of the Post-Cold Wart World." Working paper No. 9, John M. Olin Institute for Strategic Studies, Harvard University.

Part 5

The Armed Forces as Organization—
Continuity and Change

12

The Israeli Defense Forces (IDF): A Conservative or an Adaptive Organization?*

Reuven Gal

A Pessimistic Introduction

Back in 1984, in a brilliant theoretical paper, our colleague and mentor Prof. Moshe Lissak (1984) offered us a four-fold model, based on the convergence-divergence (between the military organization and its society) dimension, on the one hand, and the strengthening or weakening of linkages between the armed forces and the civilian population, on the other. According to Prof. Lissak, "one should not *a priori* assume positive or negative correlations" between these two dimensions (p.58); however, Lissak was very clear in concluding his theoretical paper by reminding us that "social and political stability of societies depend so heavily on the quality of relations between civilian and military elites" (p.59).

More particularly, in his attempt to illustrate "healthy" civil-military relations, Moshe Lissak (1980; 1985) had initially proposed that Israel and its military could be placed at the more "convergent," or rather "permeable," end of the first dimension and at the more "constructive" (i.e., strengthening) pole of the second one. Apparently, this was the case at the time. In a more updated analysis, Lissak, this time with Dan Horowitz (1990), casts some doubt whether this was still the case in the early 1990s.

The increased "price" of national security following the Yom Kippur War, the lack of societal consensus during the Lebanon War, and the continuous internal struggle concerning the Israeli-Palestinian conflict—these are the urgent reasons, claim Lissak and Horowitz, for a serious re-analysis of the relationship between the Israeli society and the Israeli Defense Forces (IDF).

In 1994, Charles Moskos and James Burk published their article *The Postmodern Military*, in which they categorized military organizations by their pace of change and adaptations from "pre-modern" to "early-modern" to "later-modern" and finally to "post-modern" phase. Assessed by the Moskos-Burk model, the IDF is an ambivalent case: While some areas of tactical and technological applications already reflect a passage towards "late-modernism" (and in isolated instances began to do so as early as the mid-1970s), in others the pace of change has failed, so far, to move this organization from its "modern" and even "early modern" phases. In fact, several critics (e.g., Gal 1996; Gordon 1997; Wald 1992) have claimed strongly that the IDF has conserved its basic premises— regarding national security, structure of force and officers' development—since the mid-1950s until today without any significant innovation.[1]

Evidently, in spite of dramatic changes occurring both in the Middle East and around the world, and even with the vast strategic, technological, economic and social transformations of recent times— the leaders of the Israeli military have failed, so far, to generate the necessary "second-order change" (Watzlawick, Weakland and Fisch 1974) that will create the equivalent transformation for the IDF as it enters the twenty-first century.[2]

To state the truth, this pessimistic evaluation should be tempered two-fold: First, one should not get the wrong impression that the Israeli military at the close of the twentieth century is, indeed, identical to its predecessor of the 1950s. Obviously, the IDF is a highly modernized fighting machine, which has proven its combat effectiveness in numerous encounters throughout the last few decades. The most significant change that *did* occur, took place in the three to five years following the (1973) Yom Kippur war: The total mobilization capability of the IDF increased during that period by almost 70 percent due to major transformations in manpower, equipment and complexity (Gal 1986). But this change, like a few others that could be found throughout the years, was mostly *quantitative* change; others were mainly changes of restructuring. One such example was the establishment of a central Headquarters for the Ground Forces ("MAFHASH"), which occurred (after a long and tiring period of hesitation) in 1983 and replaced the autonomous individual H.Q.'s for each of the land services. None of these changes, however, was of the "second-order" type.

The second qualification sounds terminological but it is actually fundamental in understanding the unique nature of the IDF. It pertains to the tendency to confuse the ability of an army to *improvise* in the chaos of battle with its ability to innovate and adapt to new eras. Indeed, "since the early stages of this training, the Israeli... commander is encouraged to improvise solutions to problems and to operate immediately upon situations when they alter unexpectedly from the proscribed plan" (Gal 1986: 130). This typical characteristic of IDF's combat officers has been frequently generalized to the IDF as a fighting organization: "At virtually every level of war (except, perhaps, the highest, that of strategy) the IDF has demonstrated throughout its history a proclivity for the dashing, the unusual, the creative solution to military problems" (Cohen, Eisenstad and Bacevich 1997: 77). But not all IDF's officers necessarily praise these creative and improvising capacities: Some, like Maj. Gen. Yossi Ben-Hanan, Commander of the National Defense College, harshly criticized the IDF as "an unprofessional army which relies excessively on improvisation and intuition" (Rabin 1994). Improvisation, thus, not only belies professionalism, but certainly does not reflect *adaptation* of a whole organization to a changing world.

In the following paragraphs, I will examine the IDF's capability at the close of the century to adapt and prepare itself, in some major areas, for the coming decades. While the range of theaters is very broad, I chose to focus on three main areas—strategic, structural and doctrinal.

Areas of Attempted Adaptations

A note of caution should be made here: At this point in time (Fall 1999), it is impossible to assess the precise nature of any trend or process occurring in Israel's society and military. The unprecedented condition of fluctuating events—the "shaky" peace process between

Israel and the Palestinians; the unstable political atmosphere within Israel, with Rabin's assassination still haunting the collective, national memory; and the shifts in the relationship between military and general society in Israel—all prevent an accurate determination of the longevity of the following phenomena.

Threat Assessment

Israel has traditionally faced three distinct categories of military threat (Cohen 1992). The first, usually referred to as "internal security," consists of insurgent activities and "guerrilla" attacks on civilian as well as military targets, emanating from within the state borders or from regions (mostly, the West Bank and the Gaza Strip) behind the lines held by IDF troops. A second category consists of perimeter (i.e., cross-border) incursions. In addition to large-scale attacks launched by the conventional armies of her immediate Arab neighbors, these could also take the form of "low-intensity" raids conducted by groups of Palestinian marauders stationed in Arab lands. The third category of military threat is remote, and consists of potential long-range aerial or missile bombardment launched by foes with whom Israel shares no geographic boundaries. The Gulf War in 1991 and the repeated crises with Iraq perfectly exemplify this category of a remote threat.

Although the scope of both the "internal-security" and the "remote" threats has increased substantially during the past decade, in comparison to the decline in the "perimeter" category (largely caused by the treaties with Egypt and Jordan), the overall perception of military threat has not been significantly altered over the years. In spite of repeated disillusioning events—the Yom Kippur War, Lebanon in 1982 and thereafter, throughout the six years of the "Intifada" and, lastly, in the face of Iraqi missile attacks in 1991—Israel's national security doctrine has not been seriously reconsidered, and its basic tenets have remained primarily unchanged. Admittedly, this failure should be blamed not solely on the military authorities: Back in 1987 (*before* Israel had faced the Intifada and the Gulf War), a special Knesset (Israel's Parliament) sub-committee concluded, after a serious investigation, that "the Government of Israel, which is responsible for Israel's national security, has not conducted any substantial and comprehensive discussion regarding the national security policy and its applications... We have not found any integrative and long-term thinking, examination or decisions. It is critical that such examinations be conducted at the national level, and not only by the military or the defense system" (quoted in Schiff 1998).[3]

Whomever's responsibility it was, the subsequent result is that the IDF has been maintaining, throughout the last decades, the same stable set of strategic policy that emphasized deterrence, preemptive attacks, short wars and forceful military-constabulary occupation of troublesome territories.

Force Structures

Ever since its establishment in 1948, the IDF has adhered to an essentially militia force structure comprised of three components. The vast majority of troops are either compulsory conscripts (male and female, drafted for periods of up to three years from age 18) and reservists (principally males, liable for further terms of roughly one month's duty per year until middle age). These two components rely on a nucleus of a permanent (career) corps, which has never exceeded 10 percent of the total force (Gal 1996). This three-tier structure of the IDF primarily reflects a military assessment of Israel's prospective force requirements. Basically, the system was historically predicated on two hypotheses. One was that Israel's principal military commitments were essentially "perimeter" in nature—a circumstance which, given the short distances involved, even in emergencies allowed for a minimal timelag between the mobilization of reservists and their participation in combat. A second was that the vast majority of troops would be able to attain the required state of battle readiness during their conscript service, and retain those standards thereafter by means of short annual refresher courses and training exercises.

Both assumptions are now deemed questionable. First, the growing complexity and sophistication of Israel's new battle systems requires efficient maintenance and multi-faceted coordination, which dictates a greater degree of specialization and professionalization than the traditional militia structure allowed. Furthermore, the need for increasing professionalization also stems from the changes that (as noted above) developed as a result of Israel's impending threats. A semi-skilled framework, primarily geared to supplying the manpower required for intermittent outbreaks of intense activity in close proximity to the borders, is unlikely to constitute an appropriate instrument for the conduct of either remote missions or persistent constabulary operations. Both are thought to demand an order of skills—and stamina—that few compulsory conscripts can attain, and even fewer reservists maintain.

Ironically enough, these new demands and their consequences have been identified not only by civilian scholars (e.g., Gal 1996; Gordon 1997; Wald 1992) and analysts (e.g., Ben 1995), but also by the IDF senior leaders themselves—but with very few consequences. Beginning in the late 1980s with Lt. Gen. Dan Shomron and followed repeatedly by his successor Lt. Gen. Ehud Barak, these two Chiefs of Staff announced publicly the need to transform the IDF into a "leaner and smarter force." Barak even declared this dictum as a "strategic decision" soon to be implemented!

The frustrating reality is that these declarations were never implemented. Though several minor changes in manpower policy have taken place recently (such as small reductions in overall reserve call-ups and a trimming of several months in the length of female conscription), the major transformation required to make the IDF "leaner and smarter" is not yet to be seen in the near future. In fact, a retiring senior officer (Maj. Gen. Shalom Hagai, former head of IDF's Logistic Division) conceded that the IDF generals "are not built" to accomplish such changes, as Shomron and Barak had envisioned. Such a transformation, suggested Hagai, could only be achieved by bringing external (i.e., civilian) experts, who would not only identify IDF's faults, but will also provide the remedies and solutions (Golan 1995).

Training and Education

Currently, there is a growing awareness in Israel of the need for higher professionalization and "academization" of IDF's personnel.[4] Strong criticism and warnings about unfortunate consequences are frequently voiced. Both military and civilian analysts (the latter being themselves retired senior officers) are in agreement that optional training at the much-needed technological—"high tech"—levels cannot be obtained through regular conscription (Gordon 1997); that leadership development in general (Gal 1987), and officer training

(Gal 1996) and staff-and-command training (Muli 1996) in particular, are anachronistic in the way they are still conducted in the IDF; and that the entire manpower policy of the IDF is not yet geared towards the challenges of the future (Goren 1993; Wald 1992).

The IDF has had a weak and partial response to this wide fault-finding. One example is the recent expansion of pre-draft training courses provided to conscripts equipped with technological skills and attributes. Another is the provision of greater incentives, professional and material, for talented junior and senior officers to contract for extended periods of career service, particularly when they possess proven technological expertise. Similarly, several changes have been introduced recently in the Staff and Command School in the direction of an advanced two-year course (instead of the former one-year only), with an emphasis on academic studies. One experimental program along these lines is the *"Ofek"* ("horizon") program, which puts mid-rank officers on a fast track toward the Lieutenant rank, while providing an opportunity for an academic degree. Indeed, today some 40 percent of IDF's lieutenants have some advanced academic education (Cohen 1995)—a figure substantially higher than fifteen years ago, but still far from an all-academic officers corps, as is the case in most Western armies.

The Need for a Paradigm Shift

The partial changes described so far are typically "first-order" changes, aiming towards more (e.g., training) or less (e.g., length-of-service) of *the same* and presenting merely reactions to "nagging" demands. But in order to achieve a real *revolution in military affairs*,[5] it is required, I believe, that the IDF become capable of making a *paradigm shift*. This term implies, according to its founder Thomas Kuhn (1962), that a new school of thought arises with answers to questions that the old school had considered uninteresting, irrelevant, or unimportant. This is no easy task!

There are many reasons why the IDF finds it so hard to introduce significant (i.e., "second-order") changes, and why it has not yet reached the ability for making a paradigm shift. Eliot Cohen and his colleagues have listed some of the reasons as the "sources of Israeli military conservatism." Cohen's list includes: The influence of *"batash"* (the Hebrew acronym for daily and routine security and border-guard operations); anti-intellectualism among IDF senior-officers; scarce resources in the face of escalating costs of techno-logical advances; and over-tasked senior leadership (Cohen, Eisenstad and Bacevich 1997). While basically in agreement with this diagnosis, I personally believe that the IDF is paying the price now for its long years of double-edged experience: On the one hand, the frequent and diverse wars and military operations have made the IDF, indeed, one of the most combat-experienced armies in the world. On the other hand, however, this preoccupation with routine and ever-pressing missions has prevented military leadership from asking questions that might look "uninteresting, irrelevant, or unimportant." The IDF, thus, has become a POW of its old paradigm. While it can quickly improvise and provide immediate solutions to emerging battle field challenges, the Israeli Military, at the same time, "does not like changes"—as its former Chief-of-Staff (Lt. Gen. Lipkin-Shahak) has recently described it. Perhaps the best nomenclature for this unique characteristic is the term Cohen and his associates choose to describe Israel's military culture: "conservative innovation" (Cohen, Eisenstad and Bacevich 1997).

An Optimistic Conclusion

While the IDF has not yet made its "big jump," it is evidently stretching its limbs at the present time to prepare for the inevitable. There are many indications of this. Never before have there been so many "think tanks," "focus groups," "brain-storming meetings" and so on, which are now flourishing within the IDF. One typical example is the special project, appropriately entitled "*Aviv Neurim*" ("Youth Spring"), which was initiated by the former Chief of Staff Lt. Gen. Lipkin-Shahak. This project was given the task to examine "in all possible directions" new ways to prepare the IDF for the next generation.

In a recently published volume of *Ma'arachot*, the IDF's official periodical, the goal of this project is reframed as to "transform the organizational culture of the IDF and to improve the command capabilities of commanders at all levels" (Moufaz 1998). Furthermore, this process is supposed to be characterized by "new rules of the game" (Ze'evi 1998)—the military nomenclature for a new paradigm. Other task-groups are likewise presently active in seriously examining training doctrines, organizational structure and strategic policies. And, quite unprecedented, much of these "brain storming" activities are conducted with the help of civilian experts, some even foreigners.

This recent atmosphere of receptivity on the part of IDF authorities is probably created as a result of dual effects—first, that the IDF has recognized that it is not as invincible or infallible as it had once thought itself to be; and second, that the IDF is healthy and strong enough to afford to make significant changes. And so, while conservatism has not yet completely vanished, it seems that the current IDF leaders are committed to inaugurate a paradigmatic transformation—perhaps even before the close of this century.

Notes

*   Some parts of this paper were modified from Gal, Reuven and Stuart Cohen, forthcoming, "Israel: Still Waiting in the Wing." In *The Postmodern Military*, edited by Charles Moskos, John Williams and David Segal.

1.   The emphasis is on the term "significant"—following Stevenson (1997).

2.   The leading daily in Israel, *Ha'aretz*, published a special volume dedicated to defense issues on the eve of the Jewish New Year (*Ha'aretz/Rosh Hashana Magazine*, September 20, 1998). The leading article, by Ze'ev Schiff, was entitled: "IDF of the Year 2000 Must Change."

3.   This sub-committee was comprised of a few members of the Knesset's Foreign and Security Committee, and was headed by Dan Meridor.

4.   For example, Maj. Gen. Ben-Hanan, in Rabin (1994).

5.   RMA—a term more and more used by military students, such as Charles A. Stevenson, Eliot E. Cohen, James R. Balker and others.

References

Ben, A. 1995. "To Rebuild the IDF." *Ha'aretz*, April 4.

Cohen, E.A., M.J. Eisenstad and A.J. Bacevich 1997. "Israel and the Revolution in Military Affairs." Unpublished manuscript, January 13, Johns Hopkins University.

Cohen Stuart 1995. "The IDF: From a "People's Army" to a "Professional Military." *Armed Forces and Society*, 21: 237-54.

Cohen, Stuart 1992. "Israel's Changing Military Commitments, 1981-1991: Causes and Consequences." *Journal of Strategic Studies*, 15: 330-50.

Gal, Reuven 1996. "For a Review of the Current Model of the Israeli Officer." *Ma'arachot*, 346: 24-25. (Hebrew).

Gal, Reuven 1987. "Military Leadership for the 1990s: Commitment-Derived Leadership." Paper presented at the Military Leadership Conference held at the U.S. Naval Academy, Annapolis, MD, June 10-12.

Gal, Reuven 1986. *A Portrait of the Israeli Soldier*. Westport, CT: Greenwood Press.

Golan, M. 1995. "A Non-Smart Army." *Globes*, May 17.

Gordon, S. 1997. *The Bow of Paris*. Tel-Aviv: Sifriyat Hapoalim. (Hebrew).

Goren R. 1993. "Good Men in the Middle of the Road." *Ma'ariv*, October 18.

Horowitz, Dan and Moshe Lissak 1990. *Trouble in Utopia: The Overburdened Polity of Israel*. Tel-Aviv: Am Oved Publishers. (Hebrew).

Kuhn, Thomas 1962. *The Structure of Scientific Revolutions*. Chicago: University of Chicago Press.

Lissak, Moshe 1985. "Boundaries and Institutional Linkages Between Elites: Some Illustrations from Civil-Military Relations in Israel." *Research in Politics and Society—A Research Annual*, 1: 129-48.

Lissak, Moshe 1984. "Convergence and Structural Linkages Between Armed Forces and Society." In *The Military, Militarism and the Polity—Essays in Honor of Morris Janowitz*, edited by M.L. Martin and E. Stern McCrate, 50-62. New York: The Free Press.

Lissak, Moshe 1980. "The Defense Establishment and the Society in Israel: Boundaries and Institutional Linkages." Paper presented at the IUS Conference, Chicago, October.

Moskos, Charles and James Burk 1994. "The Postmodern Military." in *The Military in New Times: Adapting Armed Forces to a Turbulent World*, edited by James Burk. Boulder, CO: Westview Press.

Moufaz, Shaul Maj. Gen. 1998. "'Aviv Neurim'—a Revolution in Organizational Culture in the IDF." *Ma'arachot*, 358: 1. (Hebrew).

Muli, Lieutenant Colonel 1996. "The School of Command and Staff: A Military Academy." *Ma'arachot*, 347: 47-48. (Hebrew).

Rabin, Eitan 1994. "Retiring General: IDF 'Unprofessional' Army." *Ha'aretz*, August 11.

Schiff, Ze'ev 1998. "An Old Policy in a New Reality." *Ha'aretz*, January 9.

Stevenson, Charles 1997. "Dynamics of Military Innovation." Paper presented at the Biennial Conference of the Inter-University Seminar on Armed Forces and Society, Baltimore MD, October 24-26.

Wald, E. 1992. *The Gordian Knot: Myths and Dilemmas of Israeli National Security*. Tel Aviv: Yediot Aharonot. (Hebrew).

Watzlawick, Paul, J. John Weakland and Richard Fisch, R. 1974. *Change*. New York: Norton.

Ze'evi, A. Brg. Gen. 1998. "'Aviv Neurim'—The Main Theme and Its Application." *Ma'arachot*, 358: 5. (Hebrew).

13

Organizational Complexity, Trust and Deceit in the Israeli Air Force*

Luis Roniger

Introduction

This paper examines how a military ethos is enacted into concrete mechanisms of organizational management and how fragile these mechanisms become with changes in operational scale and public morality. The analysis deals with the use of trust in the army and the development of overtrust and trust failure. Discussing the development of a large-scale case of deceit and corruption in the Israeli Air Force, it shows that the sliding of crime into scandal can be seen as part of contrasted strategies to maintain or change the matrix of military accreditation in Israel.

The Israel Defense Forces have been one of the most trusted institutions in Israel. In any society, the level of public confidence in institutions is related to the maintenance of expected performance and commitment. Public confidence in the armed forces, for example, is related to their professional performance in military confrontations, as well as to a public concern with defense and the perception of the armed forces as committed to the attainment of security. In Israel, where national values still run high and the reluctance to fight for the country is low, public trust in the army has been unsurprisingly very high. Studies conducted in the 1980s and early 1990s on institutional trust in Israel have found that a large number of people have a high level of trust in the army, while those with a low level of trust are few. The former group has systematically constituted over four-fifths of the samples, sometimes reaching the almost incredible figure of 95 percent; the proportion of those not having confidence have not exceeded 5 percent in the whole period under consideration. In Peres and Yuchtman's study (1992) the army was reportedly the most highly trusted institution; in Roniger's (1993), the IDF is placed second among twenty institutions, closely following the State comptroller, who was highly involved in the control of corruption.[1] The high levels of trust reportedly bestowed upon the IDF are greater than the trust placed by the citizens of former West Germany, France, Great Britain and the U.S. in their national armies (Dalton 1988; Dogan 1993).

Against the background of these very high levels of trust among the Israeli public, it is of special analytical interest to study the inherent fragility of those mechanisms of organizational management that rely on trust projection within the army.

The limits of delegated trust will be brought to the fore through analysis of a case of deceit and scandal in the Israeli Air Force (IAF), the biggest ever to rock the armed forces of the country. Analysis begins by outlining the place of trust in the framework of military organizational management. Although trust is cherished as a value and a source of flexible decision-making, deceit is generated under conditions of organizational complexity when individuals manipulate cultural premises and change their professional role into one of middlemen following an agenda of self-interest. There then follows a discussion of the structural and value components of trust violation in the IAF and the creation of scandal to deal with the institutional implications involved in deceit. Finally, discussion addresses the resulting institutional changes and the beginning of public review on the limits of honoring military accreditation in contemporary Israel.

The Place of Trust as a Mechanism of Organizational Management in the IDF

The IDF is a huge bureaucratic organization based on formal criteria, yet in many respects it depends upon informal "buddy" networks and relationships. As in other formal organizations, trade-offs have developed in the IDF between hierarchical controls and supervision on the one hand and reliance upon trust and self-regulation on the other. Since organizational authority is usually enforced through task delegation, controls are necessarily combined with reliance on trustworthiness in the implementation of organizational goals.

The assessment of trustworthiness has been based on bonds of mutual trust, formed through shared and often harsh experiences. These experiences range from the recruiting stage into the "total institution" atmosphere of the basic training camp to battle experience. These bonds are further strengthened as youngsters go through these experiences within age group cohorts. An informal atmosphere is encouraged by the army itself, as unit "cohesion and morale" are considered important to the functioning of army units in peacetime and for integrity and commitment in war. In contrast to armies such as that of Canada, where morale is based on the individual motivation to serve, in Israel great emphasis has been placed on committed vocation, trust in teammates and unit cohesion (Evonic n.d.; Henderson 1985; Gal 1986).

In the stages leading up to the creation of the IDF, and during its later development, several such "buddy" networks crystallized. These have survived the passing of decades, have been considered a source of committed relationships and trustworthiness, and have had sometimes a significant effect on Israeli public life. The pre-state "Palmach," aligned with the Haganah and mainstream forces of the Jewish community, created one such network (Oring 1981). Other networks include members of the pre-state dissident paramilitary organizations, the Irgun and the "Stern Group," the "101 unit" of the 1950s, and the paratroopers of the 1960s and 1970s.

Informal networking and dealings were part of the ethos endorsed by society in general and the political leadership in particular. Informality, effectiveness, institutional bypasses, innovation and the search for unconventional and a-legal solutions were part of the set of everyday principles that came to be valued by the population since pre-state times. During a decade-long period, a worldview crystallized according to which ends sanctify means and a ritualistic obedience to rules is scorned. These attitudes sustained the "*shita*," the organizational codes shaped by Mapai and its allies in the pre-state days and inherited by the state (Horowitz and Lissak 1989: ch. 2).

The above could be a source of dynamism and organizational flex-ibility but could equally work against the routine principles of mod-ern bureaucratic institutions, such as rationalization and the func-tional division of labor. These principles also permeated the armed forces, one of the institutional pillars of the state.

While controls and verification procedures have not been lacking in its hierarchical structure, the army has tended, for historical and functional reasons, to rely heavily on trust relations. Such reliance has had what can be termed both positive and negative institutional effects. Thus, for instance, on the one hand, as expectations of trust pervade relations within the structure of command, the performance of duties may take on an aura of voluntary behavior, adding flexibil-ity to decision-making. However, such reliance may impair judge-ment in connection with promotions to high rank, with excessive reliance on the trustworthiness claimed and reported by the candi-dates' patrons and friends, which leads to uncontrolled leaps and advancement. This ambivalence and conflicting potentiality come clearly into relief in the case of the Israel Air Force, since tradition-ally it has combined high professionalism and autonomy with a heavy reliance on trust relations.

Organizational Management and Trust Within the Air Force

Being a relatively small and highly specialized functional unit, the IAF has managed to create internal cohesion and a strong sense of elite identity, due to selectivity and top level performance. The IAF attained early a high degree of autonomy vis-à-vis the rest of the army, and this was reinforced following its contribution to the "blitz" victory of the Six Day War in 1967. Its social and ecological boundaries (e.g., segregated bases) have kept it as an enclave within the army. Special uniforms, good facilities on their bases, and par-ticular rituals and jargon further reinforce the sense of being a closed and quasi-aristocratic "family" within the army (Weizman 1975; Cohen and Lavi 1990; Edelist 1991). Jokingly, previous chief of staff, Dan Shomron, has been quoted as defining the IAF as "an alien though friendly army" (Bin Nun 1991: 7).

It is the combat pilots who both epitomize and sustain the sense of selectivity and top level performance in the IAF and it is norma-tive practice to appoint the commander of the IAF from their ranks.

The *elan* of being the "best" has filtered down through all ranks of the IAF, even to soldiers doing menial jobs, and has been a major factor in reinforcing the willingness of youngsters to join the IAF, which makes it possible to maintain highly selective criteria for recruitment to this corps.

In terms of internal management, the IAF has developed strict guideliness and supervision procedures. But, in parallel, it elevated trust to a central organizational principle and basis of a behavioral code typical of a familistic framework.[2]

The reliance on trust is not accessory but rather imperative organizationally due to the IAF's inner segmental structure. This structure comprises two distinct career and mobility paths: one, open to pilots, beginning with field commands and leading ultimately to the chief command of the force; the other, technical, beginning with technological and mechanical studies and leading ultimately to the command of the technical and procurement divisions within the IAF.

This segmentation is characteristic to the IAF where, unlike domains that follow a technological approach comprising both operation and servicing (e.g., tanks), the role of the IAF pilots is conceived in terms of expertise in the operation of the planes in the air, not in their servicing on the ground. In principle, the mechanical work is done under close surveillance. In practice, the pilots have to rely on the trustworthiness and competence of the mechanics. The pattern of organizational complementarity of two segmented sectors necessitates an expectation of trustworthiness as a prerequisite for successful cooperation in actions ranging from routine to complicated combat flights. Finally, trust enables the IAF to develop flexibility and rapidity in decision-making, qualities considered a functional necessity on the battlefield and a valued resource in peacetime.

Trust and trustworthiness, in addition to excellence, are cherished as ultimate values in the IAF. While this reinforces delegated accreditation as long as the mechanism operates as expected, once a failure of trust is verified its vulnerability increases. Thus, on the one hand, as the corps developed an image of trustworthiness, routine supervisory controls by external bodies were reduced, leaving it to the IAF to maintain its autonomous structure. Soldiers were expected to behave according to the IAF organizational norms, and its status as a structure generating trust was enhanced. The monitoring of performance and unit morale and confidence was adopted by the IAF in the 1980s as a means of encouraging competition for excellence among the IAF units.

However, reliance on trust facilitated domains of de-bureaucratization and opened a wider "grey area" of discretion in the use and abuse of authority, free of bureaucratic hierarchical controls. In particular the realm of arms procurement proved to be highly vulnerable to trust failure, short-circuiting the dominant pattern of accreditation.

Equipping the IAF

In the early 1950s, David Ben Gurion, Israel's first Prime Minister, established clear-cut procedural rules for arms procurement. According to these rules, a division of labor was to be observed between the technical and the financial aspects of procurement. The financial dealings were to be entrusted to the civilian sector (i.e., the Ministry of Defense), in order to keep the IDF out of the reach of commercial interests and influences (Greenberg 1991).

This principle of separation of functions has nominally been in effect in regulating arms procurement deals up to the present day. In practice, several developments hampered its effective implementation in the 1970s and 1980s, especially in procurement for the air force.

First, the decisive role of Israeli aircraft in shaping the course of the 1967 Six Days War ensured the IAF a central place in the later strategic development plans of the army. With the passing of time and rapid technological developments in Western military equipment, this centrality came to mean an increased share of the overall military budget, which itself has increased significantly since the 1970s (Mintz 1985; Horowitz and Lissak 1989: ch. 6). It can be assumed that this centrality became even more crucial as military expenditure was reduced in relative terms during the 1980s.[3]

Second, the entrusting of huge financial dealings to the often poorly paid officials of the Purchase and Production Division of the Ministry of Defense led to the proliferation of inside information being leaked and of bribery. As early as the 1970s the disclosure of such cases produced tougher bureaucratic controls (Schiff 1975). The resulting real or perceived bureaucratization of the civilian apparatus for the regulation of arms procurement had major consequences later on. IAF officers felt the formalization and proliferation of procedures hindered the effective advancement of military interests and needs.

Third, in the 1980s the nature of arms procurement changed. Instead of payments to Israeli companies in local currency, an arrangement developed in which a more substantial part of U.S. aid was designed to be used for contracts with American companies within the U.S. and for subcontractual arrangements with Israeli companies. This development prompted an increasing number of retired high-ranking army officers to go into business as military subcontractors and brokers, using their expertise and connections as major assets. This arrangement also reinforced the role of representatives of the Israeli Defense Ministry in the U.S., who were supposed to coordinate the financial aspects of the contacts, bids, and contractual links with the U.S. firms and the sub-contracting firms in Israel. Sectors of the IAF resented this financial control, seeing it as a further indication of a growing bureaucratization that was likely to slow down vital military decisions.

In addition, the growing technical complexities of military equipment made it essential that IDF officers maintain regular and continuous contacts with the firms producing such sensitive and expensive equipment. Contacts became highly important to ensure that technical standards and requirements were closely adhered to in accordance with IDF priorities. The interest of companies in gaining inside information on the technical aspects of IDF future procurement added to the pressures. In practice, technical expertise became a crucial element in the decision-making process and, conversely, financial issues often became intermingled with what were supposed to be solely technical consultations. It is clear that under existing regulations, the continuous contacts between IDF officers and arms manufacturers require a high level of civil accountability on the part of army officers, who might otherwise bypass Ministry of Defense officials or heavily influence the decisions of the latter on bids.

Thus, several structural forces were at work against the routine functional division of labor instituted by the government. As the significance of technical competence grew, trustworthiness was taken for granted. This, in turn, facilitated the potential misuse of position by IDF officers, who could transform their roles from gatekeepers, committed and accountable to formal and substantive rules, to middlemen guided by their own agenda.

The change could go unperceived as long as it was justified in terms of values traditionally held by Israelis, namely, effectiveness, celerity, initiative, innovation and a search for unconventional solutions. An example of this was the situation in the 1980s when companies were highly interested in gaining inside information on the technical requirements of future IDF procurement. This they did through their contacts with IAF officers. For their part the officers may have been willing to bypass the bureaucratic overload and delays, portraying the leaking of pertinent information as an instrumental way of saving valuable time while accomplishing the aims of the IAF in the best possible way.

That the informal bypassing of procedural rules was not used for private ends was part of the unspoken behavioral code assumed to be common among those trusted as sharing the basic experiences, commitments and values of the IAF. The fragility of such reliance and the far-reaching institutional implications a failure of trust could trigger became evident after the disclosure in December 1990 of the arms procurement network led by IAF Brigadier-General Rami Dotan. Between 1983 and 1990 this network succeeded in stashing away a multi-million sum derived directly or indirectly from military funds, U.S. aid to Israel, and bribes.

Corruption, Deceit, and Scandal in the IAF

The case involved a flagrant breach of the law by a high-standing officer, head of the Equipment Branch of the IAF. The investigation of the multi-million dollar fraud network followed accusations made by an Israeli entrepreneur and former official of the Ministry of Defense procurement office in New York, who was backed by the ministry's Director General. The investigation revealed that Brigadier-General Dotan was involved in a wide range of white collar crimes on a level that by local standards was considered tremendous. Among these crimes were: (a) the leaking of secret information concerning the Israel Defense Force's planned acquisitions in the U.S. to acquaintances, who acted as middlemen and handed over the information to companies which could then tailor their offers to win bids; (b) the abuse of official position and authority by pressuring companies to work through specific agents and subcontractors, bypassing the Israel Ministry of Defense New York representatives in charge of arms procurement with an eye to private gains; (c) extortion or forcing of payment through the misuse of power (Dotan became involved in various financial ventures, and received payments or commissions from several American firms, most notably General Electric); (d) the misappropriation of funds entrusted to the official's supervision and management, and their diversion to accounts outside public control (most of these funds were deposited in the Swiss bank accounts of dummy corporations, while others were kept for the unit's emergency needs in a covert pocket money fund, the existence of which was known to only a small number of IAF officers); and (e) the embezzlement of public monies for private ends, for example, the purchase and remodeling of a very expensive Tel Aviv apartment through a fictitious firm registered in Belgium.

After reaching a plea bargain, expressing remorse and in return for his confession, full testimony and return of the embezzled monies, the officer was prosecuted in a short and mostly closed military trial, discharged from service with the rank of private and sentenced in March 1991 to prison for thirteen years. Nine officers were also discharged from service for having obeyed the convict's orders and for withholding evidence from the investigators. In Israel, the comptroller of the Defense Ministry, together with a specially established public commission reviewed the case and suggested reforms in arms procurement procedures. By January 1992, as a result of its own assessment of the case in the U.S., General Electric had fired or disciplined over twenty of its executives. Later on, in July 1992, General Electric admitted corporate responsibility for the diversion of $40 million in U.S. military aid and the bribery of Dotan. It also reached a court settlement in the suit filed by the Justice Department along with that of a former employee, who managed GE's jet engine operations in Israel from 1984 to 1988 and had decided to suit in a "whistleblowing" legal case. Parallel to this, investigations were carried out both by federal agencies and Congress. In addition to the Justice Department, the State Department, the Pentagon, and a Congress subcommittee (Energy and Commerce, chaired by Michigan Rep. John Dingell) focused on various aspects of the case relating to the bypassing of foreign aid regulations, the involvement of

American companies in the payment of bribes and covert commissions, and the institutional levels in Israel responsible for the actions of the embezzlement network. When Israel refused to accept the State Department's request to interrogate Dotan in person, the whole fabric of U.S.-Israeli cooperation was called into question. Prime Minister Itzhak Shamir's government justified its refusal on the basis of Article 117 of the Israeli Penal Code, which forbids local officials to provide foreign countries with information received on the job. In the U.S., voices in the media joined Mr. Dingell in chastising federal agencies for not monitoring the funds distributed under the Foreign Military Aid program. He demanded that Israel comply with the terms of the ancillary agreements, calling the government to hold Israel in default of its annual aid, and eventually to cut off parts of it. With the changes in the Israeli political scene in late 1992, understandings were reached between the U.S. and Israel on this issue.

Beyond the formal definition of the case in terms of breach of the criminal law and attribution of legal responsibility, the case was also seen as a paradigmatic instance of public corruption. According to R. Tilman's classical (1970) definition, public corruption involves a series of actions kept hidden from the public, where the perpetrators operate through a cash nexus, and a frame of reference with mandatory values is transformed into a free-market orientation, within which values can be manipulated (see also Simmel 1978: 384ff; Heidenheimer 1988).

The development of a fraud and embezzlement case into a major Israeli scandal should not be taken for granted, and calls for an explanation. Many acts of public corruption, especially but not only in the civilian sector, were disclosed particularly since the 1970s. The Israeli public has long been accustomed to disclosures and claims of corruption to an extent where further disclosure, while important in itself, does not necessarily carry an overpowering, shocking significance (Kahane 1975; Ben-Yehuda 1990: 225ff).

Nonetheless, the fraud was portrayed in military circles and in the media as an emblematic case of deviance in the Durkheimian sense and it generated an emphatic outcry. Well before the case reached court, Brigadier-General Dotan was labeled an outcast, and discourse went far beyond the terms of embezzlement. Moreover, the scandal generated public debate on the advantages and limits of the honoring of delegated trust. Occurring in one of the strongholds of institutional trustworthiness, the debacle functioned as a marker of cumulative and previously unacknowledged changes in public morality. It also generated the momentum for a review of previously unquestioned assumptions about the military accreditation of the IDF in general and the IAF in particular.

Public corruption and public scandal are not coterminous, to use the expression of Marengo (1988). Accordingly, we should attempt to analyze how corruption turned into scandal, that is how the structural background and perceptions of the case caused corruption to look scandalous, and what management strategies of this scandal were forwarded by social actors concerned with the broader implications of the case.

Structural Elements and the Politics of Public Morality

The literature in the social sciences suggests that scandals tend to develop where there is ambiguity in administrative regulations (Rose-Ackerman 1999; Moodie 1989; Glees 1989). In that sense, the case under consideration seems atypical. It was initiated as a structured rebellion against a clear definition of standards. For various reasons, as indicated above, a view has crystallized in Israel that ends sanctify means, improvisation is held in high regard and ritualistic obedience to rules is scorned. It was not the lack of procedural rules that generated a scandal. Rather, it was the inability of such rules to prevent the impact of structural and cultural processes that provided opportunities for a betrayal of trust that imperceptibly slipped beyond the "acceptable" parameters of public (im)morality.

It is in this context that the shift in reactions exhibited by the high-ranking army commanders is highly instructive. At first, when the allegations had not yet been substantiated, the top commanders accepted the officer's testimony and claims of innocence at face value. The investigation they had reluctantly initiated was extremely superficial. Indeed, in spite of the accusations made by an entrepreneur formerly connected with arms procurement, Dotan was promoted to a delicate position (head of the IAF Equipment Branch) from where it became even easier for him to carry on his illegal transactions.

Significantly, the embezzlement was discovered by accident, as a result of a falling out between two international money launderers (two "outsiders") connected with the officer. The natural attitude of the IAF staff was that the case was an internal affair that should not be dealt with "outside the IAF family." Trust in Brigadier-General Dotan's integrity was clearly stated by the head of the IAF, Major-General Avihu Bin-Nun, to confirm continuing IAF confidence in its officers. However, once it became clear that he was indeed involved, the atmosphere changed dramatically. It was as though this individual was an evil to be exorcised by everyone connected with the IAF, and he was deemed worthy of treatment reserved for traitors. On December 28 1990, Beni Peled, ex-chief of the Air Force, explained:

> [His friends] felt deceived as much as I did. We trusted Dotan. We thought we had left behind people whose loyalty to the objectives and targets of the IAF was total. If we have been complacent up till now, it is only because we trusted.

In the case under consideration, the violation of trust had both personal and structural-institutional consequences. On the one hand, it engendered personal feelings of anger and betrayal, evident in the personal disappointment voiced by Major-General Bin Nun and other high-ranking IAF officers who had misjudged Dotan's character, and also by Dotan's plea during the trial for Bin Nun's personal forgiveness. On the other hand, such mismanagement of trust could generate broader images of untrustworthiness and lead to eventual attempts to redefine the IAF's delegated autonomy. It is at this second level of institutional trustworthiness and of the army's relationships with the polity and society, that the debate that ensued after the disclosure of corruption offered the public contrasting strategies for disaccreditation and reversal of disaccreditation. In some sectors, the case was perceived as being so extremely dangerous that a sense of moral crusade was generated. The case generated reactions which had generally been absent from similar cases in the past. They included the highly emotional discourse of treason to the Nation and to one of its most select institutions, the IAF. The use of this imagery is especially striking against the background of ongoing change in Israeli society, which has moved away partially from early collective ideals that predicated self-abnegation (Roniger and Feige 1992). To contextualize these reactions it is necessary to look more closely at the politics of public morality conducted by military and political forces within the context of the prevailing patterns of military accreditation.

Most government and political circles reacted with caution, and the investigating team operated with a legal-technical approach. Defense Minister Moshe Arens, who ordered a thorough examination of IAF procurement procedures, expressing commitment to increase internal scrutiny, acted with a genuine interest in preventing the scandal from hampering foreign aid. Prime Minister Shamir called on the public to wait for the complete investigation and trial before pronouncing judgement. Colonel Zvi Krochmal, head of the military investigating team, commented in April 1991, after the trial:

> Dotan is not a traitor [but] a common criminal. What he resented most were the suggestions that he had damaged the IAF. I am convinced that, although he conducted many illegal transactions for the sake of money, he did not damage the equipment of the Force.

Within the army, the closer the speaker stood to the Air Force the more highly emotional the reactions became. This led to a transposition of cultural meanings. Whereas in espionage cases it is usually claimed that greed is a major motive underlying treason, here, in a case of embezzlement clearly motivated by greed, Dotan was publicly accused of high treason. Symbols of purity and pollution were evoked, and a sense of moral scandal was voiced by many who claimed to be defending conventional morality.

As soon as the case became public, on December 20, 1990, Chief of Staff Dan Shomron said: "we shall stop at nothing before we eradicate this filth from our midst," adding that most of the personnel in the Israeli procurement network do an excellent job, totally untainted by the allegations surrounding Dotan and his associates. He also noted by way of an excuse that he himself had valued Dotan as "a most serious professional in the field." Chief Military Prosecutor, Brigadier-General Strassnoff, commented in a television interview on February 21, 1991, that he considered the case to be extremely serious. When referring to appropriate measures to be taken, he used the metaphor of "the need to eradicate thorny bushes from the flowering garden of the IAF." On December 20, 1990, ex-Air Force head and former Defense Minister (and currently President of the State of Israel) Ezer Weizman said the alleged actions of Dotan and his accomplices were "treachery of the first order" that "had caused us more damage than if we were to lose a squadron of aeroplanes." Weizman thought that

If it were possible, we should tar and feather Dotan, as they used to do in the
United States. If that were feasible, I would be more than glad to do it myself.

Two other former heads of the IAF, Mordechai Hod and Beni
Peled, were equally emphatic. Hod thought Dotan deserved a life
sentence, and Peled affirmed that if the IAF were autarchic, Dotan
would have been tried by before a military tribunal and have been
sentenced to death (December 28, 1990).

While the views voiced focused on defining the domains affected
by corruption, they were in fact referring to the effectiveness of
military accreditation. In this context, there were two distinct views
visions being expressed by contrasting metaphors. On the one hand
the corrupted officer was portrayed as the "tip of the iceberg," while
on the other he was described as the "rotten apple in the box." Various issues were implied in this debate: Could this officer have remained isolated or had a part or the rest of the force also been tainted?
Could one officer, with perhaps a small network of associates, be
corrupt, while the rest of the unit — the suspected "iceberg"— retained its "purity"? Could the Israel Air Force's claim of being an
enclave of excellence within the army still be credited and its autonomy within the IDF maintained?

From both perspectives, the head of the IAF Equipment Branch
was an outcast. His actions were seen as a case of extreme and
undisputed deviance from essential value commitments. In the
eyes of army men and politicians alike, his deeds were not only
clearly illegal but also inexcusably immoral. He was represented
as having betrayed a fundamental value commitment. It was thus
impossible for him to find moral support or create any sort of
coalition to reverse the stigma or rehabilitate himself publicly.
Once the facts of the case were ascertained, there were none of
the conventional rationalizations familiar from other cases of corruption, such as having acted for the sake of a higher cause, or having been victimized by the agents of social control (Ben-Yehuda
1990: 225-50).

The issue at the center of the debate went far beyond the discovery of fraud and embezzlement. Whereas an endorsement of the "rotten apple" metaphor implied the recognition of delegated autonomy,
the image of "the tip of the iceberg" highlighted the need for a review of accreditation and an increase in external controls. Discussion touched upon the basic matrix of relations between the army

and the civilian administration, and between the IAF and the rest of the army.

The Macrosociological Implications of Trust Failure

The boundaries between the Israeli army and Israeli society in general have been amply documented as comparatively permeable ("fragmentary" in the technical sense). The army has much influence on political and economic decision-making, yet is itself supervised by the civilian authorities, especially the Ministry of Defense. The system's ability to continue this working relationship in the context of protracted regional conflict, without turning into a Laswellian model of a garrison state, is a basic question which has long been examined by Israeli scholarship (Kimmerling 1985a; Lissak 1985; Peri 1983).

The attribution and validation of trust are an important factor in explaining the persistence of the intersystemic structure of military accreditation. On one level, the attribution of trust by the civil society is sustained by the dense networks linking the top army officers and other Israeli elites. These connections between top army officers and members of the elites are both occupational and social. Relations are further strengthened by the fact that Israeli officers retire from service at a relatively early age (mid-40s) and embark on new careers in politics or in the civilian and military-industrial sector. As a result, while still in the army they have a vested interest in maintaining good relations with potential future colleagues or employers, and, conversely, many top positions in Israeli politics and business are occupied by young ex-officers with close connections in the higher echelons of the army (Peri and Lissak 1976; Lissak 1985; Maman and Lissak 1990).

With the growing specialization and professionalization of the army and the IAF in particular, bonds of trust were extended to cover the functioning of a huge organization. Expectations of trustworthiness were simultaneously essential to the smooth running of the intersystemic relationship, and a primary factor for enabling abuse. The system, which worked well in small scale units, became subject to abuse once the unit became increasingly complex and impersonal.

One of the motifs most cited in the public outcry surrounding the scandal was a sense of personal betrayal. Dotan was trusted and he betrayed. But this focus should not obscure the institutional implications. An officer had been promoted through the army ranks because he was trusted both technically, as an expert, and more broadly, as a man of integrity. Then, his high rank itself became a proof of trustworthiness. At this stage the promotion system entered a vicious circle. The large size of the corps made it difficult to substantiate trust in a multiplex way, for instance, through shared experiences and cooperation as "in the old days." In turn, the inability to validate integrity and committed goodwill beyond performance created opportunities for the abuse of trustworthiness.

In other words, the establishment of trust allowed the system to expand, yet the expansion and increasing complexity created loopholes and opportunities for abusing the system. The IAF was able to handle multimillion dollar deals with foreign suppliers and to absorb advanced technology because of the delegation of trust (by the IDF and ultimately by the polity), yet this growth in responsibility was not accompanied by the parallel development of new control systems. The fraudulent officer was misusing a system based on trust and credibility, and the fact that he could do this for a lengthy period, with large sums of money, and exhibit status symbols bought with embezzled money, revealed the fragility of the current mechanisms of organizational management. The importance of trust as a basic correlate of the pattern of military accreditation, continuously and ritualistically reasserted, explains the outcry surrounding the breach of trust of an officer who was detached from his expected locus of commitment in the army networks. The whole structure that, on the basis of trust, delegated responsibility to the army to run its affairs autonomously temporarily lost its credibility, and its "taken-for-granted" attribute was questioned.

The heads of the IAF tried to close the confidence gap and to meet the "legitimacy deficit" by stressing the uniqueness of the case, i.e., by evoking the "rotten apple in the box" metaphor. They felt strongly that only by redrawing internal boundaries in terms of purity and pollution could they avoid losing their delegated autonomy within the IDF and within society in general. Those who were critical of that solution questioned the pattern of accreditation of the Air Force and called for the implementation of more impersonal and universalist criteria of control.

Distrust begets distrust as much as trust generates trust (Banfield 1958; Bradach and Eccles 1989: 107-8). The disclosed misdoings were considered to be a betrayal of colleagues and, moreover, an act of treason against the country because they implied a breakdown of trust with institutional implications. The deceit practiced by high-ranking army officers threaten the attribution of trust to the army by disseminating images of untrustworthiness wider than those directly associated with the case at hand.

Conclusions: The Limitation of Military Accreditation in Contemporary Israel

A debate has opened on the limits of honoring military accreditation in Israel. The case analyzed functioned as one of the first markers of cumulative changes in contemporary public morality. It generated momentum for a review of previously unquestioned assumptions about the carte blanche the army traditionally was granted by state authorities and the majority of the population in connection with its image of professionalism and the centrality of "state security" considerations in this society.[4] Subsequently, in 1993, the debate gained new momentum as parents of soldiers killed in training and operational accidents spoke through the media (with an emphasis unknown before) of their distrust of army committees of inquiry. They demanded that inquiries should be taken out of the hands of the army so that senior officers with overall responsibility should be brought to trial. In the 1990s, the army itself redefined its attitude to the public and became more media-oriented, beginning to volunteer information on subjects that had traditionally been kept hidden from the eye of civil society, for example, suicide rates and the character of special commando units in the occupied territories to name but two.

As institutions delegate trust (e.g., onto military agents), the repositories of trust derive their trustworthiness from the image of responsibility granted to them in connection with an abstract image of institutional trustworthiness (Giddens 1990: 83-92; Weber 1946: 302-22). Trustees are expected to share organizational norms to enhance the status and image of trustworthiness of the institution. From an individual perspective, this involves compliance with organizational norms in all behavior, from role performance to integrity and committed goodwill.

The crucial but fragile mechanism at work in the maintenance of trust is accountability. Trustees should remain accountable within the framework of the broader images of institutional trustworthiness (O'Loughlin 1990). But, here again, individuals can operate differently: trustees may not be willing to cling to the accepted image of trustworthiness or may manipulate it. If apparent trustworthiness is deceptive and reliability proves unfounded, institutional de-stabilization may follow. Because of the inherent fragility and potential breakdown of this form of institutional accreditation, supportive means of corroboration of trustworthiness are required. These take the form of routine and more structural measures. On the routine plane, any claim to trustworthiness is validated time and again and interpersonal pressures may be brought to encourage compliance with expected standards of trustworthiness. Structurally, the inherent fragility of the organizational reliance on trustworthiness may lead to an increased formalization and a spiraling growth of formal mechanisms of accountability, with relatively large inputs of social labor in tasks of supervision, verification and control.

Israeli military institutions seem to be entering such a stage in their accreditation. On the local scene, interventionist attitudes have been voiced; these would limit delegated autonomy, replicating moves in the U.S. to limit the licensing of professional autonomy and liability (Esquith 1987). Externally, the U.S. implemented in 1994 a policy requiring all recipients of U.S. military assistance to purchase equipment through the Pentagon. This decision gave the U.S. government greater control over foreign arms deals but added an estimated 4 to 10 percent to the cost of defense purchases.[5]

In Israel the armed forces are subject to increasing controls. In routine performance, however, controls will not be able to replace trust even under changed institutional arrangements. Despite failures in the validation of trust, it is to be expected that trust will remain a substantial force, since as an intrinsic part of relationships it can encourage commitment.

This analysis suggests two dimensions as central for further research on the matrix of military-civilian relationships in Israel. The first concerns bureaucratization, and the second, value tensions in society. Sociologists have studied continuities and changes in Israeli society from many analytical perspectives. To name but a few, in terms of the transformation of the old institutional mold from pre-independence times (Eisenstadt 1985; Horowitz and Lissak 1989); as a change in civilian culture (Liebman and Don-Yehija 1984; Roniger and Feige 1992); in terms of collective identity, as a movement from universalist to more particularist models and structures (Kimmerling 1985b; Moore and Kimmerling 1995); in terms of political revamping and the entrance of new ethnic and religious elements into the mainstream of sociopolitical life (e.g., Cohen 1983; Ben Ari and Bilu 1987; Ben Rafael and Sharot 1991; Peres and Yuchtman 1992); from a global, geopolitical perspective, as continuities and shifts in the social construction of the Arab-Israeli conflict, the status of the occupied territories and the relationships with the Palestinians (e.g., Smooha 1989; Lustick 1993); and in terms of gender domination (Bernstein 1987; Swirski and Safir 1992).

The present study focuses on a separate dimension of transformation; namely, the dialectical relationship of bureaucratization and parallel disenchantment with public life. A growing dissatisfaction with the bureaucratic setup of arms procurement prompted actions that gradually slid into white collar crime. The disclosure and ensuing scandal as well as the other developments discussed above have set in motion a process that leads to further bureaucratic regulation. This in turn can generate further disenchantment with public life within military circles. Analysis also provides indications of value tensions inherent in the move towards a post-ideological stage in a society still grounded in and regarding itself as legitimated by collective commitments. In earlier decades, and more selectively even in present times, cases of public corruption were interpreted — and until recently dismissed by large sectors of society — as by-products of the implementation of a collective project, e.g., national development or the pursuit of sectorial well-being. Ideology no longer legitimatizes public corruption, which becomes more opaque and subject to legal sanction; and yet, this decline of ideological value singles out crucial institutional arenas as especially prone to be domains in which corruption slides into emblematic scandal. It would seem that in contemporary Israel cases of routinized white collar crime and opaque corruption can become scandals with macrosociological implications when they compromise the hold of a state of mind that stills considers the army one of the most valued frameworks of institutional trust in this society.

Notes

*   This article stems from a research project supported by the Shaine Foundation and the Israel Foundations Trustees (1990/92). I am grateful to Michael

Feige and Amos Cividali who collaborated in research. Moshe Lissak, Nachman Ben-Yehuda, Baruch Kimmerling, Reuven Kahane, Eyal Ben-Ari and Edna Lomsky-Feder contributed insightful comments on various drafts. Thanks are due also to the Truman Institute for the Advancement of Peace for editorial assistance.

1.   Published and unpublished studies reflect these high levels of confidence among the Israeli Jewish population. Sources include reports in Hebrew of the IDF Research unit for 1980 and 1984; surveys conducted by the Israeli Institute for Military Research among youth in 1988 and 1991 (Mayseless 1994, Carmeli 1994, Mevorach 1994) and among those who ended military service in November 1990; country-wide studies for 1990 (Peres and Yuchtman 1992) and for 1991-92 (Roniger 1993). In these studies, the range of high trusters varied between 83 percent and 95 percent; no more than 12 percent were ever recorded at the middle point of the scale, and low trusters have not exceeded the 5 percent.

2.   This form of trust is reminiscent of behavioral codes often mentioned in studies of kinship, of ethnic trading and retailing networks in the ancient and medieval world. It also resembles the ethos of familial solidarity in the modern world. Zucker has labeled it "characteristic-based trust," that is trust based on shared background characteristics (Zucker 1986). I would like to draw attention to the constructed character of this form of trust, which spans both primordially and non-primordially patterned relationships. For comprehensive discussions of trust, see Luhman 1979; Barber 1983; and Fukuyama 1995.

3.   For example, in terms of the GNP, in the 1970s defense spending constituted between 1/4 and 1/3 of the GNP; in the 1980s it decreased to between 1/5 and 1/4 of the GNP (*Supplement to the Israeli Statistical Journal*, 35, 1986: 3-12 [Hebrew]).

4.   The issue of "state security" has had broader implications for the rule of the law and for impairing universalist adjudication in Israel and in the occupied territories. See Hofnung 1991.

5.   This decision reflected strains in U.S.-Israeli cooperation in the 1980s and early 1990s, as attested by the Pollard affair, the allegation of a payoff by two former Israeli officers to an assistant secretary of the U.S. Navy to win a Pentagon contract, the Israeli role in the Irangate, and the alleged attempt to bypass cooperation agreements and transfer U.S. technology to third parties as part of the sale of Israeli technology.

References

Banfield, Edward 1958. *The Moral Basis of a Backward Society*. Glencoe, IL: Free Press.

Bernstein, Deborah 1987. *The Struggle for Equality*. New York: Praeger.

Barber, Bernard 1983. *The Logic and Limits of Trust*. New Brunswick, NJ: Rutgers University Press.

Ben Ari, Eyal and Yoram Bilu 1987. "Saint Sanctuaries in Israeli Development Town: A Mechanism of Urban Transformation." *Urban Anthropology*, 16, 2: 243-72.

Ben Rafael, Eliezer and Stephen Sharot 1991. *Ethnicity, Religion and Class in Israeli Society*. Cambridge: Cambridge University Press.

Ben Yehuda, Nahman 1990. *The Politics and Morality of Deviance*. Albany: State University of New York Press.

Bin Nun, Avihu 1991. "To Quit Flying." *Bitaon Heil Haavir*, 183: 2-7 (Hebrew).
Bradach, Jeffrey and Robert Eccles. 1989. "Price, Authority and Trust: From Ideal Types to Plural Forms." *Annual Review of Sociology*, 15: 97-118.
Carmeli, A. 1994. "Attitudes of Israeli Inductees Toward Military Service (1991 Cohort)." In *The Military in the Service of Society and Democracy,* edited by Daniela Ashkenazy, 37-42. Westport, CT: Greenwood Press.
Cohen, Erik 1983. "Ethnicity and Legitimation in contemporary Israel." *The Jerusalem Quarterly*, 88: 111-24.
Cohen, E. and Z. Lavi. 1990. *The Sky is Not the Limit.* Tel Aviv: Maariv Press. (Hebrew).
Dalton, Russell 1988. *Citizen Politics in Western Democracies.* Chatham: Chatham House.
Doig, Alan 1989. "The Dynamics of Scandals in British Politics." *Corruption and Reform*, 3, 3: 323-30.
Dogan, Mattei 1993. "Comparing the Decline of Nationalisms in Western Europe: The Generational Dynamics." *International Social Science Journal*, 136: 177-98.
Edelist, Ran 1991. "The IAF: Reconnaissance Flight." *Iton Tel Aviv*, March 29. (Hebrew).
Eisenstadt, Shmuel Noah. 1985. *The Transformation of Israeli Society.* Boulder, CO: Westview Press.
Esquith, Stephen 1987. "Professional Authority and State Power." *Theory and Society*, 16, 2: 237-62.
Evonic, I. N. n.d. "Motivation and Morale in Military Non-Combat Organizations." Canadian Armed Forces, Central Recruiting Zone.
Fukuyama, Francis 1995. *Trust — The Social Virtues and the Creation of Prosperity.* London: Hamish-Hamilton.
Gal, Michael 1986. "Unit Morale: From a Theoretical Puzzle to an Empirical Illustration — An Israeli Example." *Journal of Applied Social Psychology*, 16, 6: 549-64.
Giddens, Anthony 1990. *The Consequences of Modernity.* Stanford, CA: Stanford University Press.
Glees, Anthony 1989. "Political Scandals in West Germany." *Corruption and Reform*, 3, 3: 261-76.
Greenberg, Itzchak 1991. *The Ministry of Defense and the General Staff. The Controversy on the Management of the Defense Budget 1948-1967.* Tel Aviv: Ministry of Defense. (Hebrew).
Heidenheimer, Arnold et al. 1988. *Political Corruption.* New York: Holt, Rinehart and Winston.
Henderson, D. 1985. *The Human Element in Combat.* Washington, DC: National Defense University Press.
Hofnung, Menachem 1991. *Israel: Security Needs versus the Rule of the Law.* Jerusalem: Nevo. (Hebrew).
Horowitz, Dan and Moshe Lissak 1989. *Trouble in Utopia.* Albany: State University of New York Press.
Jackson, John 1970. *Professions and Professionalization.* Cambridge: Cambridge University Press.
Kahane, Reuven (Ed.) 1984. *Patterns of Corruption.* Jerusalem: Akademon and Center for the Documentation and Research of Israeli Society. (Hebrew).
Kimmerling, Baruch 1985a. *The Interrupted System: Israeli Civilians in War and Routine Times.* New Brunswick, NJ: Transaction Books.

Kimmerling, Baruch 1985b. "Between the Primordial and the Civil Definitions of the Collective Identity: Eretz Israel or the State of Israel?" In *Comparative Social Dynamics — Essays in Honor of S.N. Eisenstadt*, edited by Erik Cohen et al., 262-83. Boulder, CO: Westview.

Liebman, Charles and Eliezer Don-Yehiya. 1984. *Religion and Politics in Israel*. Bloomington: Indiana University Press.

Lissak, Moshe 1985. "Boundaries and Institutional Linkages between Elites: Some Illustrations from Civil-Military Relations in Israel." *Research in Politics and Society*, 129-48.

Luhmann, Niklas 1979. *Trust and Power*. New York: Wiley.

Lustick, Ian 1993. *Unsettled States, Disputed Lands*. Ithaca, NY: Cornell University Press.

Maman, Daniel and Moshe Lissak 1990. "The Impact of Social Networks on the Occupational Patterns of Retired Officers: The Case of Israel." *Forum International*, 9: 279-308.

Marengo, F.D. 1988. "The Linkage Between Political Corruption and Political Scandal." *Corruption and Reform*, 3: 65-79.

Mayseless, O. 1994. "Attitudes Toward Military Service among Israeli Youth (1988)." In *The Military in the Service of Society and Democracy*, edited by Daniela Ashkenazy, 32-6. Westport, CT: Greenwood Press.

Mevorach, L. 1994. "Public Attitudes toward the IDF (November, 1989)." In *The Military in the Service of Society and Democracy*, edited by Daniela Ashkenazy, 27-31. Westport, CT: Greenwood Press.

Mintz, Alex 1985. "Military-Industrial Linkages in Israel." *Armed Forces and Society*, 12, 1: 9-28.

Moodie, Graeme 1989. "Studying Political Scandal." *Corruption and Reform*, 3, 3: 243-6.

Moore Dalia and Baruch Kimmerling. 1995. "Individual Strategies of Adopting Collective Identities. The Israeli Case." *International Sociology*, 10, 4: 387-408.

O'Loughlin, Michael 1990. "What is Bureaucratic Accountability and How Can We Measure it?" *Administration and Society*, 22, 3: 275-302.

Oring, Elliott 1981. *Israeli Humor. The Content and Structure of the Chizbat Tradition of the Palmah*. Albany: State University of New York Press.

Peres, Yochanan and Ephraim Yuchtman-Yaar. 1992. *Trends in Israeli Democracy*. Boulder, CO: Lynne Rienner.

Peri, Yoram 1983. *Between Battles and Ballots: Israeli Military in Politics*. Cambridge: Cambridge University Press.

Peri, Yoram and Moshe Lissak 1976. "Retired Officers in Israel and the Emergence of a New Elite." In *The Military and the Problem of Legitimacy*, edited by Gwyn Harris-Jenkins and Jacques Van Doorm, 175-92. London: Sage Studies in International Sociology.

Roniger, Luis 1993. "Patterns of Trust in Israel." Rose-Ackerman, Susan 1999. *Corruption and Government*. Cambridge: Cambridge University Press. Research report, Israel Foundations Trustees.

Roniger, Luis and Michael Feige. 1992. "From Pioneer to Freier; The Changing Models of Generalized Exchange in Israel." *Archives Europeennes de Sociologie*, 33: 280-307.

Schiff, Zeev 1975."The Need for Rotation." *Haaretz*, May 19. (Hebrew).

Shapiro, Susan 1987. "The Social Control of Impersonal Trust." *American Journal of Sociology*, 93, 3: 623-58.

Simmel, Georg 1978. *The Philosophy of Money*. London: Routledge and Kegan Paul.

Smooha, Sammy 1989. *Arabs and Jews in Israel*. Boulder, CO: Westview Press.

Swirski, Barbara and Marilyn Safir (Eds.) 1992. *Calling the Equality Bluff*. New York: Pergamon.

Tilman, Robert 1970. "Black Market Bureaucracy." In *Political Corruption*, edited by Arnold Heidenheimer, 62-4. New York: Holt Rinehart and Wilson.

Weber, Max 1946. *From Max Weber*, edited by Hans Gerth and Charles Wright Mills. New York: Oxford University Press.

Weizman, Ezer 1975. *Yours the Sky, Yours the Earth*. Tel Aviv: Maariv Press. (Hebrew).

Zucker, Lynne 1986. "Production of Trust: Institutional Sources of Economic Structure, 1840-1920." In *Research in Organizational Behavior*, edited by Barry Staw and Larry Cummings, 53-111. Greenwich, CT: JAI.

Epilogue

Uniqueness and Normalization in Military-Government Relations in Israel*

Moshe Lissak

Introduction

In the present paper I shall discuss, first of all, the nature of the interrelationships that have developed over the course of the years between the security sector, and first and foremost the Israel Defense Forces, on the one hand, and the various civilian sectors, particularly the political system, on the other. Second, this paper shall attempt to examine the future tendency of these interrelations, assuming that towards the 2000s there will be realized a period of diplomatic agreements between Israel and its neighbors. In this context, we shall also relate to the charge concerning the "militarization" of Israeli society.

Anyone examining the history of military-governmental and military-societal relations in Israel will not find it difficult to find, alongside certain characteristics shared in common with other armies, more than a few unique features. The latter are particularly striking in comparison with the patterns that took shape in other democratic societies. The uniqueness of military-societal relations in Israel is reflected in numerous ways in the political, economic, social and cultural realms. However, the best way to describe these is by an analysis of the nature of the boundaries and situations of encounter which took shape over the years between the security establishment in general, and the IDF in particular, and the various civilian sectors. Nevertheless, both the unique characteristics and those shared in common by Israel and other societies were never static or stagnant. The models that took shape during the formative years of the state gradually changed. One may imagine that the pace of change will accelerate if a post-war era indeed takes shape in the Middle East.

Several questions occur in this context: What will be the direction of change? Will the unique characteristics (positive and negative) lose their character? Will Israel get onto the track of "normalization" in this area as well; that is, will the similarity between Israel and Western democratic societies grow stronger? What are the most likely scenarios? Finally, and most interesting, what will be the implications of the changes in the IDF on Israeli society, and especially on its political culture, that has to no small degree focused around the ethos and myths of security?

Situations of Meeting Between Armed Forces and Civilian Structures

The concepts of "uniqueness" and "normalization" can only be defined in a comparative perspective, the most appropriate sociological and historical context for such a comparison being the Western democratic societies. In those societies there have taken shape over a long period of time, but particularly since the Second World War, models of military-societal and military-government relations that suited their political and democratic culture. Nevertheless, notwithstanding the fact that all these societies belonged to the democratic political culture, different patterns and secondary patterns of military-society and military-government relations emerged there.

Within the framework common to all the democratic societies—namely, complete subjugation of the armed forces to the political leadership—various different varieties of interrelationships were found between the political leadership and the leaders of the military. Thus, for example, the identity of the supreme commander of the armed forces and the legal basis in whose name he acts differs from one democratic society to another. At times one is speaking of one man, as in the United States—namely, the president. The president acts by virtue of the power invested in him by his country's constitution. In other cases one is dealing with a cabinet of senior ministers, as in the English case. This cabinet acts in the name of the supreme sovereign of the kingdom—the king or queen of the United Kingdom. There are yet other cases in which the government as a whole acts by virtue of basic laws that were passed by the Parliament. Such is, of course, the case in Israel.[1]

The variety of military-society relations in democratic societies stands out even more sharply with regard to the nature of the situations of meeting between the armed forces and the civilian frameworks. Unlike the case of the identity of the supreme commander, here the options are extremely varied. This statement is true both on the highest levels of both hierarchies—i.e., the military and the political—as well as on the intermediate and lower levels. On the level of the elites, there are various structures of consultation, exchange of information, and decision-making. Thus, for example, in the American model, the National Security Council. This institution is anchored in the law and regulations and is an integral part of the structure of the American government (Huntington 1985: 337, 384, 395-6; Jackson 1965: 30-42). There are likewise various different arrangements, more and less established, in whose center there stand advisors for national security matters, whose main function is to mediate between the heads of the political system and those of the military and the intelligence community. In the case of contacts among the non-elite populations, one is speaking of a substantive difference in the number of encounters, in the nature of the social networks, and in their number (Lissak and Maman 1996; Maman and Lissak 1990).

Notwithstanding the difference and the variety, the common denominator of all the democratic societies is a certain degree of vagueness in the nature of the boundaries and a certain flexibility in the division of functions between the two areas. This vagueness is dynamic and constitutes a constant object for redefinition of the boundaries between the political and military systems. The discussion of this subject brings in its wake constant arguments between the two sides—i.e., the security system and the civilian elites. Another public debate, and one with possibly greater implications, is connected with the need to obtain, particularly following the Cold War, a higher degree of coordination between quality manpower, which is in great demand on the civilian markets, and the development of military technology (Boëne, this volume). Similarly, the location of future fronts of conflict and the redefinition of the potential threats lie in the center of the debate concerning the doctrine of national security in democratic societies. A decision one way or the other is likely to have implications for the very nature of the military profession in the 2000s (Boëne, this volume). Among the officer class and outside it there are greater demands for rethinking of its nature and goals. This thinking has been accelerated in light of several changes in the tasks that are taken on themselves increasingly, primarily by armies in democratic societies. Thus, for example, we may observe a great increase in the number of military units serving under the flag of the UN (or of NATO). The functions of such units is in imposing peace or in insuring its existence after agreement between the warring parties (Last 1995/96).

This constitutes a historical turning point in the original functions of the military profession (Cafirio 1988; Diehl 1993). Under such circumstances, the officer class also needs to acquire skills in the diplomatic area and basic understanding of social, cultural and religious issues. In the past, such qualities were not acquired in military training schools. The exceptional cases in which such talents were required, post facto, were during periods of extended occupation of countries that were conquered in war (including following the Second World War). On the other hand, the dramatic increase in the number of acts of terrorism and their transformation into a central means of combat in political conflict involving fundamentalist elements, has influenced and will in the future influence, to no small degree, the nature of the military profession, albeit in a different way from that brought about in wake of the adoption of tasks of intermediacy between ethno-national groups. The changes will take place in the structure of the military, in the size of the deployment of forces, in the method of training of soldiers and officers, etc. But even if the direction is different, one may conjecture that, here too, the general tendency will be in the direction of a less strict distinction between the military structure and the civilian one, compared with what is usually accepted today. There is no doubt that the classic division of roles is being challenged, and in its place there is emerging a new model whose nature and character are as yet unknown (Boëne, this volume).

Models of Boundaries Between the Military Sector and Civilian Sectors

During its earliest years, the sovereign State of Israel established its own unique tone that deviated to a certain degree from the model accepted in other democratic societies. Nevertheless, this deviation gradually grew smaller, as the common denominator between it and other democratic states expanded. What then was the uniqueness of Israel, and how did its normalization find expression?

Our point of departure will again be that of the nature of the boundaries and situations of meeting between the military and the civilian sectors. At the outset, we shall briefly discuss the "theory of boundaries." From a theoretical-conceptual viewpoint, the boundary between the military sector and the civilian sector needs to be discussed in a broader framework—namely, that of different institutional boundaries in general. This subject is one of the classic subjects that engages modern social scientists. Thus, for example, sociologists and political scientists like to describe the social history of societies undergoing social change as, among other things, the history of processes of structural and functional differentiation (Lissak 1985). These processes generally bring in their wake changes in the location and nature of the boundaries among various institutional areas, such as that between the political and the economic or between the political and the cultural. The speed of change of these boundaries will of course depend upon the strength of the pressures placed upon them, on the one hand, and their power of resistance, on the other. With all their much vaunted conservatism, even professional armed forces did not evade the need to respond to external and internal pressures and to redefine the nature of the boundaries of the military system, the division of tasks, and the characteristics of situations of meeting between it and civilian systems.

With regard to the nature of the boundaries between the two systems, one may utilize the classification proposed by Luckham (1977), who distinguishes among three kinds of boundaries between the military and the civilian systems: (a) integral (impenetrable) boundaries; (b) permeable boundaries; (c) fragmental boundaries. Luckham also suggests certain central characteristics for distinguishing among different kinds of boundaries: (1) the degree of supervision of the military systems over its people on different levels in their contacts

with the non-military environment; (2) the degree of blurring between goals and frameworks of the military sector, on the one hand,
and of the civilian sector, on the other.

On the basis of these characteristics, a military system may be defined as having an integral boundary if there exists the following condition: "The interchange between persons holding roles at various levels of the military hierarchy and the environment are under the control
of those with responsibility for setting the operational goals of the
armed forces, that is the higher command." Boundaries are permeable "to the extent to which there is complete fusion both in respect of
goals and of organization between the possessors of the means of violence and other social groups." The boundary is fragmental if "in a
military with distinctive military format and goals—the interactions
of holders of military roles with holders of civilian roles escape the
control of the military elite in a way that impairs its freedom to inter-
act with the political and social environment as a single entity in a
consistent way" (Luckham 1977: 17-8).

This condition implies that certain sections of the boundaries tend
to be integral, while others tend to be more permeable. This typology is admittedly highly schematic and abstract, but one may develop various speculations on the basis of these characteristics.
Moreover, these general characteristics may be translated into more
specific and operative characteristics. In order to clarify these concepts, one needs to refer to a central aspect of this issue, to which
Luckham hardly relates—namely, the issue of *institutional linkages*
or *meeting points* between the military and various sectors of the
civilian system. These meeting points exist even if the boundaries
are integral; all the more so if the boundaries are completely permeable or fragmental—that is, alternately permeable and integral. Situations of meeting are frequently institutionalized in one way or another, but in part they are also non-established and somewhat esoteric.
In every society there are some meeting points, on different levels.
Only in a small number of meeting points is there a high level of
integrity of the boundaries, while in the majority the boundaries are
more permeable or fragmental.

Several questions arise in this context, which may be adequately
answered only by means of empirical research. Thus, for example,
one needs to ask who are the participants in a meeting (politicians,
officials of executive authorities, representatives of the public, etc.)?

What is the degree of institutionalization of the situations of meeting? What is the status of the participants in the meeting and, particularly, who has the right of veto in case of disagreements: the civilian partner or the military one? What are the rules of the game (competitive, conflictual, or consensual)? In what domain is the meeting conducted (civilian or military)? Finally, the central question that needs to be asked: what is the degree of ideological-normative agreement in relation to the nature of the boundaries and their various attributes? In other words, what is the extent of agreement in the area of national security doctrine and of the central social ethos?

It would be impossible to answer all these questions. Nor can we engage in an extensive, systematic comparison with the norms accepted in other democratic societies. For that reason, we shall refer to certain central features of the Israeli "case." As a point of departure for our discussion of this subject, we may state that the boundaries between the two sectors within Israeli society are more varied and complex, while the situations of meeting between representatives of the security establishment and the outside population are more numerous than might be thought at a superficial glance. We might also mention that the concept "security system" in this context refers to those serving in the standing and regular army, the reserves, employees of the Ministry of Defense, and those working in the intelligence community. This paper will nevertheless relate only to the IDF, and not to those belonging to other systems of the defense establishment.

Explicit testimony of the large number of meeting points and of the need for defining the interrelations among their participants lies in the fact that one may identify at least seven central meeting points, each one of which carries over into other states. These seven areas are: the political network; the economic network; the educational and cultural networks; the professional network (research and development); social network; public opinion and communication network; and the symbolic network.

Within the framework of these different areas, there are dozens of meeting points on a reasonable level of institutionalization that may be relative easily identified.[2] It should be noted that in the majority of meeting points (albeit with differing frequencies), the participants in the meeting are members of the military staff in active duty, and not civilians working in the security apparatus. In other words, the explicitly military (uniformed) component is in practice the dominant participant in these meeting points, at least from the quantitative viewpoint. The prominence of the military component nevertheless varies from one area to another.

A Nation in Uniform

Even from this schematic presentation, we learn that the interrelations between the military and the civilian sectors are conducted on a variety of meeting points, both on the individual and institutional level. The most striking meeting point on the micro level is that of compulsory military service, as dictated by the Law of Security Service. It is superfluous to elaborate the impact of this service upon the life of the individual—beginning with the limitations, during a certain period, upon his freedom of movement and expression (including the period of reserve duty), the taking of personal risks, and economic implications (Lissak 1984). Simultaneously, military service is also accompanied (or at least was in the past) by various social rewards, including status and, most important, an honorable ticket of entry into civilian society (Horowitz 1982).

The permeability of the boundaries between the military and the civilian realms is not limited to the life of the individual. In terms of the overall influence of military-societal relations, the permeability of boundaries is important specifically on the institutional level, relating as it does to the functioning of the political-democratic system under circumstances of ongoing violent conflict. As we shall see below, notwithstanding the partial permeability of the boundaries, there did not develop in Israel a "garrison state" (Laswell 1941), but what is referred to in the literature as "a nation in arms" or, in Israeli parlance, "a nation in uniform" (see Ben-Eliezer, this volume). The term implies a society in which there is extensive participation of civilians in the military effort and partial permeability of the boundaries separating between the two sectors. The model of "a nation in arms" assumes the existence of a permeable boundary and a wide variety of meeting points between the military and civilian sectors, far more so than in the model of the "garrison state." As a result, one may discern phenomena of partial "civilianization of the military" or "militarization" of the civilian sector, leading to a weakening of the relative strength of the military-security apparatus relative to the civilian sector. Nevertheless, the boundaries are not permeable in an overall or comprehensive way, so that the boundary lines between the two systems must be defined as fragmental, to use Luckham's terminology. The significance of this is that, as against those areas in which the military sector is quite independent, there also exist areas in which there is relatively high civilian involvement. At the same time, there are also areas in which the accepted rules of the game grant legitimacy to involvement of the Army in civilian activity. By contrast, in other areas the involvement of the military is seen as improper or undesirable. From the viewpoint of institutional analysis, there is particular interest in arrangements facilitating extensive involvement of the military sector in the shaping of foreign and defense policy. One is speaking here of such activities as national intelligence evaluation, participation in negotiations on cease-fire agreements and peace treaties — from Rhodes through the implementation of the Oslo agreements — and the administration of the military government until 1967 in Israel and after 1967 in the territories conquered during the Six Day War (Horowitz and Lissak 1987: 42-6).

In order to examine the institutional significance of phenomena of civilianization of the military, on the one hand, and militarization of the civilian sector, on the other, one needs to refer to the approach of role expansion of the military (Lissak 1976). The advantage of this approach is that it allows one to raise questions regarding the area of national security, in which military involvement is seen as legitimate. The expansion of the military may have two different expressions, examples of which may also be found in Israel. The role expansion of the military may be expressed in the increased involvement of the military sector in everything that pertains to policy determination and decision-making regarding matters of national security. However, it is also implemented directly by certain activities of the Army, such as in the areas of education and socialization of soldiers. Over the course of years changes have taken place, some of which point toward role expansion and others towards role contraction. Thus, for example, extensive role expansion took place in the economic or military-industrial area following the Yom Kippur War, whereas in recent years this area has been greatly reduced (Cohen 1996; Mintz 1984, 1985; Peri, this volume).

In other areas as well, which go beyond the explicitly military, there have occurred over the course of years alternating phenomena of role expansion and contraction. Thus, for example, changes have taken place in the scope of the Army's educational activity, largely as a result of the approaches and conceptions of IDF priorities by various chiefs of staff. Chief of Staff Raphael Eitan significantly expanded the activities of providing basic education to "marginal youth" (Gal 1982). By contrast, the cultural and entertainment activities for the Army as a whole were noticeably reduced under his leadership. The large Army troupes were disbanded and the activity of Gadna (the Youth Brigade) was also reduced (Dayan 1978). At the beginning of his term as chief of staff, Ehud Barak also sought to restrict, if not to entirely eliminate, several "civilian" activities of the IDF; thus, he cut back considerably the activities of the Army Radio station, Galei Tzahal, and closed several of the periodicals published by the Ministry of Defense and the IDF. Due to heavy public pressure, Barak's success in these steps was extremely limited.

One of the most sensitive areas of contact between the military and the civilian sectors, particularly from the political viewpoint, is that of the media. The general tendency during the 1950s and 1960s was towards role expansion: the setting up of the Army radio; the establishment of an IDF magazine (*Ba-Mahaneh*) and various other magazines; a book publishing house (*Ma'arakhot*); and performances of Army troupes before civilian audiences. All these reflected the extensive scope of the communication activities of the IDF, a significant portion of which was directed towards the civilian market. During the 1970s and 1980s, the scope of this role expansion was reduced relative to its civilian counterparts. This phenomenon may be explained against the background of budget limitations, on the one hand, and of a decline in the prestige of the Army, particularly following the Yom Kippur War, and of an erosion in the national consensus on the subject of security, on the other hand. I refer especially to the issue of the legitimacy of optional or elective wars. The erosion in consensus, particularly following the Lebanon War, led to particular sensitivity regarding the messages involved in the explanatory activity of the IDF in those media subject to its influence. As a result, these messages became more vague in their meaning. All this also involved a simultaneous reduction in the involvement of the security establishment in the area of civilian communications (press and television) by means of military censorship. This reduction was only carried out until after much controversy between the press and the censorship. Generally speaking, these controversies led to a softening of censorship under pressure of the media. This was particularly noticeable after the Yom Kippur War, when the press repeatedly violated the instructions of the censorship (Nussek and Limor 1995; Lahav 1993). The critical attitude of the press and its willingness to defy the IDF reached their peak during the Lebanon War, when the military correspondents of the daily papers became the vanguard of criticism of the goals of the war and the manner of its conducting. Another striking example was the "Bus Line 300" incident (Hoffnung 1991: 271-6), which took place during the period of the *Intifada*. There were numerous other violations of censorship during this period.

Thus, further changes now took place in the already vague boundaries between these two sectors. Certain boundary lines, such as those in the security-political area, became more permeable, while others, as in the cultural area, became less so.

Situations of Meeting Between IDF and Civilian Sectors

All these processes took place within the framework of rules of the game that were more or less accepted, both by the civilian and military elites and by civilian public opinion. The system of norms that took shape in this context may be described as the dominant security-political culture in Israel, that remained unchallenged over the course of years. Only with the weakening of the national consensus regarding the use of military force as a solution to political problems, and the decline in the prestige of the security establishment that occurred in wake of the Yom Kippur War and the Lebanon War, did voices of protest begin to be heard against at least some of these norms.

An active role in this criticism was played by researchers from the academic world. Their critique of the security establishment covered a broad area: beginning with discussion of the striking faults in Israeli democracy as a result of the close connection between security policy and governmental, legal and social arrangements, and ending in the attempt to marshal as much evidence as possible concerning the militaristic nature of Israeli society (see Ben-Eliezer, this volume; Helman, this volume). The faults in Israeli democracy, especially those stemming directly or indirectly from Israel's need to deal with questions of existential and ongoing security, found expression in a series of studies. A typical example of a survey of faults in Israeli democracy is that of Avner Yaniv (1993). Among the shortcomings, all of which are attributed to the mutual interrelations between security problems and the functioning of the governmental and social systems, Yaniv enumerated the inferior civilian status of Israeli Arabs, and even more so that of the Arabs of the "territories," who completely lacked even the civilian status of Israeli Arabs (Smooha 1993); restrictions on "the right of the public to know" concerning various activities of the IDF, and also regarding certain security conceptions (Negbi 1985; Lahav 1993); the partial politicization of the IDF, particularly during the period of Ben-Gurion, and the faults in the judicial system and in the quality of the rule of law in light of the existence of a dual, if not triple, system of law in the "territories" (Hoffnung 1991: 281-92; Shamgar 1982).

It would therefore seem that the institutional boundaries between the political establishment and the military establishment were not one-sided. Alongside involvement in fashioning security policy, at various periods there were manifestations of penetration of partisan politics into the realm of the Army. The dangers involved in politicization of the IDF and the means of preventing it were among the first subjects that troubled the Israeli political system during the stage of transition from settlement to state. The Israel Defense Forces Act (Ostfeld 1994: 109) was intended to fix the obligation of military personnel to swear an oath of loyalty to the State of Israel, and to its laws and duly constituted authorities (State of Israel 1948). The Knesset, that was perceived as the "duly constituted authority," was directed to prevent any partisan involvement in the military that might result from the association of Army personnel with political movements and parties, a connection whose roots lay in the tradition of the Yishuv period. The application of this principle resulted in two severe political crises during the early months of the existence of the state — the Altelena crisis, and that surrounding the breaking up of the Palmah (Brenner 1978; Gelber 1986: 225-6; Nakdimon 1978; Shapira 1985: 50-7). The breaking up of military units having particularistic movement identifications did not eliminate all manifestations of politicization of the IDF. In truth, Ben-Gurion himself took political criteria into consideration in the promotion of senior officers in the IDF (Peri 1983: 61-7). Another expression was the existence of channels of communication having a political tone between professional officers and politicians. These phenomena were greatly weakened during Levi Eshkol's administration as Prime Minister and Defense Minister (1963-1967).

Another expression of the relation of senior officers to the political establishment was the "drafting" of officers to leadership positions within the parties upon their retirement from the IDF. This phenomenon, despite ups and downs in their scope and intensity, continues to the present day. In the 1996 elections this stood out perhaps more than ever in the past. This tendency of explicitly ideological parties from all ends of the political spectrum to place retired IDF officers on their Knesset lists indicates that they saw this as a means of acquiring public legitimacy and attraction, relying upon the prestige of the IDF to symbolize the commitment to Israel's security.

Thus, the phenomena of involvement of the military in political-security decisions, on the one hand, and of connections between parties and military personalities, on the other, reflected the two aspects of the issue of permeability of boundaries between the political and the military system. At times, the same phenomenon operated in both directions. The involvement of former senior officers in political decisions served, on the one hand, as a channel of influence on policy by means of the professional knowledge and doctrines they brought with them from the IDF and, on the other hand, as a tool used by the political system to reduce its dependence upon the professional advice of active military men.

The arrangements based upon military involvement in the decision-making processes, and the phenomena of civilianization within the military establishment, are likewise expressed in the nature of the mutual inter-relations between the military and civilian elites, a phenomenon that has been described as "partnership among elites" (Peri 1983, 1981). Several different meanings may be ascribed to this definition, not all of which are unique to Israel, expressing tendencies that have already been observed by students of military-state relations in other democratic states after the Second World War. The phenomena of close official and unofficial contact between senior military staff and individuals and groups in the political elite led, in most Western democracies, to proximity and convergence between the elites (Janowitz 1965; Moskos 1971; Van Doorn 1975). Nevertheless, there still exists a difference between Israel and other Western states vis-a-vis the extent and intensity of cooperation between the elites, resulting from those factors that transformed Israel into the representative of the model of a "nation in arms." These phenomena give to the military elite, or at least gave it in the past, a more important status in the society and greater political involvement in matters of security than in other democratic countries. The most recent example of this is the significant military involvement in discussions with representatives of the Palestinian Authority regarding the implementation of the Oslo agreements. At the same time, there is a strong tendency among the civilian elites toward involvement in what takes place in the area of national security, especially through their extensive contacts with the military elites within the framework of shared social networks.

These social networks are not limited to the connections between officers in active duty and functionaries in the political and administrative establishments with whom they come into contact during the course of their duties, but are also expressed in the context of social contacts creating a circle of informal acquaintances (Lissak and Maman 1996; Maman and Lissak 1990). One of the results of these contacts is that officers who are close to the end of their military service—that is, at the zenith of their military careers—adopt, at least in part, civilian points-of-view. As a result of the relatively early retirement age, many officers are assimilated into the civilian realm following their demobilization, while simultaneously continuing to maintain contacts with military personnel on active duty. Moreover, at times these demobbed officers continue to hold senior command functions in the reserves. The complex social networks contribute to the convergence of positions between the military and civilian elites, serving as a source for exchange of information, and helping to develop mutual understanding for closing of gaps and creating a shared conceptual system easing contact among the elites.

Nevertheless, the cooperation between the military and civilian elites does not indicate full permeability of the boundaries between them. It is based upon rules of the game that define, on the one hand, the legitimate realms of involvement of the military in activities of a civilian nature and, on the other hand, the dimensions of professional autonomy of the military, which serve as a restriction against excessive involvement of the civilian sector in what occurs in the IDF.

If the social networks greatly helped to create a security-political culture with a great common denominator between the two sectors, at least until 1973, we have seen that the erosion of this common denominator greatly increased following the Yom Kippur War. Moreover, when in wake of the Oslo agreements it appeared that Israel was about to enter a post-war period, in exchange for what was described by the opposition before the 1996 elections as an act of sacrificing vital territorial assets, a new phenomenon emerged. I refer to what is known as "the phenomenon of motivation" within a given political culture—that is, an oppositionary organization which, in the final analysis, wishes to reshape the political culture from which it emerged, and thereby Israeli society as a whole, in a revolutionary manner. The most radical expression of this revolutionary tendency was the assassination of Yitzhak Rabin.

Is Israel a Militaristic Society?

Already in its early years, and especially its formation after the Six Day War, the unique political-security culture of Israel have led to attempts to describe Israeli society as a militaristic one. The arguments of those thinking in this way may be summarized in the words of Ben-Eliezer (1994, this volume), who argues that the concept of "a nation in uniform" is no more than a camouflage for "militarism." According to him, "a nation in uniform" is a cultural means utilized by the political and military elites in order to provide justification for political goals and to recruit the entire population for war. "'A nation in uniform' is based upon a blurring of the boundaries among individual, family, society, nation and state; upon the creation of a militaristic culture for the society as a whole, and not only for the Army; and upon the fashioning of a reality which negates the distinction between war-time and peace-time. This blurring of boundaries is the tool by means of which one makes wars" (1994:61). According to Ben-Eliezer, this culture originated already during the period of the *Yishuv*, was particularly cultivated by the *Palmah*, and took on shape and was institutionalized during the early years of the state. The most striking expressions of this hegemonistic culture are, in his opinion, in addition to glorification of the Army and of war, the demonization of the enemy and the banalization of the conflict, as well as phenomena of nationalism, machoism, rituals surrounding the fallen, and the industry of remembrance of those who fell in the wars of Israel. Ben-Eliezer (1994, this volume) compares these cultural characteristics to the cultures that characterized in their day Japan, Prussia, Jacobin France, and Czarist Russia. Baruch Kimmerling (1993), who shares this point of view, defines Israeli militarism as "cognitive militarism," that is, as a situation in which military considerations almost always take priority over political, economic or ideological ones. The broad public accepts the military mind set as something self-evident, without thinking too much about its long-term implications.

However, the conclusions of those who attribute characteristics of an explicitly militaristic society to Israeli society are mistaken, and one may note numerous errors in their conceptual framework (see Lissak 1996). The main misconception in the approach of these sociologists lies in their ignoring the central feature of militaristic societies, which is not included in the definition of either Ben-Eliezer or Kimmerling. In this context, one needs, first of all, to distinguish between *necessary* conditions, without which it is impossible to describe any society as militaristic, and *sufficient* conditions. The presence of sufficient conditions, in addition to the necessary conditions, enables one to make more or less definitive judgments regarding the militaristic nature of a given society. Among the necessary conditions one may include, first of all, the existence of a highly offensive security doctrine. I refer to a military doctrine that is part of an overall security doctrine, one fundamentally based upon an explicit desire for territorial expansion unrelated to self-defense or to a tactical or strategic need to make a surprise attack. Nevertheless, such a policy, even if continued over a protracted period of time, is not a sufficient indicator of the unequivocally militaristic nature of the society in question. It must by rights be accompanied by normative-ideological justifications and ethoses, largely unrelated to the objective strategic situation of the country. The connection of these justifications to the concrete threats posed by the enemy, who does not conceal his wish to obliterate the country, must likewise be very tenuous.

But beyond that, even if all these conditions do exist, relying upon this dimension alone to prove the thesis regarding the militaristic nature of any state is a short-sighted and one-dimensional perception. A broader perception of the problem requires one to examine whether some further vital and necessary condition also exists — namely, a situation in which the military establishment constitutes a source of inspiration and authority (formal and/or informal) for policy-makers. Even more important, the Army must also constitute an overall *point of reference* for the society as a whole, or at least for central groups within it.

In such a society one will find a great degree of glorification of war, of heroism and of extreme chauvinism; in other words, the Army will in this case constitute the supreme formative factor and regulator of social norms in the majority of institutional areas: political, economic and, of particular importance, cultural and life-style. This regulation will be predominantly one-directional; furthermore, there will be a nearly complete absence of powerful counter cultures in civilian society. Such overall regulation is extremely rare, albeit it existed to a considerable extent in Prussia and in imperial Japan.

It should nevertheless be noted that there may also be partial and limited regulation of only one or two areas. This will generally be the political-partisan area, within which various manipulations are performed, including those related to internal and external security policy. Such a situation does not necessitate significant overflow into other areas—economic, cultural or social—which continue to be conducted in a more or less autonomous manner, on the basis of their own normative operative principles. If such is the case, it is highly doubtful whether the society in question may be called militaristic.

It should also be mentioned in this context what was stated in the above discussion concerning military-civilian relations in Western democracies. A certain blurring of the boundaries between the military and civilian establishments is not rare, and has existed in various historical contexts (Lissak 1967, 1985; Luckham 1977). Such blurring may also be found in modern societies, particularly since the Second World War. The old formal distinctions, that wish for example to distinguish between various executive authorities, civilian and military, do not always stand up to the test of reality. Such is the case in Israel. The fact of such blurring per se is not a sufficient condition for the creation of militarism, as was thought by several authors who contributed to exaggeration in this matter, such as Ben-Eliezer and Helman. The matter also depends, as mentioned, upon the power of the regulatory power of the military and other characteristics mentioned above, such as the nature of the social networks.

In addition to ignoring this central characteristic, many of the studies attempting to describe Israeli society as militaristic suffer from additional weaknesses, the most important of which is the lack of sufficient historical perspective and the failure to discern the pendulum-like motion between tendencies of a militaristic coloration and those of a totally opposite tendency. An additional weakness of these researchers is that they refrain from comparative study of other factors likely to have greater impact than military-societal relations on the shaping of civilian society, such as the size of immigration, social composition, economic growth or decline, changes in the political culture, and various other factors.

The overall conclusion that follows from these remarks is that: first of all, one may not rely upon analysis of foreign and defense policy alone, as does Ben-Eliezer. This is particularly important in light of the fact that in the Israeli case it was not uniform or consistent during the period under discussion. Second, even if a certain degree of "militarization" does exist in the political area, this does not necessarily imply significant and consistent penetration into other areas. Third, the military elite is not all of a piece. One needs to test and examine various gradations that are to be found within it. It is a fact that, for example, following their retirement from active duty, senior IDF officers place themselves along the entire breadth of the political spectrum, from extreme left to extreme right.[3] All this leads to the conclusion that there exist strong "antibodies" in Israeli society that act, whether covertly or overtly, against the militarization of Israeli society.

Paradoxically, the source of the antibody of greatest importance lies in the ideological-political fissures that have characterized Israeli society from its earliest days. I refer to a pluralism of which there are few other examples, even if one limits one purview to the Jewish sector alone; more so if we add to this the Arab sector, which is likewise deeply divided. Ideological splits as such do not always necessarily constitute antibodies to the development of militarism. It is important to understand the different sides to the controversy and over what they disagree. But beyond that, it is important to understand the structures of the split(s): is one dealing with a dichotomous, polar split between two camps having totally different political cultures, the transition between which is almost impassable psychologically, culturally and socially? Such a situation is loaded with explosive force likely to degenerate into acts of violence and loss of control over public order (Horowitz and Lissak 1990: 33-9). In such a case, the involvement of the armed forces is a not unlikely possibility. On the other hand, when the split is not dichotomous or polarized, the likelihood of military involvement is quite low, even when there are sharp expressions of verbal and even physical violence. From a structural viewpoint, a non-polar division implies that, notwithstanding the existence of extreme camps on the right and left, there are any number of intermediary shades to be found within the political-ideological map, among which there exists partial overlap. In other words, the ideological-political distance between those camps that lie between the radical groups is not that great. In such a case, there is a good possibility of dialogue and adherence to agreed rules of the game. The creation of such a situation is greatly helped when the intermediate camps also constitute the majority of the population.

Such has been the situation within Israel, at least since the electoral turnabout of 1977. Prior to that time, the ideological-political split was less sharp, as one large party, Mapai (the forerunner of today's Labor party), enjoyed political hegemony. This party served as an axial party, without which it was impossible to compose any alternative government (Horowitz and Lissak 1990: chapter 4). Following the Labor Party's return to power in 1992, it appeared as though the ideological-political split, focused upon the execution of the Oslo agreements, might develop symptoms of an unbridgeable, dichotomous split. That this did not happen was evidently due to the sobering of the body politic with regard to verbal and physical violence—with the exception of extreme religious groups such as "*Kakh*," "*Kahana Hay*" and messianic and Kabbalistic groups — in wake of Yitzhak Rabin's assassination. Hence, the basic conditions needed for bringing the Army into the sharp public debate were never created. Despite everything, Israeli society continues to function according to the previous formulae; that is, a diversified political map with partial overlap among the camps. Moreover, the balance of electoral forces forced all past governments, as it will evidently do also in the future, to set up a coalition government. In terms of our present subject, this means that, at least since 1967, there has not been a firm consensus within Israeli society regarding basic issues of national security. The number of security conceptions and ethoses and the manner of their use in public discourse are almost as great as the number of parties and ideological streams. Hence it is mistaken to speak of one "ethos," as do Ben-Eliezer and Helman.

Another antibody, related to the former, and of no less importance, refers to the pluralism within the IDF itself, particularly among the senior staff. Such a situation did not exist during the 1950s and 1960s. It developed in later years, and there does not appear to have been any withdrawal from the tendency toward conceptual heterogenity within the IDF regarding the issue of the types of solution to the Jewish-Arab conflict. While it is true that heterogeneity among the senior staff does not reflect exactly the heterogenity among the civilian public, it is nevertheless sufficiently broad so that the General Staff and senior officers of the IDF hardly formulate a unified position on any substantive military-strategic issue. There are numerous examples of this: for example, the divergent views among the senior officers regarding the manner of conducting the war in Lebanon (1982-1985) and the solution of the conflict with the Palestinians and with Syria. We should again mention the fact that retired IDF officers who have chosen a political career are to be found in almost every shade of the political spectrum. We might summarize this subject with the statement that Israel never developed a military caste with its own separate security-political doctrine (Peri and Lissak 1976). In other words, the military never became the supreme arbitrator of the "system of beliefs" in Israel, because it had no clearcut belief system of its own (Peri 1989).

It seems quite likely that these tendencies were halted by another phenomenon—namely, the existence of numerous situations of meeting between the IDF and civilian frameworks. While such meeting points exist in every society, they stand out very strongly in Israel, and some of them are unique to it. These situations of encounter are ubiquitous, some of them institutionalized and some not; they are to be found, as mentioned above, also on the senior levels of both the military and civilian hierarchies.

In terms of our subject, this situation does not indicate, as suggested by Ben-Eliezer and Helman, a symbiosis between the IDF and the political elite, which holds a uniform, militant security outlook. The situation is quite different and far more complex: in these meeting points a dialogue takes place, at times harmonious, but quite often one in which one may observe sharp differences of opinion, that are in almost all cases resolved according to the decision of the Prime Minister and the government. A striking example of this was the criticism directed by central political figures, including Prime Minister Netanyahu, during the last months of 1996, against excessive involvement of Army generals in the diplomatic process. This criticism reached its height in the condemnation of contacts between some of these generals with opposition leaders.

A completely different example of an antibody, which may be noted only in passing, is that of the system of relations between Israel and the Jewish diaspora, especially that in the United States. The American Jewish community, with the exception of a certain portion of its Orthodox component and certain organizations participating in the President's Council, belongs to the most liberal-democratic wing of the American political map. The support of and identification of this community with Israel since its founding certainly had many roots: religious, ethnic, cultural, the Holocaust of European Jewry, etc. There is nevertheless no doubt that the democratic political culture that developed in Israel was a source of pride and identification. One may assume that, had the Jewish community of the United States (or in Western Europe) seen Israel as a militaristic society, a significant portion of the Jewish intellectual elite, as well as the heads of Jewish organizations, would have found themselves in an embarrassing and frustrating position with regard to their relations with Israel. One might mention, finally, that a similar set of considerations also holds true regarding the relations between Israel and various other democratic states of the world.

Summary and Future Trends

We asked several questions in the introduction regarding future trends in military-societal and military-governmental relations in Israel, and on the possibility that several of its unique characteristics may lose their unique character in a post-war age. Before dealing with this question, even in an oblique way, let us summarize the spirit of the things which were said above. Israel, which has been involved in an ongoing violent conflict, never behaved like a society under siege. The basic democratic patterns of life, and the life routine of its citizens, at least within the Jewish sector, were inconsistent with the moods of constant emergency situations. Israel did not become a "garrison state." Paradoxically, one may say that it was precisely the partial involvement of the military sector in those activities of the civilian sector pertaining to the realm of national security that facilitated the guarding of the democratic rules of the game and of civilian life routine. The processes of partial convergence of the military system with the civilian system, through both partial "militarization" of the civilian sector and by controlled "civilianization" of the military sector, prevented the Army from becoming a separatist caste in value conflict with the civilian elites. Nevertheless, those same characteristics which reduced the probability of de facto domination of the military over the civilian systems and its transformation into the supreme regulator of the "belief system," also made the policy-making system in Israel vulnerable to a certain degree to the manipulations of the security establishment or portions thereof.

All this took place during the "Age of Wars." What will happen if and when a post-war age begins? To answer this, we need to return to the central question asked in the introduction: will several of the unique features of the above-described interrelations between the security and civilian complexes remain as they were? And if change occurs, as it doubtless will, what will be the direction of that change? Will Israel proceed along the path of "normalization" (according to the standards accepted in Western democratic states)? What are the possible scenarios in this context, and what will be the implications of one or another change for Israeli society, and especially for the ethoses and myths of security? Due to absence of information, we do not intend to engage here in a detailed prognosis. What may be done, and even this with a certain degree of caution, is to note certain possible directions of development assuming the meeting of certain conditions.[4]

The first point of departure to be considered in any preliminary discussion of this issue is that the very fact of entering into a post-war era was and will entail a sharp ideological-political struggle. The violent and extreme nature of this conflict found dramatic expression in the recent past in the assassination of Yitzhak Rabin. Schematically, there are two possible scenarios that are likely to characterize the completion of the peace process with the Palestinians and the Syrians: the former a more "pessimistic scenario," and the latter what may be called the "optimistic scenario."

The pessimistic scenario is likely to be realized insofar as the movement towards final arrangements with the Palestinians and the Syrians involves, in the final analysis, extensive evacuation of settlements. It will likewise relate to compromises of one sort or another regarding the status of Jerusalem. Such things can occur even after the Likud victory in the 1996 elections, and all the more so if the Likud is defeated in the forthcoming (May 1999) elections. In such a case, the probability of acts of violence on the part of radical groups or individuals from the extreme right, particularly those belonging to the fundamentalist-religious wing, is high, and may include further attempts at assassination of figures in the political leadership or in the opposition. The danger entailed here is not only that of actual occurrence of violent acts or of expressions of mutual hatred, but also of the deep scars that they will leave upon Israeli society. These scars will heal, if at all, only over an extremely long course of time. Until that happens, the political atmosphere will be extremely ugly and the split between the political camps will widen.

The more optimistic scenario may be realized if the final arrangements include a reasonable solution to the issue of the settlements, to the status of Jerusalem, to the permanent borders, and to the security arrangements, that a large portion of the public will accept and be able to live with. Such arrangements will be able, in no small measure, to neutralize the destructive potential inherent in the present situation. In such a case, the danger to the democratic rules of the game will lessen and the scars left by the peace arrangements will be less traumatic and capable of healing after a relatively shorter period of time.

The second point of departure vis-à-vis the discussion of the issue of security ethoses and Army-society relations in a post-war period refers to the changes taking place at present in the IDF itself, which will receive further momentum in the near future. These changes are reflected, first of all, in the quantitative reduction of the security apparatus and its manpower in light of a real decrease of the defense budget. The order of forces in the IDF is becoming smaller and its economic-security offshots (i.e., the military-defense industries) are reducing the scope of their activity. All these, and other factors (such as an excess of certain types of manpower), will require structural changes in the numerical balance between the professional standing army, and the regular Army and reserve divisions; second, we are witnessing significant changes in the scope of the "civilian" functions of the IDF (such as education and culture); third, an increasing focus upon the standing Army and upon a more selective regular army. The reduction of the civilian tasks of the IDF is likely to have far-reaching implications for the nature of the boundaries between the IDF and the civilian realm, both on the institutional level and on the symbolic-cultural level. In light of these expected developments, the governments of Israel will need to confront several difficult dilemmas.

A. One dilemma is the degree to which conscription to the regular army will become more selective and elitist. A more selective classification constitutes a danger to the social-national goals of conscription to the IDF — i.e., the providing of opportunities to broader sections of the population to experience Army service (see Lomsky-Feder, this volume). This experience has thus far been the most honorable "entrance ticket" to civilian society. In the absence of such an entrance ticket, it will be necessary to change the goals of military service and to strengthen the motivation for service (see Horowitz and Kimmeling 1974; Lomsky-Feder 1992; Helman, this volume).

B. Insofar as the burden of reserve duty will fall primarily on selective groups of combat soldiers and officers, a decision must be made between giving appropriate rewards to those carrying this burden, or running the risk of increasing evasion of duty within the reserve system.

C. Regarding the permanent Army, there will be need for a choice among several alternative means of preserving a suitable order of forces for the standing Army, in light of the sharp competition for quality manpower with the civilian markets. The debate that has already been taking place for several years regarding the issue of the salaries, perks and retirement conditions for the regular Army does not forebode well. The strengthening of the historical tendency, away from role expansion and toward inward withdrawal, while contracting its civilian roles, is likely to harm particularly the educational activities of the IDF, such as the work of soldier-teachers; special programs for disadvantaged youth (as in the program of "Raful's youth");[5] the scope of activity of *Galei-Tzahal* (the Army radio station);[6] the activity of the *Gadna* Youth Corps[7] and the *Nahal* (units stationed on kibbutzim);[8] the *hesder yeshivot*;[9] and the projects intended to absorb new immigrants. Nevertheless, one may not ignore the fact that certain quasi-civilian functions are also likely to become strengthened. Thus, for example, the framework of the academic *atudah* (allowing postponement of regular military service until after the B.A.) and university education (the *Talpiyot* project)[10] are likely to receive support and backing beyond that which they enjoyed in the past. This will be done in order to attract particularly high-quality manpower to remain in the permanent Army for a longer period of time, thereby assuring the quality of the officers in a situation of harsh competition with the civilian marketplace. The demand for frameworks of national service for those exempt from Army service may also be strengthened. Such a framework may even include Arab-Israeli citizens who may, for various reasons, wish to become integrated within Israeli society by obtaining this entrance ticket thereto, even if one of lower status than that represented by service in the IDF.

All these potential processes, insofar as they are realized, will doubtless lead to substantial changes in the security ethos of the State of Israel, the beginnings of which may already be felt today. It is not yet clear what shape these will in fact take in the final analysis, but one may reasonably conjecture that the tendency will not be in the direction of strengthening the militaristic tendency, but towards considerable loosening of what has unjustly been called the "religion of security." One may likewise conjecture that the great symbiosis between the military and the civilian sectors will continue to shrink. Only a severe regression in the national security situation may revivify and strengthen the militaristic direction.

Notes

*   This article has been written before October 2000. The Hebrew version of this article has been published in the book, *Basic Issues in Israeli Democracy* edited by Raffi Cohen-Almagor. Tel-Aviv: Sifriat Poalim, 1999.

1.   On the special problems deriving from the fact that the government as a whole functions as the commander-in-chief is discussed elsewhere in this volume by Bar-Or, particularly from the view-point of relations between the Prime Minister, the Minister of Defense, and the IDF General Staff (Bar-Or, this volume).

2. In addition to the above-mentioned situations of meeting, there are likewise situations of meeting within the security apparatus, such as between military personnel and civilians, or with other security personnel, such as with the Police and Civil Defense.
3. Typical examples are, on the one hand, Gen. (res.) Rehavam Ze'evi, leader of the extreme right-wing party Moledet, and, on the other, the late Maj.-Gen. (res.) Mattityahu Peled, who was Knesset delegate from an ultra-Left party.
4. On the anticipated organizational changes in the IDF, see Gal's paper in this volume.
5. "Raful's Youth" (*na'arei Raful*) — a special educational project within the Army that attempts to help youths from disadvantaged classes to become integrated within the IDF. So called because it was established by Lt.-Gen. (res.) Rafael Eitan ("Raful").
6. The IDF radio station, which is the most popular one in Israel.
7. *Gadna* (*Gedudei No'ar* –"Youth Corps") — a joint organization of the IDF and the Education Ministry, functioning as a framework in which youth receive preparation for military service.
8. *Nahal* (*No'ar Halutzi Lohem*–"Pioneering Fighting Youth") — an IDF brigade in which part of the period of military service is carried out in manning and strengthening agricultural settlements.
9. *Yeshivat Hesder*. Soldiers drafted into these units divide their time between service in regular Army units and the study of Torah in special yeshivot.
10. The *Talpiyot* project is a special project intended for preparing soldiers with unusual intellectual qualifications in the natural sciences within university frameworks.

References

Bar-Or, Amir "The Link Between the Government and the IDF During Israel's First 50 Years: The Shifting Role of the Defense Minister." (in this volume).

Ben-Eliezer, Uri "From Military Role-Expansion to Difficulties in Peace Making: The Israel Defense Forces 50 Years On." (in this volume).

Ben-Eliezer, Uri 1994. "'A Nation in Arms' and War; Israel in its First Years." *Zemanim* 13, 49: 51-64. (Hebrew).

Böene, Bernard "Western-Type Civil-Military Relations Revisited." (in this volume).

Brenner, Uri 1978. *Altalena; A Political and Military Study*. Tel-Aviv: Hakibbutz Hameuhad. (Hebrew).

Caforio, Giuseppe 1988. "The Military Profession: Theories of Change." *Armed Forces and Society* 15, 1: 55-70.

Cohen, Stuart 1996. "The IDF and Israeli Society: Towards Scaling Down." In *Israel Towards the Year 2000: Society, Politics and Culture*, edited by Moshe Lissak and Baruch Kenei-Paz, 215-32. Jerusalem: Magnes. (Hebrew).

Dayan, David 1978 *Yes, We Are Youth! History of the Gadna*. Tel-Aviv: MOD Press. (Hebrew).

Diehl, Paul 1993. *International Peace Keeping*. Baltimore–London: Johns Hopkins University Press.

Gal, Michael 1982. "Integration of Soldiers from Weak Populations in the IDF; Summary and Considerations." *Ma'arakhot* 283: 36-44. (Hebrew).

Gal, Reuven "The Israeli Defense Forces (IDF): A Conservative or an Adaptive Organization?" (in this volume).

Gelber, Yoav 1986. *Why Did They Dismantle the Palmah?* Tel-Aviv: Schocken. (Hebrew).

Helman, Sara "Citizenship Regime, Identity and Peace Protest in Israel." (in this volume).

Hoffnung, Menahem 1991. *Israel: Security Needs vs. the Rule of Law.* Jerusalem: Nevo. (Hebrew).

Horowitz, Dan 1982. "The Israel Defense Forces: A Civilianized Military in a Partially Militarized Society." In *Soldiers, Peasants and Bureaucrats,* edited by Roman Kolkowicz and Andrzej Korbonski, 273-302. London: George Allen and Unwin.

Horowitz, Dan and Baruch Kimmerling 1974. "Some Social Implications of Military Service and the Reserve System in Israel." *European Journal of Sociology* 15: 262-76.

Horowitz, Dan and Moshe Lissak 1990. *Trouble in Utopia; The Overburdened Polity of Israel.* Tel-Aviv: Am Oved. (Hebrew).

Horowitz, Dan and Moshe Lissak 1987. "Democracy and National Security: an Ongoing Confrontation." *Yahadut Zemanenu* 4: 27-65. (Hebrew).

Huntington, Samuel 1985. *The Soldier and the State: The Theory and Politics of Civil-Military Relations.* Cambridge, MA: Harvard University Press.

Jackson, Henry 1965. "The National Security Council." In *The National Security Council,* edited by Henry Jackson, II-XVI. New York: Praeger.

Janowitz, Morris 1965. "Armed Forces in Western Europe: Uniformity and Diversity." *Archives Europeenes de Sociologie* 6, 2: 225-37.

Kimmerling, Baruch 1993. "Patterns of Militarism in Israel." *European Journal of Sociology* 34: 196-223.

Lahav, Pnina 1993 "The Press and National Security." In *National Security and Democracy in Israel,* edited by Avner Yaniv, 173-96. Boulder–London: Lynne Rienner.

Last, David 1995-96. "Peacekeeping Doctrine and Conflict Resolution Techniques." *Armed Forces and Society* 22, 2: 187-210.

Laswell, Harold 1941. "The Garrison State." *The American Journal of Sociology* 46, 4: 455-68.

Lissak, Moshe 1996. "'Critical' Sociology and 'Establishment' Sociology in the Israeli Academic Community: Ideological Struggles or Academic Discourse?" *Israel Studies* 1: 247-94.

Lissak, Moshe 1985. "Boundaries and Institutional Linkage Between Elites: Some Illustrations from Civil-Military Relations in Israel." *Research in Politics and Society; A Research Annual,* 1: 129-48.

Lissak, Moshe 1984. "Convergence and Structural Linkages Between Armed Forces and Society." In *The Military, Militarism and the Polity; Essays in Honor of Morris Janowitz,* edited by Michel Martin and Ellen Stern, 50-62. New York: Free Press.

Lissak, Moshe 1976. *Military Roles in Modernization: Civil-Military Relations in Thailand and Burma.* Beverly Hills–London: Sage.

Lissak, Moshe 1967. "Modernization and Role Expansion of the Military in Developing Countries: A Comparative Analysis." *Comparative Studies in Society and History* 14, 3: 233-55.

Lissak, Moshe and Daniel Maman 1996. "Israel." In *The Political Role of the Military,* edited by Constantine Danopoulos and Cynthia Watson, 223-33. Westport, CT: Greenwood.

Lomsky-Feder, Edna "The Meaning of War Through Veteran's Eyes: A Phenomenological Analysis of Life Stories." (in this volume).

Lomsky-Feder, Edna 1992. "Youth in the Shadow of War—War in the Light of Youth: Life Stories of Israeli Veterans." In *Adolescence, Careers and Culture,* edited by Wim Meeus et al., 393-408. The Hague: De Gruyter.

Luckham, A. R. 1977. "A Comparative Typology of Civil-Military Relations." *Government and Opposition* 6: 5-35.

Maman, Daniel and Moshe Lissak 1990. "The Impact of Social Networks on the Occupational Patterns of Retired Officers: The Case of Israel." *Forum International* 9: 279-308.

Mintz, Alex 1985. "Military-Industrial Linkages in Israel." *Armed Forces and Society* 12, 1: 9-27.

Mintz, Alex 1984. "The Miliary-Industrial Complex." in *Israeli Society and its Defense Establishment*, edited by Moshe Lissak, 103-27. London: Frank Cass.

Moskos, Charles 1971. "Armed Forces and American Society: Convergence or Divergence?" In *Public Opinion and The Military Establishment*, edited by Charles Moskos, 271-92. Beverly Hills, CA: Sage.

Nakdimon, Shlomo 1978. *Altalena*. Jerusalem: Idanim. (Hebrew).

Negbi, Moshe 1985. *A Paper Tiger; The Struggle for Press Freedom in Israel*. Tel-Aviv: Sifriyat Poalim. (Hebrew).

Nussek, Hanoch and Yechiel Limor 1985. "Military Censorship in Israel: A Temporary Compromise between Conflicting Values." *Kesher* 17: 45-62. (Hebrew).

Ostfeldt, Zeava 1994. *An Army is Born: Main Stages in the Buildup of Israel's Army Under the Leadership of David Ben-Gurion*. Tel-Aviv: MOD Press. (Hebrew).

Peri, Yoram "Civil-Military Relations in Israel in Crisis." (in this volume).

Peri, Yoram 1989. "The Impact of Occupation on the Military: The Case of the IDF, 1967-1987." In *The Emergence of a Binational Israel: The Second Republic in the Making*, edited by Ilan Peleg and Ofira Seliktar, 143-68. Boulder, CO: Westview.

Peri, Yoram 1983. *Between Battles and Ballots: Israeli Military in Politics*. Cambridge: Cambridge University Press.

Peri, Yoram 1981. "Political-Military Partnership in Israel." *International Political Science Review* 2, 3: 303-315.

Peri, Yoram and Moshe Lissak 1976. "Retired Officers in Israel and the Emergence of a New Elite." In *The Military and the Problem of Legitimacy*, edited by Gwyn Harries-Jenkins and Jacobus Van Doorn, 175-92. London: Sage.

Shamgar, Meir 1982. "Legal Concepts and Problems of the Israeli Military Conquest — The Initial Stage." In *Military Government in the Territories Administered by Israel, 1967-1980: The Legal Aspect*, edited by Meir Shamgar, 13-60. Jerusalem: Faculty of Law, Hebrew University.

Shapira, Anita 1985. *The Army Controversy, 1948; Ben-Gurion's Struggle for Control*. Tel-Aviv: Hakibbutz Hameuhad. (Hebrew).

Smooha, Sammy 1993. "Part of the Problem or Part of the Solution: National Security and the Arab Minority." In *National Security and Democracy in Israel*, edited by A. Yaniv, 105-28. Boulder–London: Lynne Reinner.

State of Israel 1948. *The Official Gazette*. Jerusalem: State of Israel. (Hebrew).

Van Doorn, Jacobus 1975. *The Soldier and Social Change*. Beverly Hills–London: Sage.

Yaniv, Avner 1993. "An Imperfect Democracy." In *National Security and Democracy in Israel*, edited by Avner Yaniv, 227-30. Boulder–London: Lynne Reinner.

About the Editors and Contributors

Daniel Maman is Lecturer at the Department of Sociology and Anthropology, The Hebrew University of Jerusalem. His fields of research include economic and organizational sociology, elites, civil-military relations, and networks analysis.

Zeev Rosenhek is Lecturer at the Department of Sociology and Anthropology, the Hebrew University of Jerusalem. His fields of research include state-society relations, ethnicity and citizenship, and labor migration.

Eyal Ben-Ari is Professor at the Department of Sociology and Anthropology, The Hebrew University of Jerusalem. He has carried out research on Japanese culture and society. He has also done research on the Israeli army, military leadership in the industrial democracies, and United Nations peace-keeping forces.

Bernard Boëne, Professor, Director General for Academic Affairs, ESM de Saint-Cyr, France.

James Burk, Professor of Sociology at Texas A&M University, USA.

Yoram Peri, Senior Lecturer at the Department of Communication and Journalism, The Hebrew University of Jerusalem, Israel.

Uri Ben-Eliezer, Senior Lecturer at the Department of Sociology and Anthropology, University of Haifa, Israel.

Stuart A. Cohen, Professor at the Department of Political Studies and BESA Center, Bar-Ilan University, Israel.

Dafna Izraeli, Professor of Sociology, Department of Sociology and Anthropology, and Chair of the Interdisciplinary Graduate Program in Gender Studies. Bar-Ilan University, Israel.

Edna Lomsky-Feder, Lecturer at the School of Education, The Hebrew University of Jerusalem, Israel.

Sara Helman, Lecturer at the Department of Behavioral Sciences, Ben Gurion University, Israel.

Amir Bar-Or, The Ben-Gurion Research Center, Ben-Gurion University, Israel.

Henning Sørensen, Director of Institute for Sociological Research (ISF), Denmark.

Reuven Gal, Director, Carmel Institute for Social Studies, Israel.

Luis Roniger, Associate Professor at the Department of Sociology and Anthropology, The Hebrew University of Jerusalem, Israel.

Moshe Lissak, Professor at the Department of Sociology and Anthropology, The Hebrew University of Jerusalem, Israel.

Name Index

Subject Index

Printed and bound by CPI Group (UK) Ltd, Croydon, CR0 4YY

22/10/2024

01777628-0010